Smart and Sustainable Power Systems

Operations, Planning, and Economics
of Insular Electricity Grids

Smart and Sustainable Power Systems

Operations, Planning, and Economics of Insular Electricity Grids

EDITED BY

João P. S. Catalão

UNIVERSITY OF BEIRA INTERIOR, COVILHA, PORTUGAL

CRC Press
Taylor & Francis Group
Boca Raton London New York

CRC Press is an imprint of the
Taylor & Francis Group, an **informa** business

CRC Press
Taylor & Francis Group
6000 Broken Sound Parkway NW, Suite 300
Boca Raton, FL 33487-2742

First issued in paperback 2020

© 2015 by Taylor & Francis Group, LLC
CRC Press is an imprint of Taylor & Francis Group, an Informa business

No claim to original U.S. Government works

ISBN-13: 978-1-4987-1212-5 (hbk)
ISBN-13: 978-0-367-73815-0 (pbk)

Visit the Taylor & Francis Web site at
http://www.taylorandfrancis.com

and the CRC Press Web site at
http://www.crcpress.com

I dedicate this book to Carla and Alex,

for inspiring me with their love and energy.

Contents

Preface

To foster the transition toward a sustainable energy future, islands represent a priority test bed for pioneering technologies such as smart grids, alongside renewable energy sources and electrical energy storage devices. The smart grid initiative, integrating advanced sensing technologies, intelligent control methods, and bidirectional communications into the contemporary electricity grid, provides excellent opportunities for energy efficiency improvements and better integration of distributed generation, coexisting with centralized generation units within an active network.

A large share of the installed capacity for recent renewable energy sources already comprises insular electricity grids, since the latter are preferable due to their high potential for renewables. However, the increasing share of renewables in the power generation mix of insular power systems presents a significant challenge to efficient management of the insular distribution networks, mainly due to the variability and uncertainty of renewable generation.

More than other electricity grids, insular electricity grids require the incorporation of sustainable resources and the maximization of the integration of local resources, as well as specific solutions to cope with the inherent characteristics of renewable generation. Insular power systems need a new generation of methodologies and tools to face the new paradigm of large-scale renewable integration.

This book deals with modeling, simulation, and optimization of insular power systems to address the effects of large-scale integration of renewables and demand-side management, from forecasting to operations and planning and proposing efficient methodologies, tools, and solutions toward the development of a sustainable and smart grid. Comprehensive testing and validation of the mathematical models are provided using real-world data.

This book covers several important topics, such as insular power systems, forecast techniques and models, renewable energies, uncertainty and variability, power flow calculations, probabilistic and stochastic approaches, scenario generation, scheduling models, short-term operation, reserves and demand response, electric price signals, competitive operation of distribution networks, and network expansion planning.

João P. S. Catalão
University of Beira Interior

Acknowledgments

This book would not have been possible without the exceptional contributions of several globally-recognized experts in the power systems field.

I deeply thank Professors Tasos, Javier, Gianfranco, Claudio, and Radu, as well as all my current students and postdoctoral scholars, especially Nikos, Ozan, Shafie, and Gerardo, for the great and *singular* work carried out and for their support and friendship. Thank you very much.

A special thank you also goes to Professor Matias, for always having the amazing capability to surpass life's difficulties with a smile on the face, embodying the best of the human character.

I thank the EU Seventh Framework Programme FP7/2007-2013 under grant agreement No. 309048, the Portuguese Science and Technology Foundation, and FEDER through COMPETE, under Projects FCOMP-01-0124-FEDER-020282 (Ref. PTDC/EEA-EEL/118519/2010) and PEst-OE/EEI/LA0021/2013, for financial support.

Finally, I thank the staff at CRC Press, particularly Nora Konopka and Laurie Oknowsky, for their immense competence and kind attention during all phases of the book preparation. Many thanks to all!

João P. S. Catalão
University of Beira Interior

Editor

João P. S. Catalão received a master's degree from the Instituto Superior Técnico, Lisbon, Portugal, in 2003, and a PhD degree and habilitation for full professor ("Agregação") from the University of Beira Interior (UBI), Covilhã, Portugal, in 2007 and 2013, respectively.

Currently, he is a professor at UBI, director of the Sustainable Energy Systems Lab, and researcher at INESC-ID. He is a senior member of IEEE. He is the primary coordinator of the EU-funded FP7 project SiNGULAR (Smart and Sustainable Insular Electricity Grids Under Large-Scale Renewable Integration), a €5.2-million project involving 11 industry partners. He has authored or coauthored more than 320 publications, including 100 journal papers, 200 conference proceedings papers, and 20 book chapters, with an *h*-index of 22 (according to Google Scholar). He has supervised more than 25 postdoctoral, PhD, and MSc students. He is the editor of the book entitled *Electric Power Systems: Advanced Forecasting Techniques and Optimal Generation Scheduling* (Boca Raton, FL: CRC Press, 2012), translated into Chinese in January 2014. His research interests include power system operations and planning, hydro and thermal scheduling, wind and price forecasting, distributed renewable generation, demand response, and smart grids.

Professor Catalão is an editor of *IEEE Transactions on Smart Grid*, an editor of *IEEE Transactions on Sustainable Energy*, and an associate editor of *IET Renewable Power Generation*. He was the guest editor-in-chief for the special section "Real-Time Demand Response" of *IEEE Transactions on Smart Grid*, published in December 2012, and he is currently guest editor-in-chief for the special section "Reserve and Flexibility for Handling Variability and Uncertainty of Renewable Generation" of *IEEE Transactions on Sustainable Energy*. He is the recipient of the 2011 Scientific Merit Award UBI-FE/Santander Universities and the 2012 Scientific Award UTL/Santander Totta.

Contributors

José M. Arroyo
University of Castilla–La Mancha
Ciudad Real, Spain

Miguel Asensio
University of Castilla–La Mancha
Ciudad Real, Spain

Anastasios G. Bakirtzis
Aristotle University of Thessaloniki
Thessaloniki, Greece

Emmanouil A. Bakirtzis
Aristotle University of Thessaloniki
Thessaloniki, Greece

Adriana Carillo
Italian National Agency
    for New Technologies,
    Energy and Sustainable
    Economic Development
Rome, Italy

Gianfranco Chicco
Energy Department
Politecnico di Torino
Torino, Italy

Valeria Cocina
Energy Department
Politecnico di Torino
Torino, Italy

Javier Contreras
University of Castilla–La Mancha
Ciudad Real, Spain

Ozan Erdinç
Yildiz Technical University
Istanbul, Turkey

Pedro Fonte
Department of Electrical Energy
    and Automation
Lisbon Superior Engineering Institute
Polytechnic Institute of Lisbon
Lisbon, Portugal

Ehsan Heydarian-Forushani
Isfahan University of Technology
Isfahan, Iran

Evaggelos G. Kardakos
Aristotle University of Thessaloniki
Thessaloniki, Greece

Andrea Mazza
Energy Department
Politecnico di Torino
Torino, Italy

Pilar Meneses de Quevedo
University of Castilla–La Mancha
Ciudad Real, Spain

Cláudio Monteiro
Department of Electrical and
    Computer Engineering
School of Engineering
University of Porto
Porto, Portugal

Sergio Montoya-Bueno
University of Castilla–La Mancha
Ciudad Real, Spain

José I. Muñoz
University of Castilla–La Mancha
Ciudad Real, Spain

Gregorio Muñoz-Delgado
University of Castilla–La Mancha
Ciudad Real, Spain

Rafael Nebot-Medina
Canary Islands Institute of
    Technology
Las Palmas De Gran Canaria, Spain

Nikolaos G. Paterakis
University of Beira Interior
Covilhã, Portugal

Radu Porumb
Department Electrical Power
    Systems
University Politehnica of Bucharest
Bucharest, Romania

Angela Russo
Energy Department
Politecnico di Torino
Torino, Italy

Gianmaria Sannino
Italian National Agency for
    New Technologies, Energy
    and Sustainable Economic
    Development
Rome, Italy

Bruno Santos
Smartwatt Intelligence Department
Porto, Portugal

Tiago Santos
Smartwatt Intelligence Department
Porto, Portugal

George Serițan
Department of Measurements,
    Electrical Apparatus and Static
    Converters
University Politehnica of Bucharest
Bucharest, Romania

Miadreza Shafie-khah
University of Beira Interior
Covilhã, Portugal

Christos K. Simoglou
Aristotle University of Thessaloniki
Thessaloniki, Greece

Carina Soares
University of Beira Interior
Covilhã, Portugal

Filippo Spertino
Energy Department
Politecnico di Torino
Torino, Italy

Ion Trîştiu
Department Electrical Power
    Systems
University Politehnica of
    Bucharest
Bucharest, Romania

Stylianos I. Vagropoulos
Aristotle University of Thessaloniki
Thessaloniki, Greece

1

Overview of Insular Power Systems: Challenges and Opportunities

Ozan Erdinç and Nikolaos G. Paterakis

CONTENTS

ABSTRACT This chapter aims to provide an overview of insular power system structures and operational requirements. First, a general evaluation of insular power systems is realized together with an analysis of sample insular grid structures. Then, the challenges and opportunities of insular grids are discussed also considering possible future developments in technology and applications to overcome challenges and to promote opportunities. Finally, the chapter is summarized with concluding remarks.

1.1 Evaluation of Insular Power Systems

1.1.1 Overview

The term *island* basically means a subcontinental land that is surrounded by water, which implies a physical total insularity (which refers to isolation and/or dispersion) from the mainland. Thus, insular power systems correspond to electric power grid structures in physically isolated geographical areas that are mainly islands in the above sense [1].

It was stated in the Treaty of Amsterdam that "insular regions suffer from structural handicaps linked to their island status, the permanence of which impairs their economic and social development" [2] (p. 136). In other words, insular areas have limitations that need to be identified. Limited range of resources, inability to achieve economies of scale, often seasonal change of population, higher infrastructure costs, distance from the mainland, climatic conditions, and microclimates within the insular area are examples of such constraints. These limitations lead to several negative outcomes, including overseas trade dependency, economic weakness that reduces the chance of gaining access to conventional markets, and the need of oversizing infrastructure such as power systems [3]. Among the mentioned limitations, priority should be given to the fact that insular areas are among the most vulnerable places regarding climate change as the frequency of natural disasters (mostly tropical storms, typhoons, etc.) that has increased because of climate change are a more important threat for insular areas than for the mainland [4]. Climatic effects such as rising sea levels, climate variability, and abnormal climatic conditions also affect the environment of islands [3].

Thus, the general problems of insular areas can mainly be categorized as dependence on imported fuel, availability of freshwater, management of wastes, and other problems related to climatic conditions [3,5], which are affected by many factors, including the components of insular economies.

The economies of most of the insular areas rely on tourism, which is specifically an advantage driven by their added geographical value. For European islands, a research concluded that the dependency ratio of these insular economies on tourism is nearly 70%. Apart from tourism, insular economies mainly depend on agriculture and fishery, since industrial development is very limited in islands due to scarce resources, high infrastructure and transport costs, limited competitive markets for companies, and so on [3]. Nearly all the imported goods at many insular areas are shipped over very long distances, a fact that has had a serious impact on their economic sustainability. Many islands worldwide rely nearly 100% on imported fuel for many activities, such as energy production, transportation, and heating. Such dependence on imported fuel is a major issue that has economic, technical, and social results. Many critical industries such as fisheries are highly vulnerable to fuel prices [4].

Tourism is a high-quality-water- and an energy-consuming industry. Generally, tourism is mostly seasonal and leads to a significant increase in the island's population for a few months. Tourism also increases waste and the need for waste management, especially in such high-population periods during which waste is a serious problem due to limited area and high cost and scale of recycling processes. If high-quality water is not available within the own resources of an island, there are two ways to fulfil this requirement: (1) water import from another area and (2) desalination of seawater, both of which are fuel- and energy-consuming activities. As can be seen, both waste management and quality water preparation increase the need of energy, together with the fact that other daily energy needs (air-conditioning and other daily consumptions) in high-population periods are significantly greater than those during the rest of the year. This leads to oversizing infrastructure in all fields as systems sized to accommodate the requirements of the summer period are significantly underutilized in winter time.

The mitigation of dependence on imported fuel, especially for electricity production, is an important parameter for the economic sustainability of insular areas. The fuels (gasoline, oil, and liquefied petroleum gas) required for conventional energy sources are generally carried to islands by tankers, which creates both an unsustainable service model especially during peak times and a problematic strategy from an environmental point of view. Besides, energy production from fossil fuels is costly, especially due to transportation costs. Thus, the utilization of local resources mainly in the form of renewable energy systems (RESs) is a pivotal aim of many energy policies, especially during the last decade, and the structures of electric power grids have started to significantly change with the recently increasing interest in RESs. Comparatively, weaker electricity grid structures such as insular grids are more sensitive to power-quality issues such as frequency and voltage deviations, especially if the penetration level of RESs is high due to the volatile nature of such units. Insular grids with a lower inertia due to reduced number of generating units connected to the system are especially more vulnerable to large frequency and voltage deviations, which in turn reduces system reliability and security. Thus, the policies aiming to improve the penetration of RESs, especially in insular areas, are limited by insular-power-system deficiencies. As a result, insular power systems are considered as a *laboratory* for testing the impacts of new technologies and strategies [3], which underlines the need for deeper examination of insular areas and insular power systems in terms of current status, challenges, and opportunities for future technological advancements, ultimately including the *smartification* of all operations.

1.1.2 Analysis of Typical Insular Grid Structures

1.1.2.1 *System Size and Typical Infrastructure*

According to their peak power demand [megawatt (MW)] and annual energy consumption [in gigawatt/hour (GWh)], insular areas are classified into four groups [6]:

1. Very small islands (<1 MW and <2 GWh)
2. Small islands (1–5 MW and 2–15 GWh)
3. Medium islands (5–35 MW and 15–100 GWh)
4. Big islands (>35 MW and >100 GWh)

Over the last few decades, the electric power loads have significantly increased with the investments in the new transmission and distribution system construction to enable nearly all citizens in insular areas to gain access to electricity, apart from the people living in significantly rural areas, where geographical conditions prevent the possibility of installing power lines. Table 1.1 briefly describes on power loads of sample insular areas geographically located in different parts of the world.

Generally, the power system of insular areas comprises a single or a few and mostly conventional fuel-based generators, especially in small and very small islands. Thus, the inertia of the total system is significantly low, and the current status of insular power systems can be considered unreliable due to possible outages and fuel shortages for such a small number of generating options. Moreover, most of the generating units as well as other

TABLE 1.1

The Power Loads of Sample Insular Areas

Location	Population	Installed Capacity	Peak Load	Reference
Guam	~170,000	550 MW	280 MW (40,000 customers)	[4]
St. Thomas-St. John (U.S. Virgin Islands)	~53,000–~4,000	199 MW	86.3 MW	[4]
St. Croix (U.S. Virgin Islands)	~55,000	120.8 MW	55 MW	[4]
Cyprus	~860,000	990 MW	775 MW	[7]
Crete	~600,000	1085 MW	650 MW	[8]
Gran Canaria	~840,000	860 MW	552 MW	[7]
São Miguel (Azores Islands)	~140,000	160 MW	65 MW	[8]
La Graciosa (Canary Islands)	~1,000	1.3 + 1.5 MW from the main island	1.2 MW	[8]
San Juanico, Mexico	~640	205 kW	167 kW	[9]

electrical infrastructure components are old and need maintenance more than newly installed systems, an issue that reduces both reliability and economic sustainability.

The technical and nontechnical losses in insular areas are considered to have a higher percentage when compared to those of bulk power system that promotes the increase of both fuel utilization and the cost per unit of electricity [4]. Thus, the overall power system operating efficiency is significantly lower compared to that in the mainland, which places further economic burden on energy service companies and accordingly end-user customers.

1.1.2.2 Economical Framework of Insular Power System Operation

Over the last few decades, the organization and regulation of the electricity sector has substantially changed in many countries around the world, where vertically integrated companies and monopolies have been or are being replaced by competition and market structures. In favor of these profound changes, supporters of deregulation claim that competition among an increasing number of firms will permit cost and price reductions [10]. Practically, restructuring of the electricity sector entails the introduction of several new entities and agents that operate in all levels of the production-to-consumption chain [11]. Although there are several examples of successful liberalization around the world, the power systems of islands pose challenges toward the implementation of market structures and competition because of their peculiarities. The main barriers are as follows:

- In contrast with continental grids, a generator in an island cannot have significant capacity due to system security reasons.
- More reserve capacity is required than in the mainland networks on account of absence of interconnections.
- Provision of electricity is more expensive in insular power systems, owing to high fuel-transportation costs.
- Limited space, public opposition, and local factors (e.g., the island is a tourist resort) do not allow vast investments in conventional power plants.
- RESs are alternative candidates for electricity production, promising economic efficiency and sustainability. Nevertheless, network security issues and the interruptible nature of such resources limit the penetration of RESs.

Until recently, most insular power systems were managed by a public vertically integrated company that was socializing the higher cost of producing electricity in islands in order to achieve social fairness and boost the economies of the islands.

There are several models to introduce competition in the electricity market [10], a few are as follows:

- The single-buyer model
- Bilateral contracts
- Wholesale markets (pools)

The single-buyer model and bilateral contracts are considered to be the most suitable structures for noninterconnected insular power systems [12]. All insular power systems share the aforementioned limitations. However, the organization of the electricity market may differ significantly from island to island. In Malta, a small-sized island, there exists only a national corporation (Enemalta), which acts as a generator, distributor, and retailer. Here, the market structure follows the single-buyer model since independent producers may sell electricity produced to Enemalta. Furthermore, consumers have to buy energy from the same corporation since no other retailer exists [13]. In Cyprus, a noninterconnected island country, the market is based around bilateral contracts. Legislation allows 67% of the island's electricity consumers to choose their supplier and also foresees for competition in the generation of electricity. Nevertheless, the Electricity Authority of Cyprus remains the only producer and supplier of power [14]. In Canary Islands (Spain), Red Eléctrica de España acts as a single buyer using a minimum cost-generation scheduling procedure [10,15]. The island of Crete offers the opportunity of having a pool market structure given the size and demand of the load [12]. It is obvious that although the legal framework regarding insular power systems in many cases allows competition, in practice, it is not operational. In order to attract investors' interest and overcome the practical barriers in insular power systems, several alternatives such as hybrid portfolios (e.g., combining energy storage and RESs) should be considered, and possible interconnections with the nearby islands or the mainland should be investigated.

1.1.2.3   The Need for Sustainability

In general, it is evident for many islands that it is technically hard or economically unviable to have a direct interconnection with the mainland. This leads to the necessity of supplying all the energy demand by producing on site. The insular electricity grids generally rely on a single type of energy resource that is typically of conventional type and is dependent on imported fuel with price volatility. This issue is likely to hamper the economic sustainability of insular areas. The utilities in these areas suffer dramatically from the increased cost for imported fuels that in turn has a major impact on the cost of electrical energy per unit. As a real example, electrical energy prices reflected to end users within insular areas were varying between 25 and 34 cents/kWh, whereas the same cost was in the range of 10–14 cents/ kWh in the mainland for the United States in 2005 [4]. The cost of electricity for residential and commercial end users was approximately 31 cents/kWh in September 2010, 40 cents/kWh in December 2012, and 42 cents/kWh in the third quarter of 2013 for American Samoa [16]. It is clear that the price has been significantly volatile for a short period in insular areas. This was mainly caused by the higher and variable cost of fuel/barrel, due to additional costs such as transportation (as all imports are shipped over very long distances to most of the islands). Other major reasons for this cost difference include the increasing percentage of maintenance within total costs and the decreasing efficiencies of employed engines and turbines, both caused by aging of electricity production facilities, as investment in new technologies with surely significant capital costs is not regular for insular areas as opposed to the bulk power system renovations [4]. Some insular areas have their own bulk fuel facilities and can purchase fuel from the bulk market at more affordable prices. Besides, in some countries, the difference in prices between insular areas and the mainland is supported by public funds to ensure that kWh cost for the customers is the same in both the mainland and islands. An apt example of this is the case of the small islands in Sicily, Italy, where the Ministry of Industry gave the aforementioned support [17]. However, this is not a common case, since it is an additional burden on the economy of the country.

In some islands (e.g., Pacific Islands of the United States, which depended nearly 100% on petroleum for their energy needs in 2006), local wood, coconut husks, and other possible biomass products are also employed for some daily activities such as cooking instead of using electricity. However, the need for fossil fuels, especially petroleum, is also increasing because of other sectors, for example, transportation, apart from electricity production. Among the transportation facilities, commercial airlines and local airports are major fossil fuel users. Furthermore, fishing fleets and other marine crafts promote the need for fossil fuel utilization.

This increasing use of petroleum and other kinds of conventional resources not only threatens the insular economies but also significantly contributes to the environmental degradation, which can have serious results. First of all, such an increase in conventional fuel utilization increases the major impact of global warming similarly to all over the world. However, the climatic changes related to global warming will have a greater impact on insular areas compared to the mainland. Rise in sea levels, natural disasters, and drought threaten the natural life in insular areas dramatically. Life quality will certainly worsen with such an increase of greenhouse gas emissions, especially in places such as islands, where end users are more closely located to production sites compared to a mainland. Thus, both economic and environmental impacts of fossil fuel utilization in insular areas should be examined in a more detailed way, aiming to improve the sustainability of insular life.

1.1.2.4 Reliability Requirements

Given the economic and social aspects the usage of electricity entails, measures have to be taken in order to guarantee the uninterrupted and quality operation of the power system. In this respect, the concepts of power system stability and reserve adequacy should be given importance for both interconnected continental grids and insular power systems.

According to the IEEE/CIGRE Joint Task Force on Stability Terms and Definitions "power system stability is the ability of an electric power system, for a given initial operating condition, to regain a state of operating equilibrium after being subjected to a physical disturbance, with most system variables bounded so that practically the entire system remains intact" [18] (p. 1388). Three types of stabilities have been identified, namely, rotor angle, frequency, and voltage.

The rotor angle stability indicates the ability of the power system to remain synchronized under severe or small disturbances. Transient stability (ability to withstand severe disturbances) depends on both system properties and the type of faults. Generally, systems with centralized and similar-technology synchronous generators have better transient stability. Small disturbances are linked with small signal stability. This means that the system has the properties of damping electromagnetic oscillations and improved frequency response (frequency restores its predisturbance value quickly with no oscillations) [19].

Frequency should be maintained within acceptable limits around the nominal value. Off-nominal frequency has negative and potentially hazardous results such as resonances in rotating machines that tear them because of mechanical vibrations, changes in the speed of asynchronous machines, overheating of transformers and machines because of increased core loses, flickering, and so on.

To maintain the frequency at its nominal level, a balance between production and consumption of active power should be retained. In order to achieve this, an amount of active power is rendered available in order to control frequency through its variation. Despite the fact that the energy stored in the rotating masses of generators compensates energy deficits constantly due to their inertia, control schemes are required in order to guarantee appropriate values for frequency.

The typical procedure in order to control frequency comprises three levels. The first level is the primary frequency control, which is a local automatic control and is performed by generators that are equipped with a speed governor. It is designed to react instantly (several seconds), stabilizing the frequency value after large generation or load changes (outages). Frequency-controlled loads such as induction motors or frequency-sensitive relay equipped loads can also participate in this control. There are two issues regarding this control level. First, the active power provided by several generating units or demand-side resources must be replaced after some time due to technical or other restrictions. Also, the resources that contribute to this frequency control level should not be concentrated in a specific location of the grid but should be distributed instead. In this way, unexpected transit phenomena are avoided and the network security is enhanced.

Primary frequency control limits frequency deviations, but cannot recover its nominal value. For this reason, another level of control, namely, the secondary frequency control, is used in order to restore the target frequency value. This is a type of centralized automatic control and refers typically only to the generation side. In contrast with the primary frequency control, the secondary frequency control is not mandatory for all power systems. Such systems exert frequency control based on the primary frequency control and a third level of control, known as the *tertiary frequency control*, that comprises manual changes in the commitment and dispatch status of the generating units. This is a definite measure to restore both the primary and the secondary frequency reserves to their normal levels and to reestablish the nominal frequency value.

The control of a power system's voltage can be also organized into three levels. Primary voltage control is a local automatic control that maintains the appropriate voltage level at every bus. This task can be performed by the automatic voltage regulators of the generating units or by static devices such as static voltage compensators. Then, a rather rare level of voltage control is the centralized secondary voltage control that regulates the local reactive power injections. Finally, the manual tertiary voltage control reestablishes the reactive power flow through the power system. Naturally, the demand-side resources that are capable of absorbing or generating reactive power have the potential of participating in voltage control [20,21].

1.2 Challenges and Opportunities for Insular Power Systems

1.2.1 Overview

The economic and environmental issues that were examined in detail in the previous sections call for new policies to ensure the sustainable growth of insular life. To reduce the dependency on fossil fuel utilization, energy-related policy making and relevant R&D studies are of utmost importance. The energy-related R&D studies are driven by four major concerns as follows [22]:

- *Supply pressures* corresponding to risk in the security of global energy resources
- *Demand pressures* corresponding to economic growth, increasing consumer expectations and industrial demand, and limitations of existing infrastructures

- *Environmental pressures* corresponding to actions that should be taken to reduce carbon emissions resulting in global warming
- *Political pressures* corresponding to political actions in energy-key regions such as the Middle East and Russia

One of the leading solutions to partly overcome the supply, environmental, and political pressures is to increase the share of renewable and clean alternative energy production technologies in the generation mix. Generally, islands have a good RES potential. Wind and solar resources are the pivotal available resources in many insular areas. Besides, hydro, biomass (also biofuels for transportation), geothermal, and oceanic energy (wave or tidal) have found areas of application in specific islands. According to a report [3], for European islands, more than 15% of all imports are energy imports, which can be reduced by RES integration. However, the intermittency of RESs is especially a core concern for small grid structures such as insular areas [5]. Thus, the major concern for increasing RES penetration is the stability of the grid in terms of frequency and voltage control, reactive power supply, and so on [23].

Another combined social and technical problem is the limitations of available capital and human capacity to efficiently operate and maintain such technologies. There have been many cases where insular area citizens who had lived in the mainland for some time to gain training and technical skills finally chose to migrate rather than returning to their islands to take part in technical actions. If the local know-how necessity is well managed, it can also create job opportunities with high-educational requirements, which islands generally lack.

The natural climatic conditions, apart from temperature, wind speed, and solar irradiation, should also be examined. For example, industrial equipment that works well in normal conditions may fail while used outdoors in tropical environmental regions such as Pacific Islands. Besides, many insular areas are prone to tropical cyclones that make it harder to protect especially wind turbines from high winds. In a similar context, from the geographical point of view, one of the problems faced during the electrification of rural areas is the practical difficulty in maintenance after installing an off-grid system based on RESs.

Apart from the RES integration, many reports have been prepared and many actions have been taken around the world to reduce fossil fuel dependence in insular areas. Recently, demand-side actions including responsive demand strategies and energy efficiency actions took the second place, which is linked to *demand pressures*. Some general issues that have been pointed out in the reports including these facts are as follows [4]:

- Using more efficient devices in all areas, from engines to street lights
- Developing forums to share experiences of different insular areas in terms of policies, projects, programs, and so on

- Developing strategies to increase competition among fuel suppliers to reduce prices in the market
- Examining the supply-side and especially the demand-side energy issues
- Evaluating power losses within generation, transmission, and distribution systems
- Reevaluating the current maintenance practices
- Examining the power demand pattern in a more detailed way rather than the daily total energy usage profiles with advanced metering systems
- Examining the reactive power requirements within services and reducing reactive power flow within the system
- Examining the loading ratios of all transformers to reduce no-load losses by loading all transformers at a similar loading ratio
- Examining the loading of power lines and balance the loads to further balance neutral currents
- Taking actions and establishing legislations (such as criminalizing electricity theft) to reduce nontechnical losses caused by meter tampering
- Taking actions for supporting large power consumers to employ cogeneration technologies for excess power
- Providing fuel diversity with possible investments on power-production technologies based on different conventional fuel resources, such as coal, apart from petroleum
- Providing a public transportation structure to reduce personal automobile-based transportation fuel utilization

Thus, it is obvious that not only there are many challenges but also many opportunities for insular areas to more efficiently and effectively operate insular power systems that should be examined in detail.

1.2.2 Challenges and Opportunities

1.2.2.1 Generation Side

Most islands do not have any exploitable indigenous conventional energy sources (e.g., fossil fuels and hydrocarbons) and they are not connected with any other energy sources such as natural gas pipe networks. Hence, due to their location, most islands are highly dependent on external energy sources, mainly oil that is transported by ships at a very high cost. An apt example of this is the case of the Canary Islands, where 94% of the electricity generation depended on imported fuels in 2010 [24]. Similarly, the island of Cyprus uses heavy fuel oil and diesel for electricity generation almost exclusively [25].

To alleviate the negative economic and environmental effects of fossil fuel dependence, RESs are considered as a means of increasing the self-sufficiency of the insular power system. There are many autochthonous energy sources that may be used according to the specific needs and peculiarities of each insular system, in order to mitigate imported fuel dependence and to diversify the production mix.

This subsection discusses available green energy production options, their current applications in insular power systems, and future trends.

1.2.2.1.1 Solar Energy

Solar radiation may be used directly in order to produce energy either through the direct conversion of the solar energy to electrical [through photovoltaics (PVs)] or to provide energy for side applications (e.g., water heating, solar drying, and solar cooling systems). Essentially, most RESs (wind, ocean, and biomass energy) are indirect forms of solar energy [26]. Solar energy is available all around the world and may be considered as a suitable generation opportunity for remote islands. Naturally, according to the location of each island, generation potential differs from place to place. For example, all Greek islands are characterized by high solar irradiance, varying from 1500 to 1700 kWh/m². Furthermore, the annual variation of the solar potential coincides with the annual variation of the system's load demand [27], rendering it an appealing green energy option.

There are several ways to integrate PV modules. They can be installed on rooftops of buildings (several kW) or, if larger production is required, collective solar power plants (e.g., municipal), such as concentrated photovoltaic or concentrated solar power plants.

There are two major drawbacks concerning electricity generation using solar energy. First, it is still an expensive technology and subsidies are required in order to render it competitive. However, governments of leading countries take initiatives in order to reduce the relevant costs [28]. Second, as a result of its relatively low energy density, significant space is required in order to achieve adequate electricity production from solar potential. This can raise issues of limited space, as well as issues of environmental impacts. These are especially important parameters for islands, whose economies rely on tourism [24].

Several initiatives regarding the integration of renewable energy sources in islands consider vast investments in solar energy. In 2010, 112 MW solar PVs were installed in the Canary Islands, mainly concentrated in PV farms and a small portion in several premises. The Canary Islands Energy Plan aims to have 30% of the electricity produced by RES, mainly solar (160 MW) and wind (1025 MW) [24]. Most recent data (2013) regarding the RES share in the insular power system of Cyprus suggest that it stands only for the 1.2% of the total electricity production. Production of rooftop PV systems and PV parks amounts only for about 7.7% of this small share. Due to the commitment of Cyprus to comply with the European Union (EU) 2020 obligations, the country developed a program (National Renewable Energy Action Plan of Cyprus) that, among others, targets to install 192 MW of solar PVs and 75 MW of concentrated solar power by 2020 [25]. Furthermore, the island of Crete is expected to have installed 140 MW of solar energy by 2030 [29]. Already in 2010, 60 GWh were produced in Réunion Island by PV systems installed (80 MW), both stand-alone and interconnected [26]. Recently, the Hawaiian Islands of Oahu, Maui, and Kauai had significant solar resources, reaching a penetration of 10% in Oahu [30]. Finally, the example of the U.S. Virgin Islands, where PV installations are considered an economic way to reduce fossil fuel consumption, is very important to realize the potential of the solar energy, especially for the electrification of noninterconnected islands [31].

1.2.2.1.2   *Wind Energy*

It is estimated that world's wind resources have the capacity to generate 53,000 TWh of electrical energy/year, which accounts for three times the global energy consumption [32]. Because of the increasing interest (due to national and international targets) in reducing the carbon footprint, many countries have motivated the development of this type of RES during the last decade. Most remote islands are located in favorable locations with exploitable onshore and offshore wind potential. As a matter of fact, there are many examples of relevant investments in islands of different sizes across the globe. On the island of Rhodes, a Greek island, approximately 6% of the energy production comes from the 11.7 MW installed wind power [33]. The biggest Greek island, Crete, has an installed wind capacity of 105 MW, which accounts for 12.5% of the total capacity, and the 12 wind farms may instantaneously provide up to 39% of the total generated power (in 2006). However, the total licensed capacity exceeds 200 MW [29]. In 1998, Samso Island was chosen by the Danish Government as a demonstration of a 100% RES-electricity producing island. As evidence of this successful endeavor, Samso Island currently has 23 MW of offshore wind power generation and 11 MW of onshore wind power generation, although all its demand needs are produced by RES. The El Hierro Island in Spain is also subject to an ambitious target of becoming a 100% renewable energy island and currently wind power penetration reaches 30% [34]. Jeju Island in South Korea is also an example of high wind power generation penetration. There is the goal of installing 250 MW of wind power, and in 2010, 88 MW of wind power generation was already installed [35]. Finally, plans for increasing the RES penetration in many islands are set by other countries as well. The Canary Islands, the American Hawaiian Islands, and Germany's Pellworm Island are a few indicative examples.

The major challenge that needs to be addressed when planning to utilize wind energy to produce electricity is the intermittent and variable nature of this kind of production. Intermittency refers to the unavailability of wind for a considerably long period, whereas volatility describes the smaller, hourly oscillations of wind. Due to the reduced control over the wind energy production, some quality characteristics of the power system such as frequency and voltage may be affected. Also, to balance the lack of production during some periods, generation adequacy has to be reserved, leading the power system to a vulnerable state, especially in noninterconnected islands. Nevertheless, intermittency management is performed using sophisticated tools, and wind can be considered as a reliable source of energy in the long run [36]. Another concern that is associated with the operation of wind farms and can be important for small islands is the noise pollution near the wind park. This is the case for the island of Faial in the Azores, where the wind farm is shut down for several hours during the night in order to cease the disturbance of local housing [37]. Offshore wind farms seem to improve several issues related to the regular onshore wind farms such as the visual and noise impact and the efficiency because of higher speed and less turbulent wind streams [38].

1.2.2.1.3 Wave Energy

Over the past few decades, a great effort has been devoted to develop solar and wind energy generation. However, the idea of exploiting the high energy potential of the waves has drawn significant attention recently. Wave energy has been recognized as more reliable than solar and wind power because of its energy density (typically 2–3 kW/m^2 compared to 0.4–0.6 kW/m^2 of wind and 0.1–0.2 kW/m^2 of solar potential). Besides, wave energy offers several advantages in comparison with other RES. First of all, waves can travel long distances without losing much of their energy and as a result wave energy converters can generate power up to 90% of time compared to 20% and 30% for wind and solar converters, respectively. This fact renders wave energy as a credible and reliable energy source. Furthermore, there are also specific advantages that make it an appealing choice for the electrification of insular power systems. First, the resource is available in multiple locations (from shoreline to deep water). Second, the correlation between demand and resource (distance between generation and load) is high in islands. Finally, this type of renewable resource has less environmental impacts than other alternatives. This is particularly important for islands with limited space, particularly for islands whose economies rely on tourism.

The main challenge toward the large-scale integration of wave energy is the infant phase of relevant technologies. To provide high-quality power to the grid, both frequency and voltage have to be at appropriate levels. Together with the fact that the wave power is uncertain, special storage systems are needed to support the output of such plants. To exploit efficiently the wave power, especially in offshore applications where energy flux is greater, infrastructure has to withstand severe stress due to environmental conditions. Regardless of the attractive features of wave energy, the lack of funding poses a further hindrance toward the development of the required technology. Other renewable sources are more competitive since their respective markets are mature and, still, large investments are required to construct wave energy-harnessing plants.

TABLE 1.2

Wave Energy Potential in Different Islands

Location	Estimated Wave Energy	Reference
Madeira Archipelago	20 kW/m (Madeira), 25 kW/m (Porto Santo)	[39]
Hawaii	60 kW/m in winter time, 15–25 kW/m throughout the year	[40]
El Hierro	High energy: 25 kW/m, 200 MWh/m (annually)	[41]
	Low energy: 13 kW/m, 150 MWh/m (annually)	
Lanzarote	Over 30 kW/m, 270 MWh/m (annually). Seasonal variation: energetic winter and spring, mild summer	[42]
Menorca	8.9 kW/m, 78 MWh/m (annually)	[43]
Canary Islands	25 kW/m	[44]
Azores	Peak power 78 kW/m, minimum 37.4 kW/m	[45]
Vancouver Island	34.5 kW/m, 7 km from shore	[46]

Wave power varies with the location and the season and, therefore, the placing and the technology of such plants should be carefully considered. Also, the variability of the resource changes significantly according to the same parameters. Nevertheless, measurements regarding the wave energy potential near several islands around the world have provided promising results. Table 1.2 presents relevant information. Based on the previous review of wave power, it is evident that it emerges as an opportunity for ecological and smarter insular grids. Already, for 2015, the Canary Islands Energy Plan has established that 30% of the electricity generation should be supplied by RES, mainly wind and solar. This plan establishes, among others, that wave energy has to reach 50 MW of production [24].

1.2.2.1.4 *Geothermal Energy*

Geothermal energy comes from the natural heat under the crust of the Earth and is linked to earthquakes and volcanic activity; therefore, thermodynamic characteristics (e.g., temperature and enthalpy) of geothermal resources may significantly vary with the place. However, the available technology to exploit geothermal energy has evolved to adapt to the specific characteristics of a place's resources and it may be considered mature.

Geothermal energy has a potential to be used for electrical energy generation in insular power systems. For example, based on several studies, a 2.5 MW geothermal power plant is being considered to be installed on the Island of Pantelleria, Italy. It may be possible to achieve a production of 20,000 MWh/year, which stands for about the 46% of the island's consumption [47]. Also, the government of the Azores has launched an ambitious plan to achieve 75% of sustainable electricity production, on average, of all islands by 2018. Electricity of the Azores strategy, among others, includes additional investments in geothermal plants in the major islands (São Miguel) [48]. In February 2009, approximately 20.6% electricity was produced by geothermal energy on Hawaii Island, Big Island [49]. Significant geothermal power is installed in Jeju Island, South Korea, where 130.1 MW of geothermal energy contribute to the total RES generation by 15% [35].

Geothermal plants are characterized by high capital investments (exploration, drilling, and plant installation). However, operation and maintenance costs are low and thus, geothermal plants may serve as base load units [50]. Recently, several hybrid systems combining geothermal energy have attracted research interest in order to achieve a more efficient usage of this resource. Hybrid fossil-geothermal plants have been developed, but they have led to a compromise of the environmental benefits that stand-alone geothermal plants have to offer because of the increased greenhouse gases emissions. To maintain the advantage of sustainability, combining other renewable energy sources (e.g., solar and biomass) with geothermal energy production has been proposed [51].

1.2.2.1.5   Biomass

Biomass is considered as a mature and a promising form of RES. It offers the advantages of controllability and the possibility of creating liquid fuels and the flexibility to adapt to any raw material available locally (agricultural and livestock residuals, urban garbage, etc.). The major challenge is that the installation should be strategically considered near a populated area in order to guarantee the constant availability of the resource.

A recent study [29] indicates that based on agricultural residues (olive kernel, citrus fruits, etc.) and forestry material, Crete has the potential to develop a total of up to 60 MW of biomass power plants around the island. In the Hawaiian Islands, two biomass stations operate with installed capacity of 57 and 46.1 MW, respectively. Currently, two more are under construction and have a rated capacity of 24 and 6.7 MW, respectively. Especially, the 6.7 MW station that is being constructed on Kauai Island will provide 11% of the island's annual energy needs [52].

1.2.2.1.6   Small Hydroelectric Power Plants

Small hydroelectric power plants (SHEP) have small installed capacity (e.g., less than 10 MW in Europe) and do not generally use large reservoirs and, therefore, the interference with the environment is minimal. Such units exist in several islands. On the island of Crete, there exist two SHEPs, and a third one is being considered to be built [29]. In Faial (Azores), a 320 kW hydropower unit exists [37]. On El Hierro Island, a 9.9 MW of hydropower capacity is installed with pumping capability. In this way, excessive wind power is used to pump water in the upper reservoir in order to achieve energy storage and cope with the intermittency and variability of wind power generation [34].

1.2.2.2 Energy Storage Systems

On the way to the *smartification* of electrical power system, many factors should be taken into account. In this regard, a specific part of the smart grid definition of Department of Energy (DoE) should be given importance: "All generation and storage options should be considered in a smart grid structure" [53]. Thus, the role of energy storage in new smart and sustainable power grid structures cannot be neglected.

There are many roles storage systems can perform in a power grid. Fast storage units that offer short-duration (seconds or minutes) storage option can be used for sustaining power quality, including frequency regulation, whereas middle-duration (hours) storage systems can be used for load leveling, peak shaving, and actions for load pattern reshaping. On the other hand, storage units offering long duration (days) of storage are generally used for providing autonomy to the system also by leveling the output of intermittent RES with a considerable penetration within the relevant grid structure.

Energy storage technologies can be either electrical or thermal. An electrical input–output is available for electrical energy storage systems; similarly, there is a thermal input–output for thermal systems. Electrical energy storage systems can be in the form of electrochemical systems (battery), kinetic energy storage systems (flywheel), or potential energy storage system (pumped-hydro and compressed air). Recently, large-scale storage options such as compressed air energy storage and pumped-hydro-based storage systems have been utilized with a growing ratio. Such energy storage options can be utilized for storing hundreds of MWh of energy. However, the geographical dependence of such systems is a major drawback especially for insular areas as a suitable underground storage space and an inclined land with multiple water storage options in different levels are necessary for compressed gas and pumped-hydro systems, respectively. Large-scale new types of battery units are also considered as an option especially for smoothing the fluctuating power output of large power wind turbines. A 245 MWh sodium–sulfur (NaS) battery system is available in Japan for ensuring the dispatch of a 51 MW wind farm [54]. New-generation lithium-ion batteries for electric vehicles (EVs) will also be a significant option for energy storage, particularly when the penetration of EVs on road transport reaches a considerable level.

As another storage option, the flywheel technology has already been employed in different areas, especially for rapid responses to frequency deviations caused by the fluctuating nature of nondispatchable renewable energy sources. With their significantly high-power density, superconducting magnetic energy storage (SMES) and ultracapacitors can also play a similar role in short-period-rapid-response requirements. As a promising technology the hydrogen storage systems composed of electrolyzer-large scale hydrogen tank-fuel cell (FC) combinations have been given importance in the research studies realized, especially on the cost reduction and reliability improvement of the mentioned systems. The generated hydrogen can also

TABLE 1.3

Ranges for Concise Specifications of Storage Technologies

	Very Low–Very Weak	Low–Weak	Medium	High	Very High
Energy density (Wh/kg)	$0.01 < X < 10$	$10 < X < 30$	$30 < X < 50$	$50 < X < 150$	$150 < X$
Power density (W/kg)	$10 < X < 25$	$25 < X < 50$	$50 < X < 150$	$150 < X < 1000$	$1000 < X$
Storage duration	Seconds to minutes	Seconds to hours	Minutes to hours	Minutes to days	Hours to months
Self-discharge/day	$X < 0.1\%$	$0.1\% < X < 1\%$	$1\% < X < 10\%$	$10\% < X < 30\%$	$30\% < X$
Capital cost (€/kW)	–	$100 < X < 600$	$600 < X < 1500$	$1500 < X$	–
Charge-discharge cycle efficiency	–	$X < 60\%$	$60\% < X < 90\%$	$90\% < X$	–
Lifetime (years)	$X < 1$	$1 < X < 5$	$5 < X < 15$	$15 < X < 50$	$50 < X$
Technological maturity	Developing –	Developing +	Developed –	Developed +	Mature

Source: Rious, V., and Y. Perez, *Renew. Sust. Energ. Rev.*, 29, 754–65, 2014.

be utilized directly for powering FC-based hydrogen vehicles, which has also been a topic of research for more than a decade. A concise summary of the specifications of main energy storage systems is given in Tables 1.3 and 1.4 [55].

The role of energy storage systems in insular power grids is gaining more importance especially since RES investments have become more important toward reducing dependence on imported fuels to ensure sustainable growth of insular areas. RESs are highly dependent on natural conditions. Thus, the energy produced by these resources can significantly vary monthly, daily, or even instantly. This leads to the fact that the energy produced may not exactly match with the energy demanded. In order to supply the load demand in every condition, energy storage units show great potential. The excess energy produced by alternative resources is transferred to energy storage units, and this stored energy is used to supply the load demand when the main sources are nonexistent or are not sufficient. Besides, the sudden unavailability of RES units especially during high RES penetration may force the conventional units to reduce their power ratings below an allowable lower limit [55]. Here, storage systems can effectively be employed to meet this discrepancy in production and consumption to ensure the safe operation of main power units. This issue is especially significant for small systems such as insular areas where the number of available generators is tightly limited, which offers nearly zero flexibility. Energy storage units with

TABLE 1.4

Concise Specifications of Different Major Storage Technologies

	Energy Density	Power Density	Storage Duration	Self-Discharge/Day	Capital Cost	Charge-Discharge Cycle Efficiency	Lifetime	Technological Maturity
Pumped-hydro	Very weak	–	Very long	Very weak	High	Medium	Very high	Mature
Compressed air energy storage	Medium	–	Very long	Weak	Medium	Medium	High	Developed +
Battery (lead-acid)	Medium	High	Long	Weak	Weak	Medium	Medium	Mature
Battery (NaS)	High	High	Short	High	High	Medium	Medium	Developed +
Battery (Li-ion)	High	High	Long	Weak	High	High	Medium	Developed +
Hydrogen	Very high	Very high	Very long	Very weak	High	Weak	Medium	Developing +
SMES	Very weak	Very high	Medium	High	Weak	High	High	Developed –
Flywheel	Weak	Very high	Very short	Very high	Weak	High	Medium	Developed +
Ultracapacitor	Very weak	Very high	Short	Very high	Weak	Medium	Very high	Developed +

Source: Rious, V., and Y. Perez, *Renew. Sust. Energ. Rev.*, 29, 754–65, 2014.

TABLE 1.5

Different Applied Energy Storage Systems in Sample Insular Areas

Location	Type	Rating	References
Bella Coola, Canada	Hydrogen	100 kW	[9]
	Flow battery	125 kW	
Bonaire, Venezuela	Nickel-based battery	3 MW	[9]
King Island, Tasmania, Australia	Vanadium redox battery	400 kW	[9,58]
Metlakatla, Alaska	Lead-acid battery	1 MW	[9]
Ramea Island, Canada	Hydrogen	250 kW	[9]
El Hierro, Canary Islands	Pumped-hydro	Hydro 11 MW, Pumping 6 MW	[59]
Gran Canaria, Canary Islands	Li-ion battery	1 MW/3 MWh	[59]
La Gomera, Canary Islands	Flywheel	0.5 MW	[59]
Ventotene	Li-ion battery	300 kW/600 kWh	[59]

different characteristics come into action at this point to add different levels of flexibility in required points of the low-inertia insular power system.

Especially, research and examination of energy storage units for stand-alone applications is also significant [56]. With the usage of energy storage systems in such applications, the need for energy transmission lines that have a high investment cost is highly reduced, which is an important issue, especially for the most secluded areas of insular grids [57]. From this point of view, the different applied energy storage systems in sample insular areas are given in Table 1.5.

1.2.2.3 Operational and System Planning Aspects

The abundance of RES in combination with the high fossil fuel prices has led to a significant penetration level of such technologies in order to produce electricity in insular networks. Although leading RES technologies such as wind and solar are mature and are able to compete with conventional power plants, they are linked with variability due to their intrinsically stochastic nature. Therefore, the integration of such levels of nondispatchable resources in relatively small-sized insular systems poses operational and economic challenges that need to be addressed. The magnitude of the problem depends on the penetration of RES in the production mix, whereas its mitigation is reflected on the *flexibility* of the power system.

There are several issues that need to be considered in the operational practice of insular networks. First, production of RESs such as wind and solar depends on wind and irradiation values, which in turn fluctuate according to climatic changes and spatial characteristics. As a result, instantaneous, seasonal, and yearly fluctuations affect the generation output. In the case of wind power production, power quality is also affected by these fluctuations (especially by very short-term turbulences) depending on network impedance at the point of common coupling and the type of wind turbine used.

Variable RES production affects the operation of conventional generators [60–62]. Under high penetration, conventional units are likely to operate in a suboptimal commitment. Fluctuation in RES output power leads to the cycling of conventional units and shortens the life of their turbines, while causing increased generation costs. Emission reduction potential is also suppressed. Reserve needs are also increasing with the penetration of RES and especially ramping requirements (load following) because of the uncorrelated variation of wind generation and load demand. A case study for the insular power system of Cyprus [63] concluded that the available reserve is not adequate to balance the real-time fluctuations of wind, whereas higher penetration of wind power generation would further constrain the downward-ramping capability of the island's system due to the part loading of generators.

An indispensable tool in the decision-making process taking place in insular power systems is the accurate modeling of uncertainty factors using advanced forecasting and scenario-generation techniques. It has been reported [64] that commercially available modern forecasting methods are able to provide up to 80% of the benefits perfect predictability would have for wind power integration. Accurate forecasting may improve the results of unit commitment and minimize operational costs. Besides, sophisticated scheduling tools should be employed in order to accommodate uncertainty at different timescales. A multistep approach is proposed for the short-term operation of insular networks under high RES penetration [65]. It comprises a day-ahead scheduling model, an intraday dispatch scheduling model, and a real-time economic dispatch model. It also considers through appropriate forecasts and scenarios several uncertainty sources such as wind and solar power production, load variation, and unit availability. This model is developed based on the insular power system of Crete.

Apart from the aforementioned solutions, the demand-side and electrical energy storage [36,66] technologies offer a great opportunity for increasing the flexibility of insular power systems. Flexibility will enable the most efficient utilization of the RES production that may be curtailed in an inflexible power system to maintain the generation–demand balance (or to avoid congesting transmission lines) and thus decreasing the environmental and economic potential of wind power production. Given that investments in fossil fuel-fired power plants are neither motivated by policy makers and governments nor cost-effective because of the increased fuel prices in islands, it is not possible to increase flexibility by generation-side resources. A study on the power system of Cyprus [63] suggested that in order to accommodate the fluctuations of the wind power generation, load shedding and extensive wind power curtailments are necessary. The potential of the active demand-side participation in several timescales of the grid's operation is illustrated in Figure 1.1. A study concerning the demand response (DR) integration has

FIGURE 1.1
Potential of active demand-side participation into the operation of a power system.

been conducted for the insular power system of Gran Canaria in order to assess the benefits on the operation of the power system and the cost savings [67]. Also, a new type of load that can act as a means of energy storage integration, namely, the EV, is considered as a factor of flexibility in power systems. A study concerning the island of Flores (Azores) has concluded that through appropriate control schemes, the current distribution system can support the penetration of EVs that, apart from transportation, not only provides services to the system in order to accommodate the interruptible RES but also reduces distribution system losses [68].

When planning the integration of RES in insular power systems, system operators should consider the following aspects:

- RES-based power plants may either reduce or increase the transmission line capacity requirements (and subsequently active power losses) according to their location and their distance from the load.
- Voltage quality may be improved because of the capability of several RES technologies [such as doubly fed induction generators (DFIG) wind turbines] to control their reactive power. As a result, connecting renewable energy sources to weak parts of a power system may contribute to the voltage stability of the system.
- RES installations should be as much geographically dispersed as possible in order to avoid further requirements in transmission capacity.
- Demand of loads that contributes to peaks (such as air-conditioning) may be correlated with the peak production of several RES (such as solar energy). This is a fact that should be recognized during the sizing and connection procedure.

1.2.2.4 Reliability

Challenges concerning the reliability of insular power systems are a result of both their structure and the increasing penetration of RES.

Insular power systems have low system inertia and, therefore, are highly sensitive to frequency deviations. Renewable energy sources such as PVs that use an inverter interface further reduce the inertia of the system. Furthermore, lower inertia implies that after a disturbance in the network, large frequency deviations are expected since there is not enough active power to mitigate the rate of change of frequency. The fact that insular power systems, especially in small islands, depend on a few conventional fuel-fired generators further jeopardizes the security of the system. Generators tend to be large in comparison with the system size and during off-peak hours, a single generator may represent a large proportion of the system's total generation. It is then evident that the sudden loss of this unit will lead to significant frequency deviations [69].

Another problem that is caused by increasing renewable penetration in combination with the typical insular power system described above is the inability of conventional generators to reduce their output when nondispatchable RES production is high. Typically, diesel-fired generators have a minimum output limit of 30% of their installed capacity. Forcing a load-following unit to shutdown in order to retain the generation and demand balance may compromise the longer-term reliability of the power system. In order to avoid such deficit in system's inertia, RES generation is normally curtailed in order to avoid switching off synchronous generators at the expense of economic losses [70].

The penetration of RES may also affect voltage stability because power sources such as fixed-speed induction wind turbines and PV converters have limited reactive power control. Also, high system loading and inductive loads such as air-conditioning have a serious impact on the voltage stability.

Surely, operational reserves (spinning or nonspinning) are required in insular power systems as well. Apart from frequency regulation, load-forecasting error, sudden changes (ramps) in the production of RES units, and equipment forced or scheduled outages also need to be confronted. To deal with these issues adequate generation or demand-side capacity should be kept.

In order to improve both frequency and voltage stability, the following measures can be taken:

1. The reduced inertia in isolated power systems with inverter-interfaced RES can be replenished by energy storage that keeps the power balance [71]. Grid code changes have been proposed and many different applications of energy storage systems in island systems already exist [6,72,73].

2. Demand-side resources can be used in order to provide frequency stabilization reserves. According to the literature, there are two ways of controlling consumed power [74]. Either there can be a continuous control, if the load is supplied by power electronics-based power supply, or the loads can be switched on and off for specific periods. The first group of loads is rare and expensive, whereas the second is linked to the minimal inconvenience of consumers. Heating, ventilation, air-conditioning, dryers, water heaters, and pumps are loads of this group. Using frequency relays, they can be switched on or off according to the frequency shifts. This is a particularly promising opportunity for islands that primarily serve residential loads.

3. A relatively new concept that may prove indispensable is the inertial control of wind turbines. *Virtual* wind inertia is created through utilizing the kinetic energy stored in the rotating mass of wind turbines in order to respond to frequency drops. Variable-speed wind turbines may accept wide speed variations and as a result, the inertial response of wind turbines is greater (more kinetic energy can be transformed into electrical energy) than the regular synchronous generators given the same inertia value [70,75,77]. Power control of wind farms with respect to frequency can complement the frequency control schemes of the insular power system.

4. Solar and wind power plants have the ability of controlling their reactive power. DFIG and permanent magnet synchronous generators are capable of injecting or consuming reactive power. Power electronics that interface solar power plants have the same ability. Therefore, regulating voltage at the common connection point is possible [75]. Connecting wind farms and solar power plants at weak network points can thus improve the overall stability of the system.

Finally, regarding transient and small-signal stability, no major concerns are expressed for insular systems under high penetration of RES. As it has already been stated, transient stability from the system's perspective is linked with the type of controllers and the dispersion of synchronous generators and, therefore, it is not possible to draw general conclusions since each single insular power system is a unique case. However, several studies indicate that moderate presence of RES (~40%) does not seem to have severe impacts on transient stability [77]. Besides, given a specific insular network, one could suggest a maximum penetration limit for RES in order to avoid affecting transient stability. Regarding small-signal stability, it is generally not considered as a concern for insular power grids [19].

Before concluding this subsection, it is worth mentioning that the concepts of security and reliability may as well have a localized component because of geographical or other peculiarities of the specific island. For example, we will refer to the interesting case of the tropical island of Guam that has an area of 542 km^2 and 157,000 residents and is a major tourist destination, attracting 1.5 million visitors/year. Like in many other tropical islands, there is an increased need for maintenance due to corrosion, rapid growth of tropical vegetation that threatens distribution lines, and storms. However, what makes the case of Guam different is the increased population of the brown tree snakes that are the reason for at least 1658 power outages from 1978 to 1997 and most recently approximately 200 outages/year. Outages span the entire island and last for eight or more hours and cost over three million dollars. Excluding repair costs, the damage of electrical equipment, and lost revenues, the damage on the economy of the island exceeds 4.5 million dollars/year over a seven-year period [78]. This example confirms the fact that each insular power system's operational requirements are directly linked to the geographical and other local factors.

1.2.2.5 Demand-Side Management

The demand side in power systems was formerly referred to as *consumption level* due to the fact that the consumers were just consuming energy without power-producing facilities. However, as the applications of small-scale grid tied renewable RES (rooftop PVs, small wind turbines, pico-hydro systems for areas located near a small river, etc.) recently increased, the consumers have gained the chance of also selling energy back to the grid in low demand–high production periods. For instance, PV-based power production is at the highest level during noon times when generally the consumption is at the lowest due to the fact that most people are out of their home (working hours). Besides, on-site energy storage facilities and additional new-generation consumer appliances such as EVs have also changed the demand-side structure and made the demand side more flexible, able to respond to power system requirements.

The demand side of a power system surely includes different kinds of end users. These end users can be listed under three main types: residential, commercial, and industrial. The demand-side management actions that can be taken to change the load pattern considering the requirements can be basically classified into two groups: energy efficiency actions and demand-side participation programs.

Energy efficiency actions mainly focus on reducing the total energy consumed for realizing a particular amount of work. One of the widely known energy efficiency actions is to change the old light bulbs with new-generation energy-efficient bulbs to obtain a particular lighting level with a reduced amount of energy consumption. Many different actions can be taken in order to improve the efficiency of energy usage in all areas of a power system from production to consumption. Energy policies on insular areas within the last decade give specific importance to energy efficiency actions. First, it is proposed to make investments in new generating facilities instead of old-technology-based pollution-generating units that contribute to imported fuel consumption significantly, which is the most critical issue for the sustainable growth of insular areas. Apart from the generation level, investments in transmission and distribution levels such as proper sizing of assets and reduction of reactive power flows, are also proposed in order to improve system efficiency similar to such investments on the mainland. There are also policies for energy efficiency actions in the demand side to reduce consumption levels, especially for common areas. For example, the replacement of street light bulbs with more efficient ones was strongly recommended for many islands in the U.S. territories to reduce the impact of such a consumption component in a detailed report on recommendations for insular areas [4].

Demand-side participation programs generally focus on shifting the energy usage to off-peak periods to reduce the need of extra electrical infrastructure investment requirements to cover the peak load conditions. Demand-side programs have successfully been applied in industrial area for the past few decades, but new technologies also provide opportunities to apply such programs in other end-user areas. Individual residential or industrial end users have their own appliances/devices that can be classified under three main categories [79]:

1. Thermostatically controllable appliances
2. Nonthermostatically controllable appliances
3. Noncontrollable appliances

Controllable appliances provide the leading opportunity for new-generation demand-side programs. The thermostatically controllable devices include heating and ventilation air conditioners and water heaters for residential end users that can be switched from on to off position and vice versa and curtailed during peak periods of power demand, also considering the comfort settings of the residents. Besides, nonthermostatically controllable appliances can be controlled by positioning the appliance to work consuming a higher or lower power value without just totally opening or closing the appliance.

Washers and dryers for residential areas may be given as examples for such kinds of appliances. On the other hand, noncontrollable appliances are termed *noncontrollable* due to the fact that these appliances, such as televisions and lights, have a comfort-lowering impact on consumers if controlled without the permission of the end user. It is well known that there is a chance of automatic dimming for the lightening of an end-user unit, but this is considered in the concept of energy efficiency investments.

As a new type of end-user appliance/load, EVs have recently gained more importance as the electrification of the transportation sector is a major fossil fuel consumer and is a hot topic of environmental sustainability. EVs have a different structure with challenges/opportunities that should be examined in detail. As a load, the energy needs of EVs can reach to the levels of new power plant installation requirements. The recommended charging level of a Chevy Volt as a small-sized EV is 3.3 kW [80], which can even exceed the total installed power of many individual homes in an insular area.

Thus, special importance should be given to EV charging with smart solutions, especially with the consideration of high-level EV penetration. On the contrary to the charging power challenge of EVs, the opportunity of using EV as a mobile storage unit with the vehicle-to-grid concept may provide some opportunities, especially for shaping the demand without adding stress to transformers and, accordingly, to the bulk power system. Analysis of EVs is expected to have an important place in a better deregulation of electric power systems.

Demand-side management activities especially in the context of new-generation smart grid vision mainly lies on DR strategies employed in the end-user local areas. DR is a term defined as "Changes in electric usage by end-use customers from their normal consumption patterns in response to changes in the price of electricity over time, or to incentive payments designed to induce lower electricity use at times of high wholesale market prices or when system reliability is jeopardized" by DoE and is composed of incentive-based and price-based programs (time-of-use, critical peak pricing, dynamic pricing, and day-ahead pricing) [81].

The utilization of DR strategies is considered as mature for industrial customers, but this is a relatively new concept for employing in residential areas responsible for nearly 40% of energy consumption in the world [83]. There are many supporting devices/technologies for DR activities in residential areas. Especially, home energy management systems (HEMs) and smart meters have the leading role for effectively applying DR strategies.

Smart meters are new-generation electronic meters that have the capability of a two-way communication with the utility/system operator. For DR activities, smart meters can receive signals from the system operator, such as the maximum allowable level of power in a certain period (especially reducing the possible foreseen stress on the relevant transformer) or price signals determined in a dynamic way, and can share this information with HEMs. HEMs also receive information signals from smart appliances and/or smart plugs, if available, including state of the appliance and power consumption. Also, the power production information of available renewable generation facilities is received by HEMs.

For insular areas that have limited industrial facilities, such DR applications in residential and possibly in commercial (e.g., hotels for more touristic islands) areas may come into prominence in order to improve the economic and technical sustainability of insular power systems. Such flexibility in the demand side is likely to overcome the high inertial structure of the generation system in insular areas. In this regard, many projects and investments have been ongoing all over the world to provide an enabling infrastructure for DR activities in insular areas. For the Caribbean Islands, which mostly depend on imported diesel fuel for energy production and has a system loss of 27%, it was announced that 28,000 smart meters will be installed in a close future to provide a basis for future smart grid activities by the Caribbean Utilities Company, Ltd [83]. Utility Hawaiian Electric, the local operator of Hawaiian Insular Power System, declared to have 5200 customers with smart grid facilities by smart metering. Other islands in the U.S. Virgin Islands also have a similar plan of improving the number of smart-metering facilities with different deadlines between 2016 and 2018 [84]. Another project in the EU is conducted with a similar aim for Bornholm Island in Denmark [85], although many different insular areas are considered in this regard in a different research project [8]. There are many activities, incentives, and policies in different regions of the world for insular areas for reducing the dependence on imported fuel and losses in the system, where demand-side actions play, and will surely play, a leading role in the efficient and effective balancing of the production and consumption.

1.2.2.6   *Other Innovative Approaches and Possible Future Developments*

The discussion conducted in the previous sections has highlighted several opportunities to address the issues of reliability, cost-effectiveness, and the efficient operation of insular power systems. Complimentarily, there exist several other innovative approaches that can further boost the goal of sustainability in all aspects of insular power system operation.

Desalination process employed in many islands in order to fulfil the needs of high-quality water may be embodied in a general operational scheduling of the islands power-consumption portfolio. For example, excess of wind power generation, which would be in other case spilled, can be used to desalinate seawater [86]. Waste management in insular areas can also be turned into an opportunity by producing energy from waste for covering self-needs of the waste treatment procedure. Furthermore, locally available resources such as plantation can be used in order to produce biofuels that, together with the opportunity EVs offer, will reduce the dependence on imported fuels of the transportation section. Moreover, similar to the areas with low solar potential but high solar energy use such as Germany, all the available solar energy can be procured also in insular areas with high or low solar potential by the combination of all different civil, architectural, engineering methods within the concept of "energy neutralization."

The possible future developments should not just focus on the technical aspect of addressing the problems of insular power systems, but also should aim toward reshaping the way governments, energy producers, and end users consider sustainability as a keystone of their daily life.

1.3   Conclusions

Insular areas have naturally different characteristics compared to mainland a fact that is also reflected to power system structures. The insular power systems that generally rely on nearly 100% imported and fossil-based fuels lack sufficient characteristics for the sustainability of insular economies and natural life. Thus, the mitigation of the dependence on imported and polluting fuels is pivotal. A few energy efficiency measures, policies, renovation-based new technological investments, demand-side strategies, and pivotally higher levels of RES integration are proposed as viable opportunities. However, challenges such as the geographical limitations of insular areas, and the technical limitations of small-sized grid structures with low inertia, for example, insular grids, are all barriers to overcome in order to seize the opportunities for ensuring a sustainable insular power system. Many policies and reports have been announced, specifically focusing on insular power systems. Besides, many largely funded R&D projects have just been finalized or still continue. All of these efforts have resulted in specific applicable solutions and also have shown that some pre-proposed possible solutions are not applicable or not feasible to apply in real life. In this regard, a worldwide cooperation between insular areas, policy makers, and researchers is strongly required to share valuable experiences to overcome such drawbacks of insular structures. Agencies such as the United Nations and the EU that are active in every region of the world may take responsibility in this regard to ensure the sustainable continuity of insular life.

References

1. Duic, N., and M. G. Carvalho. 2004. Increasing renewable energy sources in island energy supply: Case study porto santo. *Renewable and Sustainable Energy Reviews* 8:383–99.
2. The Treaty of Amsterdam. 1997. *The Treaty of Amsterdam: Amending the Treaty on European Union, the Treaties Establishing the European Communities and Certain Related Acts.* Amsterdam, the Netherlands.
3. Warmburg, B. M. 2006. *Sustainable Energy for Islands: Opportunities versus Constraints of a 100% Renewable Electricity Systems.* MSc thesis, The International Institute for Industrial Environmental Economics, Lund University, Lund, Sweden.
4. Pacific Power Association. 2006. *United States of America Insular Areas Energy Assessment Report.* Pacific Power Association, Suva, Fiji.
5. Chen, F., N. Duic, L. M. Alves, and M. G. Carvalho. 2007. Renew islands-renewable energy solutions for islands. *Renewable and Sustainable Energy Reviews* 11:1888–902.
6. Bizuayehu, A. W., P. Medina, J. P. S. Catalão, E. M. G. Rodrigues, and J. Contreras. 2014. Analysis of electrical energy storage technologies' state-of-the-art and applications on islanded grid systems. *Proceedings of the IEEE PES T&D Conference and Exposition,* Chicago, IL.
7. Rious, V., and Y. Perez. 2012. *What Type(s) of Support Schemes for Storage in Island Power Systems?* EUI Working Papers RSCAS 2012/70, Robert Schuman Centre for Advanced Studies, Loyola de Palacio Programme on Energy Policy, San Domenico, Italy.
8. Smart and Sustainable Insular Electricity Grids Under Large-Scale Renewable Integration (SINGULAR). 2014. http://www.singular-fp7.eu/home/ (accessed September 15, 2014).
9. International Renewable Energy Agency (IRENA). 2012. *Electricity Storage and Renewables for Island Power: A Guide for Decision Makers.* IRENA, Abu Dhabi, UAE.
10. Perez, Y., and F. J. R. Real. 2008. How to make a European integrated market in small and isolated electricity systems? The case of the Canary Islands. *Energy Policy* 36:4159–67.

11. Conejo, A. J., M. Carrión, and J. M. Morales. 2010. *Decision Making Under Uncertainty in Electricity Markets*. Springer, New York.

12. Addressing barriers to STORage technologies for increasing the penetration of intermittent energy sources (STORIES). Deliverable 3.4. 2014. http://www.storiesproject.eu/ (accessed September 15, 2014).

13. Enemalta. 2014. http://www.enemalta.com.mt (accessed September 15, 2014).

14. Electricity Authority of Cyprus. 2014. https://www.eac.com.cy/EN/ (accessed September 15, 2014).

15. Energy Storage Association, 2015. *Sodium Sulfur (NaS) Batteries*. http://energystorage.org/energy-storage/technologies/sodium-sulfur-nas-batteries (accessed February 16, 2015).

16. Conrad, M. D., S. Esterly, T. Bodell, and T. Jones. 2013. *American Samoa: Energy Strategies*, The National Renewable Energy Laboratory (NREL) Report for U.S. Department of the Interior's Office of Insular Affairs (OIA). NREL, Golden, CO.

17. Sanseverino, E. R., R. R. Sanseverino, S. Favuzza, and V. Vaccaro. 2014. Near zero energy islands in the Meditteranean: Supporting policies and local obstacles. *Energy Policy* 66:592–602.

18. Kundur, P., J. Paserba, V. Ajjarapu et al. 2004. Definition and classification of power system stability IEEE/CIGRE joint task force on stability terms and definitions. *IEEE Transactions on Power Systems* 19:1387–401.

19. Potamianakis, E. G., and C. D. Vournas. 2003. Modeling and simulation of small hybrid power systems. *Proceedings of the IEEE Power Tech Conference*, Bologna, Italy.

20. Rebours, Y. G., D. S. Kirschen, M. Trotignon, and S. Rossignol. 2007. A survey of frequency and voltage control ancillary services—Part I: Technical features. *IEEE Transactions on Power Systems* 22:350–7.

21. Rebours, Y. G., D. S. Kirschen, M. Trotignon, and S. Rossignol. 2007. A survey of frequency and voltage control ancillary services—Part II: Economic Features. *IEEE Transactions on Power Systems* 22:358–66.

22. World Energy Council. 2007. *Deciding the Future: Energy Policy Scenarios to 2050*. World Energy Council, London, UK.

23. Xydis, G. 2013. Comparison study between a renewable energy supply system and a supergrid for achieving 100% from renewable energy sources in islands. *Electrical Power and Energy Systems* 46:198–210.

24. Schallenberg-Rodríguez, J. 2013. Photovoltaic techno-economical potential on roofs in regions and islands: The case of the Canary Islands. Methodological review and methodology proposal. *Renewable and Sustainable Energy Reviews* 20:219–39.

25. Fokaides, P. A., and A. Kylili. 2014. Grid parity in insular energy systems: The case of photovoltaics (PV) in Cyprus. *Energy Policy* 65:223–8.

26. Praene, J. P., M. David, F. Sinama, D. Morau, and O. Marc. 2012. Renewable energy: Progressing towards a net zero energy island, the case of Reunion Island. *Renewable and Sustainable Energy Reviews* 16:426–42.

27. Kaldellis, J. K., D. Zafirakis, and E. Kondili. 2010. Optimum sizing of photovoltaic-energy storage systems for autonomous small islands. *International Journal of Electrical Power & Energy Systems* 32:24–36.

28. Feldman, D., G. Barbose, R. Margolis, R. Wiser, N. Darghouth, and A. Goodrich. 2012. *Photovoltaic (PV) Pricing Trends: Historical, Recent, and Near-Term Projections*. Technical Report DOE/GO-102012–3839. NREL, Golden, CO.

29. Giatrakos, G. P., T. D. Tsoutsos, and N. Zografakis. 2009. Sustainable power planning for the island of Crete. *Energy Policy* 37:1222–38.

30. Wesoff, E. 2014. *How Much Solar Can HECO and Oahu's Grid Really Handle? Testing the Limits of a Large Island's Electrical Grid with 10 Percent PV Penetration.* Greentech Media. http://www.greentechmedia.com/ (accessed September 15 2014).

31. Miller, M., P. Voss, A. Warren, I. Baring-Gould, and M. Conrad. 2012. Strategies for international cooperation in support of energy development in Pacific Island Nations. Technical Report NREL/TP-6A20–53188. NREL, Golden, CO.

32. Kumar, A., and S. Prasad. 2010. Examining wind quality and wind power prospects on Fiji Islands. *Renewable Energy* 35:536–40.

33. Katsaprakakis, D. A., and D. G. Christakis. 2014. Seawater pumped storage systems and offshore wind parks in islands with low onshore wind potential. A fundamental case study. *Energy* 66:470–86.

34. Wu, Y., G. Han, and C. Lee. 2013. Planning 10 onshore wind farms with corresponding interconnection network and power system analysis for low-carbon-island development on Penghu Island, Taiwan. *Renewable and Sustainable Energy Reviews* 19:531–40.

35. Kim, E., J. Kim, S. Kim, J. Choi, K. Y. Lee, and H. Kim. 2012. Impact analysis of wind farms in the Jeju Island power system. *IEEE Systems Journal* 6:134–9.

36. Rahimi, E., A. Rabiee, J. Aghaei, K. M. Muttaqi, and A. E. Nezhad. 2013. On the management of wind power intermittency. *Renewable and Sustainable Energy Reviews* 28:643–53.

37. Suomalainen, K., C. Silva, P. Ferrão, and S. Connors. 2013. Wind power design in isolated energy systems: Impacts of daily wind patterns. *Applied Energy* 101:533–40.

38. Veigas, M., R. Carballo, and G. Iglesias. 2014. Wave and offshore wind energy on an island. *Energy for Sustainable Development* 22:57–65.

39. Rusu, E., and C. G. Soares. 2012. Wave energy pattern around the Madeira Islands. *Energy* 45:771–85.

40. Stopa, J. E., K. F. Cheung, and Y. L. Chen. 2011. Assessment of wave energy resources in Hawaii. *Renewable Energy* 36:554–67.

41. Iglesias, G., and R. Carballo. 2011. Wave resource in El Hierro—An island towards self-sufficiency. *Renewable Energy* 36:689–98.

42. Sierra, J. P., D. González-Marco, J. Sospedra, X. Gironella, C. Mösso, and A. Sánchez-Arcilla. 2013. Wave energy resource assessment in Lanzarote (Spain). *Renewable Energy* 55:480–9.

43. Sierra, J. P., C. Mösso, and D. González-Marco. 2014. Wave energy resource assessment in Menorca (Spain). *Renewable Energy* 71:51–60.

44. Gonçalves, M., P. Martinho, and C. G. Soares. 2014. Assessment of wave energy in the Canary Islands. *Renewable Energy* 68:774–84.

45. Rusu, L., C. G. Soares. 2012. Wave energy assessments in the Azores islands. *Renewable Energy* 45:183–96.

46. Robertson, B. R. D., C. E. Hiles, and B. J. Buckham. 2014. Characterizing the near shore wave energy resource on the west coast of Vancouver Island, Canada. *Renewable Energy* 71:665–78.

47. Cosentino, V., S. Favuzza, G. Graditi et al. 2012. Smart renewable generation for an islanded system. Technical and economic issues of future scenarios. *Energy* 39:196–204.

48. Pina, A., P. Baptista, C. Silva, and P. Ferrão. 2014. Energy reuction potential from the shift to electric vehicles: The Flores island case study. *Energy Policy* 67:37–47.

49. Matsuura, M. 2009. Island Breezes: Renewable energy integration from a Hawaiian perspective. *IEEE Power & Energy Magazine* 7:59–64.
50. Camus, C., and T. Farias. 2012. The electric vehicles as a mean to reduce CO_2 emissions and energy costs in isolated regions. The São Miguel (Azores) case study. *Energy Policy* 43:154–65.
51. DiPippo, R. 2015. Geothermal power plants: Evolution and performance assessments. *Geothermics* 53:291–307.
52. Biomass Magazine. 2014. http://biomassmagazine.com/ (accessed September 15, 2014).
53. U. S. Department of Energy. 2009. *Smart Grid System Report*. July. p. vi. http://energy.gov/sites/prod/files/2009%20Smart%20Grid%20System%20Report.pdf (accessed February 16, 2015).
54. Energy Storage Association. 2015. *Sodium Sulfur (NaS) Batteries*. http://energystorage.org/energy-storage/technologies/sodium-sulfur-nas-batteries (accessed February 16, 2015).
55. Rious, V., and Y. Perez. 2014. Review of supporting scheme for island power system storage. *Renewable and Sustainable Energy Reviews* 29:754–65.
56. Hadjipaschalis, I., A. Poullikkas, and V. Efthimiou. 2009. Overview of current and future energy storage technologies for electric power applications. *Renewable and Sustainable Energy Reviews* 13:1513–22.
57. Ibrahim, H., A. Ilinca, and J. Perron. 2008. Energy storage systems— Characteristics and comparisons. *Renewable and Sustainable Energy Reviews* 12:1221–50.
58. King Island Renewable Energy Integration Project. 2014. http://www.kingislandrenewableenergy.com.au/ (accessed September 15, 2014).
59. ENEL. 2012. Energy storage on islands: A sustainable energy future for islands. http://www.eurelectric.org/media/64743/Fastelli.pdf (accessed September 15, 2014).
60. Albadi, M. H., and E. F. El-Saadany. 2010. Overview of wind power intermittency impacts on power systems. *Electric Power Systems Research* 80:627–32.
61. Ackermann, T. 2005. *Wind Power in Power Systems*. Wiley, Chichester, UK.
62. Bhati, R., M. Begovic, I. Kim, and J. Crittenden. 2014. Effects of PV on conventional generation. *Proceedings of the 47th Hawaii International Conference on System Sciences (HICSS)*, Waikoloa, HI.
63. De Vos, K., A. G. Petoussis, J. Driesen, and R. Belmans. 2013. Revision of reserve requirements following wind power integration in island power systems. *Renewable Energy* 50:268–79.
64. Smith, J., M. Milligan, E. DeMeo, and B. Parsons. 2007. Utility wind integration and operating impact state of the art. *IEEE Transactions on Power Systems* 22:900–8.
65. Simoglou, C. K., E. G. Kardakos, E. A. Bakirtzis et al. 2014. An advanced model for the efficient and reliable short-term operation of insular electricity networks with high renewable energy sources penetration. *Renewable and Sustainable Energy Reviews* 38:415–27.
66. Medina, P., A. W. Bizuayehu, J. P. S. Catalão, E. M. G. Rodrigues, and J. Contreras. 2014. Electrical energy storage systems: Technologies' state-of-the-art, techno-economic benefits and applications analysis. *Proceedings of the 47th Hawaii International Conference on System Sciences*, Waikoloa, HI.

67. Dietrich, K., J. M. Latorre, L. Olmos, and A. Ramos. 2012. Demand response in an isolated system with high wind integration. *IEEE Transactions on Power Systems* 27:20–9.

68. Baptista, P. C., C. M. Silva, J. A. P. Lopes, F. J. Soares, and P. R. Almeida. 2013. Evaluation of the benefits of the introduction of electricity powered vehicles in an island. *Energy Conversion and Management* 76:541–53.

69. Lalor, G. R. 2005. *Frequency Control on an Island Power System with Evolving Plant Mix.* PhD thesis, University College Dublin, Dublin, Ireland.

70. Wang, Y., G. Delille, H. Bayem, X. Guillaud, and B. Francois. 2013. High wind power penetration in isolated power systems—Assessment of wind inertial and primary frequency responses. *IEEE Transactions Power Systems* 28:2412–20.

71. Milosevic, M., and G. Andersson. 2005. Generation control in small isolated power systems. *Proceedings of the 37th Annual North American Power Symposium*, Ames, IA.

72. Merino, J., C. Veganzones, J. A. Sanchez, S. Martinez, and C. A. Platero. 2012. Power system stability of a small sized isolated network supplied by a combined wind-pumped storage generation system: A case study in the Canary Islands. *Energies* 5:2351–69.

73. Delille, G., B. Francois, and G. Malarange. 2012. Dynamic frequency control support by energy storage to reduce the impact of wind and solar generation on isolated power system's inertia. *IEEE Transactions on Sustainable Energy* 3:931–9.

74. Khederzadeh, M. 2012. Frequency control of microgrids by demand response. *Proceedings of the CIRED 2012 Workshop Integration of Renewables into the Distribution Grid*, Lisbon, Portugal.

75. Rodrigues, E. M. G., A. W. Bizuayehu, and J. P. S. Catalão. 2014. Analysis of requirements in insular grid codes for large-scale integration of renewable generation. *Proceedings of the IEEE PES T&D Conference and Exposition*, Chicago, IL.

76. Acquaviva, V., P. Poggi, M. Muselli, and A. Louche. 2000. Grid-connected rooftop PV systems for reducing voltage drops at the end of the feeder—A case study in Corsica Island. *Energy* 25:741–56.

77. EIRGRID. 2014. *All Island TSO Facilitation of Renewables Studies Report*. http://www.eirgrid.com/media/FacilitationRenewablesFinalStudyReport.pdf (accessed September 15, 2014).

78. Fritts, T. H. 2002. Economic costs of electrical system instability and power outages caused by snakes on the Island of Guam. *International Biodeterioration & Biodegradation* 49:93–100.

79. Du, P., and N. Lu. 2011. Appliance commitment for household load scheduling. *IEEE Transactions on Smart Grid* 2:411–9.

80. Pipattanasomporn, M., M. Kuzlu, and S. Rahman. 2012. An algorithm for intelligent home energy management and demand response analysis. *IEEE Transactions on Smart Grid* 3:2166–73.

81. Borlease, S. 2013. *Smart Grids: Infrastructure, Technology and Solutions*. CRC Press, Boca Raton, FL.

82. Chua, K. J., S. K. Chou, W. M. Yang, and J. Yan. 2013. Achieving better energy efficient air conditioning—A review of technologies and strategies. *Applied Energy* 104:87–104.

83. Smart Metering in Caribbean Islands. 2014. http://www.engerati.com/article/smart-metering-helps-caribbean%E2%80%99s-energy-security (accessed September 2014).

84. Smart Metering in Caribbean Islands. 2014. http://www.metering.com/us-islands-connect-to-smart-grid/ (accessed September 2014).
85. EU EcoGrid Project. 2014. http://www.eu-ecogrid.net/ (accessed September 2014).
86. Ma, Q., and H. Lu. 2011. Wind energy technologies integrated with desalination systems: Review and state-of-the-art. *Desalination* 277:274–80.

2

Forecasting Models and Tools for Load and Renewables Generation

Cláudio Monteiro, Bruno Santos, Tiago Santos, Carina Soares,
Pedro Fonte, Rafael Nebot-Medina, Gianmaria Sannino, and
Adriana Carillo

CONTENTS

ABSTRACT   This chapter introduces state-of-the-art and present developments in short-term forecasting approaches, with a focus on specific applications in the operation of insular power systems. The chapter describes the
general information sources used for short-term forecast, including numerical weather prediction (NWP) and wave forecast systems. An introduction to
forecast techniques used in the scope of SiNGULAR is presented to understand the tools used in the forecast model application. Because of the specific
requirements of insular systems, probabilistic forecast tools will be adopted.
Forecast models will be presented for application in wind power generation,
photovoltaic (PV) solar generation, hydropower generation, and load prediction using some practical examples. Benefits of providing geographical
aggregated forecasts are also shown, with special attention to wind power.
Finally, the forecast system interface and its usage in a specific scheduling methodology developed within the scope of the SiNGULAR project is
analyzed.

2.1 Operation Forecast Framework

Variability and uncertainty always have been features of electric power
systems, which have made operation more difficult. Due to demand variations over time that cannot be perfectly predicted or due to unexpected
outages of generators (or other equipment), additional capacity from conventional generators needs to be available, in order to ensure the balance
between demand and power generation. This uncertainty is intensified
as the integration of intermittent and uncontrollable generation (as wind
or solar power) increases, due to their limited predictability. Unexpected
variations in their output, generally, lead to the necessity of higher levels
of spinning reserve for an efficient and reliable operation of the electricity
grid. Therefore, when a large amount of intermittent power is feeding a conventional electric grid, scheduling must take this into account, committing
backup controllable power sources (e.g., gas turbines for day-ahead scheduling) to cover unplanned variations of injected renewable energy.

2.1.1 Temporal Framework

To produce a simple deterministic forecast of variable p at time t for a look-ahead period $t+k$ (where k can be in range $1...n$, and n represents the complete forecast horizon), it is necessary to use a set of explanatory variables X known when the forecast is issued. This set can include measured recent-past observations of the variable of interest to predict or include the estimation of future values for a variable issued by other forecasting methods, provided at time t (for instance, forecast variables from NWP).

When we are forecasting the load or renewable energy generation, questions related to the forecast horizon, time resolution (step or interval), forecast refresh frequency, and latency of NWP may be assessed.

For instance, load forecasts are commonly divided in the literature into long-term forecasts (longer than a year), medium-term forecasts (from a week up to a year), and short-term forecasts (typically from 1 hour up to a week) [1]. The first two ranges are essential for network planning, the short-term forecast being critical for power system operation. Wind power forecast, in terms of timescale classification, is done based on forecasting horizon. However, in the literature, this classification is done in different ways. In [2], wind power forecasts are divided into immediate-short-term forecasting (8 hours ahead), short-term forecasting (for a day ahead), and long-term forecasting (multiple days ahead). In the classification done in [3], the forecasting is divided into four categories, very short term (from few seconds to 30 minutes ahead), short term (from 30 minutes to 48 or 72 hours ahead), medium term (from 48 or 72 hours to 1 week ahead), and long term (from 1 week to years ahead). Other different classifications can be found in [4]. Also, considering radiation and PV forecasts, they can be provided for the next 3–4 hours ahead (now casting), up to 7 days ahead (short-term forecasting) and for the next months or years (long-term forecasting) [5].

From the electric power system's point of view, a rough prediction from some days ahead up to a week ahead can be sufficient for maintenance planning. For this kind of horizons, NWP is the best source of information. In this case, global models could provide forecasts with sufficient temporal and spatial resolution with lower computational cost than mesoscale models. As the time horizon is reduced, higher resolution forecasts from some hours ahead up to 3 days will be necessary. For this range, mesoscale models will provide nested forecasts with better resolution. One of the main drawbacks of this type of approach is latency. Latency [6] is the delay that occurs, since the observations are assimilated by the NWP model, when data is processed until the instant at which forecasts are available. For horizons up to a few minutes or hours, the use of autoregressive models, based on the recent-past-observed values of power, can add value to the forecast accuracy. Instead of their simplicity, the performance of these models is hard to be beaten for very-short-term forecasts. In [7], a comparison of wind power forecasting model performance based on time-series techniques and NWP for a time horizon up to 48 hours is shown.

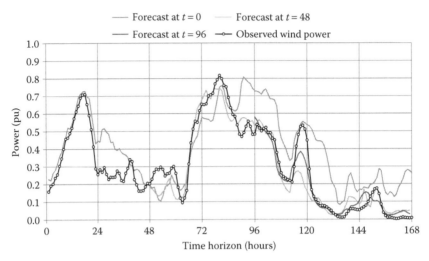

FIGURE 2.1
(See color insert.) Influence of the forecast refreshments along time. The forecasts are provided at 00 UTC and refreshed after 48 and 96 hours.

The forecast's refreshment is justified whenever new information arrives. Regarding power forecasting models, based on NWP systems, this refreshment occurs, typically, every 6 hours. The performance of power forecasting models degrades as the time horizon of the forecast increases. This situation is illustrated in Figure 2.1.

When more recent meteorological forecasts are available, power forecasts are updated, providing more accurate results. The example in Figure 2.1 represents three different forecasts provided at different times (this is done with a low refreshment rate in order to facilitate the analysis of the figure) for an ensemble of wind farms.

2.1.2 Geographical Framework

In a short-term interval, the most important factor that affects the renewable power production variability is the atmospheric conditions. Fortunately, when the output of a group of wind farms or PV power plants is considered, the variability is reduced. For example, consider two PV power plants. The more separated they are, the less sensitive they will be to simultaneous variations by the same cloud (or cloud system for large regions). For large groups of geographically distributed PV plants, the reduction of the power intermittency is even more effective.

Figure 2.2 presents two different forecasting models. An aggregated output using the most representative wind farms of a certain region is shown in Figure 2.2a, whereas Figure 2.2b shows an individual wind farm. The forecasts are provided at 12 UTC (Universal Time Coordinated) for the day ahead.

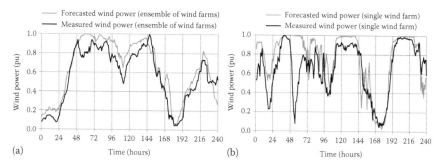

FIGURE 2.2
(See color insert.) Example of forecast (day ahead) and observed wind power of a group of wind farms (a) and a single wind farm (b) for the same period.

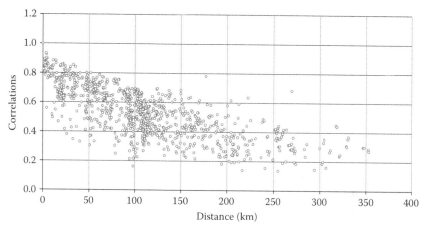

FIGURE 2.3
Correlations between power productions in function of distance between wind power plants.

The output of the forecasts of a larger geographical area (geographically dispersed farms) is smoother when compared with the output of a single wind farm. It is also possible to conclude that the accuracy of the aggregated wind power forecasts is significantly better than that of a single wind farm, so clustering could be a good approach.

This smoothing effect may be explained by means of Figure 2.3, which shows that the correlation between all the power production series of different wind farms (and therefore the primary wind resource) tend to decrease with increasing distance between them. It means that weather conditions are different along the areas where the power plants are installed. This situation will conduce to a different production profile, which is expected to lead to a smoothing effect. Considering the wind production case, it is clear that the influence of a certain wind profile in an area is not the same in adjacent areas. This situation also creates a cancelation in the forecasting errors, as also shown in Figure 2.2. It has also been verified that generally the correlation values decrease with increase of distance between wind farms, as concluded in several publications [8–10].

Also regarding PV generation, Lorenz et al. [11] analyzed the effect of aggregation of several PV power plants, which permits improvements in power PV forecasts.

2.1.3 Application Framework

Load demand and renewable power forecasts for different time horizons and different time resolutions are crucial information for the reliable planning and operation of the electric power systems.

With some details changing from country to country, there is normally nested planning (*planning* understood as the organization of resources in the future in order to accomplish some goal) for the electric grid actors. This planning is based on forecasts, which can be classified with regard to an increase in time resolution (and shorter time horizon), ending with the real-time balancing of consumption and production:

- *Years*: For power system planning (changes in networks, commissioning or decommissioning of generation units, and substations, among others).
- *From year to week*: For maintenance planning and planned outages of grid elements.
- *Week to day*: For high-level scheduling for the week and also maintenance planning.
- *Following day (day ahead)*: For trading in the day-ahead electricity markets or for scheduling unit commitments (UCs). At this level, normally, an hourly program schedule for each generator (conventional and renewable) that will operate and how much power it will/ should inject is determined.
- *Following hours to dozens of seconds (intraday, hours ahead, intra-hour)*: For trading in intraday electricity markets and for the adjustment of schedules based on current forecasts and previous schedules.
- *Following few seconds*: This shortest time horizon is left to automatic generation control.

With the long-term planning horizon, the following tasks are completed: plan interconnections between country networks, build continental transmission lines, or commission/decommission generation units. In this chapter, the time range higher than a week will not be addressed.

Some of the applications of the forecasts up to a week, day ahead, and hours ahead are listed in [4,12–14], respectively. Figure 2.4 represents a simple scheduling diagram over the period t up to a week ahead.

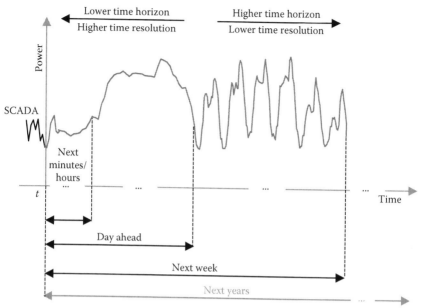

FIGURE 2.4
(See color insert.) Scheduling diagram over the period *t* up to a week.

At instant *t*, assuming that meteorological forecasts (resulting from NWP) for the next 7 days are available, it is possible to obtain a prediction of the load or net load (load forecast minus renewable generation forecast) for a high-level scheduling or maintenance planning. Since, for this horizon, a rough prediction is needed, it can be provided by a global model. This type of model resolves the whole planetary atmosphere, with a relatively coarse temporal and spatial resolution (it is expected that with increment of computational power, the temporal and spatial resolution can be increased); however, the time processing is faster than a mesoscale model (which resolves a local planetary subdomain, providing finer resolution). As seen in Figure 2.1, large deviations between the forecast and observed information for longer horizons are usual.

For day-ahead forecasts, NWP is still the best source of information. In this case, a refined resolution (both temporal and spatial) is necessary in order to provide to decision makers a more precise prediction of shorter time intervals. For this time frame, mesoscale models as weather research forecasting (WRF) with an hourly time step or steps of 15 minutes are used. They also have the capacity to provide forecasts taking into account the local characteristics of the terrain using a more detailed digital elevation model (DEM) of a terrain surface. Figure 2.5 shows graphically the DEM of the terrain surface with differences in their resolutions of 25 km (less detail Figure 2.5b) and 5 km (more detail Figure 2.5a) for the same local characteristics. The terrain surface with more detail gives better accuracy regarding the complexity of

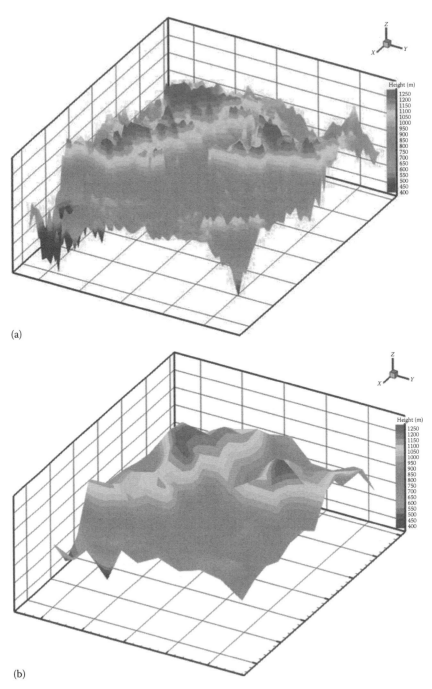

(a)

(b)

FIGURE 2.5
(See color insert.) Digital elevation model of the terrain surface with spatial resolutions of
5 km (a) and 25 km (b).

the terrain and due to that more detailed information is available for site-specific forecast. Obviously, with crescent detail, the computational efforts become excessive to some applications. In this sense, sometimes, and for some applications, a less accurate but faster models are good enough.

The update rate (of meteorological forecast) depends on the refresh cycle of newly available data and on the computational resources needed to execute the whole process. For instance, NWP are typically refreshed between two and four times a day. Currently, NWPs are being investigated to be refreshed more times (*rapid refresh*) [15].

One of the main disadvantages of this approach is related with latency. The available meteorological forecasts were obtained by a numerical model that can take up several hours to run (from data assimilation to forecast delivery), so when the forecasts are obtained, the initial presupposes may have changed, and this situation may lead to an increase in forecast deviations. Nowadays, meteorology focused on power systems is pushing the development of better atmospheric models in order to improve energy forecasts.

When recent observations are available, it is possible to increase and improve the forecast accuracy when the instant t (value to be forecast) is approaching. It is useful in the case of a system such as islands, in order to provide a rapid fine-tuning of the previous schedule of rapid conventional generators. In this case, models are based on the recent-past values (time series) of the variable to be predicted. Some models use a single value, such as persistence (normally used as reference model to be outperformed by proposed methods) and autoregressive models (where the output variable depends on its own previous values). These models are characterized by their simplicity, but their performance in the first few hours is better than the performance of models using only data from NWP. However, most of the time, data measured from SCADA is only available a few days or months later, making impossible the tuning of the forecast in the first instance, reducing the uncertainty.

Also, for this time horizon, a set of techniques based on satellite [16] and sky-camera image processing [17,18] is being exploited, mixing physical models with computer vision techniques, leading to better resolution of radiation and solar power forecasts.

Although the accuracy of renewable forecast benefits from the smooth effect, when spatial aggregation is done, the geographic component of the information is lost. This is critical when, along with scheduling, a predictive load flow analysis is needed in order to transmission system operator (TSO) guarantee the secure operation of the transmission system. In this situation, individual forecasts of load and power plant may be necessary for the management of congestion, preventing overloading. An aggregation in small clusters (by injection points and substations) may be a solution in order to keep part of the geospatial information and simultaneously reduce forecasting errors.

Moreover, probabilistic forecasts are currently a very active area of research, especially in wind power area. This type of forecast provides some uncertainty quantification, based on the supposition that past errors (which can result from several factors such as incorrect or incomplete models, incorrect starting conditions, among others) will be repeated in the future under the same conditions. Contrary to deterministic approaches of the scheduling, which define spinning reserve as a fraction of the peak load, or the same value of the largest online unit, new approaches are emerging that determine the system reserve implicitly, by incorporating explicitly the stochastic nature of the forecast uncertainty leading to more economic solutions [19,20].

2.2 NWP and Wave Forecast System

There are many types of physical models, depending on forecast horizon, spatial scale, and phenomena studied, each with their own advantages, limitations, and specifications.

2.2.1 Weather Forecast

The history of NWP dates to 1904, when V. Bjerknes tried to define the problem of prognosis as the integration of motion equations of the atmosphere [21]. It was the first reliable recognition where the future state of the atmosphere was determined by the initial state and boundary conditions, together with Newton's equations of motion, the Boyle–Charles–Dalton equation of state, the equation of mass continuity, and the thermodynamic energy equation. Bjerknes is known for his program of observation, graphical analysis of meteorological data, and graphical solution of the governing equations.

In 1922, L. F. Richardson, an Englishman, published *Weather Prediction by Numerical Process*. Richardson demonstrated how the differential equations, which reflect the atmospheric motions, could be written as a set of differential equations of variables at a finite grid [21]. At that time, there was no computational capability to solve such equations in an acceptable time period; even so, he proceeded to calculate the surface pressure prediction, at two points. The results were very poor, and they were attributed to lack of data. However, nowadays, it is known that there were problems in Richardson's scheme.

After Richardson's lack of success, many years passed without anyone attempting computing again; only after World War II, interest reemerged, with the development of digital computers and the implementation of a wide meteorological observations network. In 1948, the American meteorologist J. G. Charney showed that dynamical equations could be simplified by the systematic introduction of hydrostatic and geostrophic approximations [22] and filtering sound and gravity oscillations [23], which were the problems in Richardson's scheme. Charney's filtering approximations were the base for the equivalent barotropic model, the first successful numerical forecast, in 1950 [21,23].

The base and methodology of NWP is provided by dynamical meteorology, consisting of the use of numerical approximations to the dynamical equation, predicting the future state of the atmosphere circulation by knowledge of the present state. The present state of the atmosphere can be defined by a set of variables, which is required to initialize NWP. This is integrated in the model time prediction equations to achieve the future distribution of weather variables.

In the general circulation models, dynamic and thermodynamic processes and radiative and mass exchanges are modeled using five basic sets of equations [24]. The basic equations describing the atmosphere are as follows:

- The three-dimensional equations of motion (conservation of momentum)
- The equation of continuity (conservation of mass or the hydrodynamic equation)
- The equation of continuity for atmospheric water vapor (conservation of water vapor)
- The equation of energy conservation (thermodynamic equation derived from the first law of thermodynamics)
- The equation of state for the atmosphere
- The conservation equations for other atmospheric constituents such as sulfur aerosols that may be applied in more complex models

The accuracy of NWP depends on the accuracy with which physical and numerical processes is represented on the model resolution and, no less important, on the meteorological data that will bring initial conditions to NWP [23,25]. Meteorological data need to be modified dynamically, that is, data assimilation [23], to turn them suitable for NWP ingestion. The process of data assimilation has the following four purposes: data quality control, objective analysis, initialization, and initial guess from a short-range forecast. These components have been taken to form a continuous cycle of data assimilation, often called *four-dimensional data assimilation* [21].

Forecasts are typically mentioned as short-range (up to approximately 3 days), medium-range (up to approximately 14 days), and long-range (monthly or seasonal) outlooks. The short-range and medium-range forecasts could be considered together because their methodology is similar, and the increase of computing power is becoming less distinguishable as a separate type of forecast.

2.2.1.1 Global Models

Global NWP models are operated by National Centers for Environmental Prediction (NCEP) [26], the European Centre for Medium-Range Weather Forecasting (ECMWF) [27], and important national weather services (such as the United Kingdom, Germany, France, Brazil, Japan, and Russia). The global models are operated in UTC between two (00 and 12 UTC) and four (00, 06, 12, and 18 UTC) times a day.

The more relevant institutions in NWP are the NCEP and the ECMWF, who operate their models several times regarding the original forecast, in parallel with a lower resolution with perturbations. The NCEP ensembles have perturbed initial conditions [28] and the ECMWF ensembles are obtained with resource to perturbed physics [29].

2.2.1.1.1 Global Forecast System

The NCEP's global forecast system (GFS) is a global model with a horizontal resolution of about 35 km, with 64 unequally spaced vertical levels [26,30]. The forecasts are provided for the next 16 days. The model comprises the calculation models of divergence equation, vorticity equation, hydrostatic equation, thermodynamic equation, conservation equation of water vapor, and mass continuity equation. The model has real data from various sources as input with different spatial and temporal resolutions (airports, ships, meteorological balloons, radiosondes, satellite surface station networks, among others). This process is described in [26].

2.2.1.1.2 Global Ensemble Forecast System

There are no perfect initial conditions for NWP models ingest, and therefore the errors contained in those can be extrapolated along the simulation, leading to significant errors [31–33]. This is the case when the resolution and mesh dimensions required by the user do not have the same grid dimensions and horizontal resolution of the simulations, requiring interpolation with consequent forecast errors. Throughout the simulation, solving numerical algorithms that characterize the physics and dynamics of the atmosphere also incur errors, such as limitation of the equations order, truncation errors, and the impossibility of solving certain phenomena whose scale does not fit with the resolution, among others. Taking into account the previous considerations (input and modeling errors), it can be stated that NWP should always be regarded as having an uncertainty associated, and ensemble forecasting aims at quantifying that.

The main goal of the ensemble project is to investigate the use of meteorological ensemble forecasts for the analysis of uncertainty of the forecasts [33]. The model is run with a minor variation of the initial conditions, showing how sensitive the results are to it.

The global ensemble forecast system (GEFS) model used is the same as GFS [28]. The NCEP operates the GEFS four times a day (00, 06, 12, and 18 hours). Each time the NCEP operates the ensemble model, 21 forecasts are made, one unperturbed forecast (control forecast with the same initial conditions as GFS) and 20 perturbed ensemble forecasts (with perturbed initial conditions), predicting for the next 384 hours (16 days) [34,35].

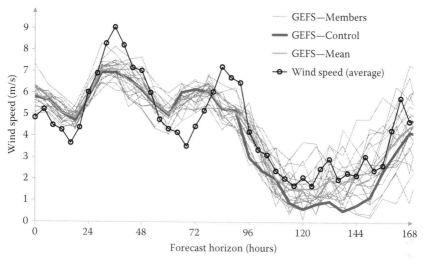

FIGURE 2.6
(See color insert.) Example of a forecast for the next 168 hours of wind speed at a height of 10 meters and mean measured value.

Figure 2.6 presents an unperturbed forecast of wind speed at a height of 10 meters (from GEFS) initialized at 00 UTC for a horizon of 168 hours with a time step of 3 hours. It also shows the 20 ensemble members and the measured values (every 4 hours).

It is possible to verify that (in this case) the measured values are not well forecast by the members of the ensemble, and the dispersion of the ensemble members increases as the forecast range increases. The limited performance of the GEFS can be explained by the limited spatial and temporal resolution of this type of global model, which does not have the capacity to provide forecasts taking into account the local characteristics and when the measured data correspond to a complex terrain. The mean of the 20 members of the ensemble seems to fit better the observed values of wind speed than the control forecast.

2.2.1.2 Mesoscale Models

The regional climate of a specific region is affected by several meteorological occurrences with distinct spatial and temporal resolution. Semipermanent controls are large-scale standards that have a well-defined behavior, mostly of static nature. Synoptic disturbances are transient meteorological occurrences shaped by semipermanent climatic conditions, being of unstable nature, and with the duration of several days. These local effects are transient occurrences with a duration of a few hours up to approximately 3 days, having a high dependency of regional features, such as topography and soil type.

Limited area models were developed for the purpose of research of mesoscale atmospheric processes. The mesoscale process includes every atmospheric event with a horizontal scale between one and a few hundred kilometers.

Mesoscale NWP needs to capture the large-scale environment, which is responsible for the formation and modification of smaller-scale weather systems; on the other hand, the grid and time intervals should be fine enough to resolve small-scale processes, representing multiscale processes in a finite domain [36,37].

The evolution of mesoscale models, since its creation, has converged on the way to better simulation resolution, with the aim to better predict storm-scale phenomena. With increased resolution, nonhydrostatic models have a tendency to prevail against hydrostatic models; for example, advanced regional prediction system (ARPS) is not accurate when the resolution is equal or smaller than 10 km because of the hydrostatic approximation [21].

Determining appropriate domain size is one important quest, as is determination of grid interval, time interval, and total integration time of the mesoscale numerical model. Those choices should take into account the following physical and numerical factors [37]:

- Spatial scales and dimensionality of the forcing and physical processes
- Timescales of the forcing and the fluid responses to the forcing
- Stability criterion of the adopted numerical scheme
- Limitations of predictability of the atmospheric phenomena
- Computational resource

Mesoscale models need four-dimensional data for initial conditions. Meteorological boundary conditions could be provided by data assimilation of observations or by global model data [4].

2.2.1.2.1 Numerical Mesoscale Model

Simulation of the regional pattern of several climatic variables, such as wind, can be obtained using the WRF and University Corporation for Atmospheric Research (UCAR) systems [38–40]. The mesoscale model is used by many institutions around the world, and its versatility ensures complete customization for a wide range of studies. This numerical modeling system consists of several modules, especially designed to assimilate observational data and simulate atmospheric conditions and to solve multiple equations that describe the dynamics and thermodynamics of the atmospheric flow in limited areas [41–43].

Simulation domains have a regular grid, where the placement of the simulation is determined relative to the geometric center, defined by latitude and longitude [38–40]. The given coordinates will correspond to the center of the main domain, with the following areas positioned relative to the lower-left corner of the domain that precedes it.

Domain dimension is determined by the number of nodes of each domain, and its spatial discretization is performed using a staggered mesh in space by Arakawa scheme C [38–40]. The variables are defined in the mass center of the mesh; however, the wind components are defined within the limits of the mesh, and thereafter interpolated to calculate the wind speed.

WRF uses for vertical characterization eta coordinates (η) [38,44]. The use of η coordinates enables representation of the lower layer of the atmosphere in the model, for each mesh point, by a horizontal step [38–40].

2.2.1.2.2 *Local Analysis and Prediction System*

Nowadays, the availability of meteorological data sites leads to their integration to improve resource evaluation. The National Oceanic and Atmospheric Administration's (NOAA) local analysis and prediction system (LAPS) [45] is a mesoscale meteorological data assimilation tool that allows us to assimilate several meteorological data (meteorological networks, radar, satellite, soundings, and aircraft), with different spatial and temporal resolutions, in order to get a result of all inputs. This tool generates an analysis with a defined time step, spatially regular, in a three-dimensional grid that represents features and processes at all atmospheric layers [46–49].

The physical boundary conditions of LAPS are provided by U.S. Geological Survey's (USGS) 30 second global elevation data (USGS and University Corporation for Atmospheric Research). This dataset consists of a DEM that provides terrain elevation information in a horizontal grid of 30 arc seconds (approximately 1 km) [50,51].

The land cover data is provided by the global land cover characteristics database (USGS, National Center for Earth Resources Observation and Science—EROS, Joint Research Centre of the European Commission). It consists of a set of global information about land coverage characteristics with a horizontal resolution of 1 km, acquired from global observations performed by the advanced high-resolution radiometer, aboard the NOAA satellite [50,52,53].

It is necessary to take into account the instrumentation and calibration of the acquisition data equipment, as well as the validation of datasets that the operator wants to input in the LAPS analysis. If the previous factors were not taken into account can lead to higher errors, especially when it refers to derived variables, such as relative humidity, in the LAPS analysis [54]. Hiemstra et al. [54] completed the LAPS analysis with several meteorological sources with distinct time and spatial resolution and used 107 independent meteorological observations. The conclusions derived from the validation show a matched trend in independent temperature and relative humidity; although the magnitude error is less accurate in diurnal changes, LAPS duplicated it. The wind and precipitation obtained with LAPS are more variable and less reliable. Knowing that the wind speeds are well represented on lands with low elevation, the following factors are mentioned as possible disparity causes: observation errors, scale problems, and limitations in radar measurements.

The use of LAPS analysis as input of numerical models has the advantage of assimilating local data to the database previously chosen; this could have low index data in its elaboration taking into account the interest area. Nowadays, there is a wide global coverage of meteorological data; however, these are not distributed in a regular way. Assimilation of surface data with other data sources with vertical levels (such as GFS, GEFS, and ECMWF) will improve the input data for numerical models. If we are looking, however, for mesoscale or microscale models that solve for the entire atmosphere, and not just the surface layer, the assimilation of surface data will not make a substantial difference [55].

The Department of Geography in Harokopio University of Athens used LAPS to generate initial conditions for meteorological models [55], such as WRF, using public data from GFS/NCEP and incorporating metars, synops, and raobs. The data sources have different time, resolution, and spatial configuration. With the purpose of evaluating the accuracy of initial conditions generated by LAPS, these were compared and verified with independent observations over Greece that were not assimilated in LAPS. The test was performed with the POSEIDON model (Prevision Operational System for the mEditerranean basIn and the Defence of the lagOon of veNice) ingesting LAPS output. It achieved results very close to the scores of the GFS model [56] for the following variables: temperature, relative humidity, mean sea level pressure, and wind speed. The results lead to the conclusion that GFS background fields seem to play a key role in the quality of analyses produced given this LAPS configuration.

The research about LAPS seems to lead a valuable and reliable choice as a data assimilation system for distributed meteorological data with high temporal resolution [54], as well as ways to improve global data for input mesoscale simulations [55].

Including an ensemble-based estimate of background error statistics and improving the weighting of the model background information in the LAPS data assimilation system are very interesting ways to improve LAPS. In the study, Ok-Yeon et al. [57] tested the integration of ensembles to reduce the forecast errors with LAPS analysis. LAPS analysis was used as meteorological input data for the WRF model's 48-hour forecast. This study showed that wind speed analyses are improved at most pressure levels by ingesting background error statistics; however, the wind direction and temperature analyses improved at many levels by ingesting background wind variance, but deteriorated at other levels.

The direct approach will affect the horizontal and temporal resolution of LAPS, especially in wind and precipitation variables [54,58]. Using LAPS analysis as input in prediction models seems to lead to forecasting improvements [55,57].

2.2.2 Wave Forecast

The forecast of sea wave conditions started in the middle of the nineteenth century. Initially, only waves generated by local winds were considered. Wave forecasts were based on empirical relations between local wind speeds and heights and periods of the waves. Then also a single wave height representative of waves propagated from other areas was included.

This approach was oversimplified as it did not consider that the sea state is determined by the superposition of waves of different heights and wavelengths, propagating in different directions. A more complete description of the sea state can be obtained using the energy spectrum that describes the distribution of energy over wave frequency and wave propagation direction. Nowadays, numerical models commonly used to simulate sea wave propagation are spectral models, based on the formulation of the transport equation for the two-dimensional wave spectrum, derived in [59].

The main source-and-sink processes determining wave energy are represented by wind forcing, dissipation due to wave breaking, and physical processes due to the interaction of waves with the bottom. Moreover, nonlinear interactions transfer energy between the spectral components. Three generations of spectral models have been developed. The first-generation models describe the evolution of parameterized spectra and do not take into account nonlinear interactions among different frequencies. In the second-generation models, some nonlinear interactions are parameterized but the spectral shape is still fixed. Since the late 1980s, the most-used wave models belong to the so-called third-generation models that solve explicitly the nonlinear energy transfer equation, without any a priori assumption on the shape of the wave spectrum. The first model of this generation to be developed was the WAM model [60], which is used worldwide for the simulation of waves in open ocean. Over the last few years, other third-generation spectral models have been implemented, characterized by the possibility of selecting a number of different parameterizations of the physical processes, including characteristics of shallow waters such as dissipation due to bottom friction and depth-induced breaking. The two models more frequently applied in shallow waters are SWAN (Simulating WAves Nearshore) [61] and WAVEWATCH III [62], which can be used both for global and regional applications.

Wave model analysis and forecasts have been produced operationally at ECMWF since 1992. Two versions of the WAM model are running, a global model and a limited area model at higher resolution. One of the limited area models covers the Mediterranean Sea at a resolution that has been continuously increased over the years.

A variety of wave forecast systems is nowadays running operatively over the Mediterranean Basin and its subbasins and has been developed by research centers and the European University often as a result of European projects. They differ not only in the wave model used and in its resolution but also in the wind fields used to force the wave models that are often derived by regional atmospheric models. The quality of the atmospheric forecast is in fact a limiting factor in the performance of wave forecasts, in particular for simulations in small basins surrounded by a complex topography, such as the Adriatic and the Aegean Basins.

For the most significant regions, successive nesting procedures are adopted, which are coarse resolution models covering the entire Mediterranean Basin to provide boundary conditions for the higher resolution models. In some cases, the nesting procedure is applied more times producing very-high-resolution forecasts, of the order of hundreds of meters, over small areas.

2.3 Power Forecasting Models

It is undeniable that short-term forecasts of load demand and renewable power production are indispensable for efficient and reliable operation of the electricity grid. Depending on the variable to forecast, time horizon, time step (granularity) of the prediction, and the different sources of information available (known up to the instant that power forecasts are provided), different techniques may be used.

According to [1], it is possible to predict the day-ahead load, with relatively low errors. Regarding renewable technologies, over the last few years, the efforts of the scientific community to present efficient and effective tools to predict the behavior of their production have been significant. Recent publications containing exhaustive reviews of wind power forecasting models can be found in [4] and [63]. The wind power forecasting has years of experience and it is especially advanced in probabilistic forecasts. Concerning solar power forecasting, [64] provides a review of the current methodologies for the forecast of solar resource. Since the power produced by a PV system is closely related, almost linearly, with the incident global irradiance, which is highly affected by the presence of clouds, a great deal of attention has been paid to satellite image processing techniques and most recently to all sky images in order to forecast the movement of the clouds. The short-term forecasting of small hydropower plants (SHP) production has also been a target of study. A recent publication [65] presents a short-term forecasting model for the electric power production of SHPs for a few days ahead, using as inputs past values of power production, as well as values of forecast precipitation from NWP. In a much smaller number, works related to wave energy forecast can be found in literature. In [66], the performance of physical and statistical approaches for longer and shorter time horizons forecasts and the combination of both approaches are explored. A probabilistic approach is studied in [67].

2.3.1 Uncertainty Forecast

Although every forecasting process implies some uncertainty around the predicted variable, up to the very recent past, deterministic forecasts (which provide a single expected value for each forecast horizon) were used in scheduling procedures. Over the recent years, the increasing penetration of noncontrollable renewable energy sources (RESs) in actual power systems awakened the interest in the development of probabilistic forecasting tools in order to obtain a predictive probability density function (PDF) for each time step ahead of the forecast horizon.

There are several probabilistic estimator models in the literature; however, they focus on wind power uncertainty forecast, ignoring other renewable sources such as solar or SHP production. The classic approach is to assume a parametric distribution. On the other hand, the KDE technique [68–70] is a nonparametric conditional density estimator, whereby no distribution parameters must be assumed. Also, in a quantile regression model [71,72] the shape of the distribution does not have to be specified. These models are the most widely used.

2.3.1.1 Kernel Density Estimation

The KDE models developed in [68–70] were used to compute a PDF as response to the necessity of probabilistic wind power density forecasts. This method presented slight improvements for wind power forecasts when compared with other forecasting models [68,69]. The KDE model developed in [68] uses biweight kernel functions and is trained offline, which means that the model does not have the capability to adapt to temporal changes in the relation between target and explanatory variables. In [69], the author adopted specific kernel functions for each used variable and introduced a forgetting factor (for a time-adaptive Nadaraya–Watson kernel density estimator) that controls the importance of old and recent data in the forecast. In [70], a conditional KDE for wind power density forecasting with exponential time decay is explored since the relation between wind power and explanatory variables evolves over time, giving more weight to recent observations. The same authors recommend the use of conditional KDE to generate density forecasts for electricity load.

2.3.1.1.1 Forecasting Model Formulation

In forecasting problems, the estimation of the future conditional density function has a very important role, since it describes the relation between both explanatory and target variables. The conditional density estimation can be seen as a generalization of regression, since conditional density estimation aims at obtaining the full probability density function $f_{y|x}(y|x)$, whereas regression aims at estimating the conditional mean $E(y|x)$ [73].

The conditional density function $f_{y|x}(y|x)$ can be estimated using the joint density $f_{y,x}(y,x)$ and the marginal density $f_x(x)$, as shown in Equation 2.1:

$$f_{y|x}(y|x) = \frac{f_{y,x}(y, x)}{f_x(x)} \tag{2.1}$$

Power forecasting problems consists of estimating the future conditional PDF of a random variable for each look-ahead time step $t+k$, given a learning set with N samples constituted by pairs of (X_N, P_N) summarizing all historical information available up to instant t. Each pair consists of a set of explanatory variables X_N and the corresponding measured value of the variable to predict. By generalizing the wind power forecast case in [68,69] to other sources, it is assumed that at instant t, a set of explanatory variables $x_{t+k|t}$ is known for each time step ahead we want to forecast, resulting in Equation 2.2, where p_{t+k} is the power forecast for look ahead time $t + k$.

$$\hat{f}_p\left(p_{t+k} \mid x_{t+k|t}\right) = \frac{f_{P,X(p_{t+k},\, x_{t+k|t})}}{f_x\left(x_{t+k|t}\right)} \tag{2.2}$$

Since joint and marginal densities are not known, a nonparametric kernel estimation of the regression function can be used [74]. Once renewable power production or load depends on several variables, a multivariate KDE needs to be used [69,74].

For a given independent and identical distributed multivariate data $(X_{1d}, \ldots, X_{nd})$ from different d variables from an unknown multivariate density function f, the multivariate KDE is given by Equation 2.3.

$$\hat{f}(x_1, \ldots, x_d) = \frac{1}{N} \sum_{i=1}^{N} \left[\prod_{j=1}^{d} \frac{1}{h_j} K_j\left(\frac{x_j - X_{ij}}{h_j} \right) \right] \tag{2.3}$$

where:
  N is the number of samples
  d is the number of variables
  K_j is the kernel function to each variable j
  h_j is the bandwidth (smoothing parameter) of each kernel around each sample X_{ij}

Using the Nadaraya–Watson estimator, the conditional density is given by Equation 2.4.

$$\hat{f}_{P_{t+k|t}}(p, x_j) = \sum_{i=1}^{N} K\left(\frac{p - P_i}{h_p} \right) \cdot \frac{\prod_{j=1}^{d} K_j\left[(x_j - X_{ij})/h_j \right]}{\sum_{i=1}^{N} \left\{ \prod_{j=1}^{d} K_j\left[(x_j - X_{ij})/h_j \right] \right\}} \tag{2.4}$$

In the case of deterministic forecast, to estimate the conditional mean $\hat{p}_{t+k} = E(p_{t+k} \mid x_{t+k|t})$, the Nadaraya–Watson estimator is also used to determine $\hat{p}_{t+k}$ as a locally weighted average (see Equation 2.5).

$$\hat{p}_{t+k}(x_j) = \sum_{i=1}^{N} \frac{\prod_{j=1}^{d} K_j\left[(x_j - X_{ij})/h_j\right]}{\sum_{i=1}^{N}\left\{\prod_{j=1}^{d} K_j\left[(x_j - X_{ij})/h_j\right]\right\}} \cdot P_i \qquad (2.5)$$

The approach of Equation 2.5 can be found in [75], which is used to provide a short-term load forecast.

The first step is to choose the kernel, and in the literature, there are several possible kernels, namely, normal, biweight, Epanechnikov, and logistic, among others, but the most important step is to choose the bandwidth h. In [68], the biweight kernel was used instead of the classical Gaussian kernel in order to decrease the computational efforts. In this sense, in [68], it is stated that the kernel function has a minor impact on the estimated quality when compared with the bandwidth selection. On the other hand, and specifically to wind power forecast in [69], the choice of the kernel function is considered as a critical issue. In this sense, different kernels were proposed to different forecasting variables. Both authors agree that behind the choice of the kernel function, the choice of the smoothing parameter is much more critical. Small bandwidth values lead to an over-fitted prediction function, whereas high values generalize too much [74].

2.3.1.2 Evaluation Framework

There are several standard error measures and evaluation criteria for point forecasts such as, the root mean square error, mean absolute error, and mean average percentage error [2]. Evaluating a probabilistic forecast is more difficult than evaluating spot predictions. There are some indicators that are widely used, namely, reliability and sharpness [68,69,76,77].

Reliability is a measure of the agreement between nominal proportions (forecast probabilities) and the ones computed from evaluation samples. Reliability diagrams plots the empirical coverage versus the nominal coverage for various nominal coverage rates. If the values tend to be equals, the diagram becomes a diagonal line (*perfect reliability*), so the results are better the nearer the diagonal the diagrams are. Alternatively, the diagrams can be drawn as a function of the deviation from the *perfect* reliability. In this case, the empirical coverage should be equal to the nominal coverage.

$$\xi_k^{\alpha_i} = \begin{cases} 1 & \text{if } p_{t+k} \leq \hat{q}_{t+k|t}^{\alpha_i} \\ 0 & \text{otherwise} \end{cases} \qquad (2.6)$$

$$n_{k,1}^{\alpha} = \#\left\{\xi_{i,k}^{\alpha} = 1\right\} = \sum_{i=1}^{N} \xi_{i,k}^{\alpha} \tag{2.7}$$

$$n_{k,0}^{\alpha} = \#\left\{\xi_{i,k}^{\alpha} = 0\right\} = N - n_{k,1}^{\alpha} \tag{2.8}$$

$$\hat{\alpha}_{k}^{\alpha} = \frac{n_{k,1}^{\alpha}}{n_{k,1}^{\alpha} + n_{k,0}^{\alpha}} \tag{2.9}$$

In Equation 2.6, p_{t+k} is the realization of the variable and $\hat{q}_{t+k|t}^{\alpha_i}$ the forecast quantile with the coverage values α_i. The indicator, $\xi_k^{\alpha^i}$, resulting from this equation, takes the value "1" if the realization of the variable p_{t+k} hits the forecast quantile with the nominal coverage α_i and "0" if missed. The number of *hits* and *misses* are counted by Equations 2.7 and 2.8, and the coverage $\hat{\alpha}_k^{\alpha}$ is calculated by Equation 2.9. With the difference between empirical and nominal proportions, it is possible to have an idea about the bias of the probabilistic forecasting methods and measure the quality of the forecasts.

Another indicator is sharpness, which evaluates the prediction independent of the observations and gives an indication of the level of the prediction's usefulness. Following Equation 2.10, the quantiles are gathered by pairs in order to obtain intervals around the median with different nominal coverage rates. Sharpness of the predictive intervals is measured by the average interval size (see Equation 2.11), where $\delta_{t,k}^{\alpha}$ is the size of the interval forecast with nominal coverage rate $1 - \alpha$ estimated at time t for lead time $t + k$.

$$\delta_{t,k}^{\alpha} = \hat{q}_{t+k|t}^{(1-\alpha/2)} - \hat{q}_{t+k|t}^{\alpha/2} \tag{2.10}$$

$$\bar{\delta}_{k}^{\alpha} = \frac{1}{N} \sum_{i=1}^{N} \delta_{t,k}^{\alpha} \tag{2.11}$$

Large intervals mean that the approach is very conservative and it is not designed to "take risks." Both reliability and sharpness have contradictory results, because good values of reliability conduced to bad values of sharpness. As example, low values of sharpness can lead to "narrow" uncertainty intervals which can result in underestimation or overestimation of the uncertainty, with consequent degradation of the reliability. On the other hand, high values of sharpness could increase the uncertainty interval and by this way improve the agreement between the nominal proportions and empirical ones. The drawback is that too high values could cover extreme prediction or even outliers. In other words, all measured values would fall inside the uncertainty interval. A trade-off between reliability and sharpness has to be accepted because improving reliability generally degrades sharpness and vice versa.

2.3.2 Wind Power Forecast

In short-term forecasts of wind power, there are fundamentally two paths: one uses physical models and the other uses statistical models; however, there are systems that use a combination of both methods, since in reality, both are necessary to forecasting success [13].

The physical approaches are composed of a large number of physical specifications, and their inputs are physical variables, such as the orography of a wind farm, atmospheric pressure, atmospheric temperature, wind machine specifications, and other types of physical data, which can have some advantages in long-term forecasting [78]. Nevertheless, statistical approaches try to find the relationships in inherent structures within the measured data, which can have advantages in short-term forecasting [79,80]. Furthermore, statistical forecasting approaches can overcome the physical methodologies, that is, in relation with the accuracy in very-short-term horizon, due to the stable efficiency in forecasting [81].

Furthermore, nowadays, these methodologies include some new intelligent and hybrid methodologies, which have been reported in the technical literature with interesting results, such as neural networks [82,83], neural networks with wavelet transform [84], adaptive wavelet transform with neural networks [85], neural networks with fuzzy systems [86], evolutionary approaches [87, 88], and hybrid methods [89]. These statistical methodologies can be more efficient and more accurate if proper inputs are selected by the predicted system [90].

Typically, wind speed forecasts at the hub height of the wind turbines are the most important explanatory variable used to forecast wind power production in the short term. Figure 2.7 presents the relationship between wind power production data of a single wind farm and meteorological forecasts of wind speed resulting from the simulation of a local NWP system where the farm is installed.

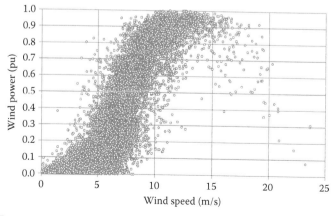

FIGURE 2.7

Example of the relation between measured power output of a single wind farm (in pu) and forecast wind speed (in m/s) (provided at 00 UTC for the next 24 hours) using a mesoscale NWP system.

Using wind speed forecasts, among other variables (e.g., wind direction and air density forecasts), as input of a point forecast model, it is possible to obtain forecasts for a single wind farm. Occasionally, it could be necessary to provide forecasts for a certain region that covers an ensemble of wind farms instead of one, as in [91]. In this case, the output of a set of wind farms could be calculated by adding the output forecast of each wind farm that is obtained individually. According to [91], this is the most accurate method to calculate the power output of an ensemble of wind farms. However, the feasibility of this type of approach depends on the number of wind farms to predict, on the quality of the data from predict, on the quality of the data from supervisory control and data acquisition (SCADA), and on the availability of data from NWP (computational complex and burden). To overcome these limitations, a set of representative wind farms may be used, followed by extrapolation of their sum to the total generation.

Figure 2.8 presents the performance of two different forecasting models. Figure 2.8a illustrates an aggregated output using the most representative wind farms, whereas Figure 2.8b illustrates an individual wind farm. The forecasts are provided at 12 UTC for the next day using artificial neural networks.

Considering the distribution of the errors in Figure 2.9, it is possible to conclude that the distribution of the forecasting errors of an ensemble of wind farms is narrower than that of a single wind farm, as expected. The uncertainty existent in aggregated forecasts is significantly lower than the uncertainty for a single farm, as mentioned before.

According to [92], there are different approaches to model the uncertainty existing in wind power forecasts, the NWP being the main source of error. Regarding the spot forecasts of NWP variables, they can be used directly as input in a probabilistic forecasting model or first converted in spot forecasts of wind power. Another approach to estimate uncertainty is use of ensemble forecasts of NWP. In this case, there are three ways to estimate the uncertainty: the ensembles may be fed directly to a probabilistic model, the

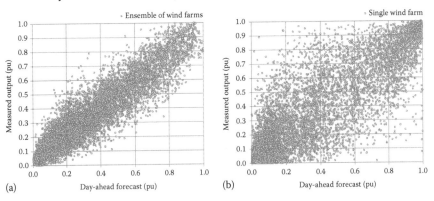

FIGURE 2.8
Relationship between forecast and measured wind power for (a) an ensemble of wind farms and (b) a single wind farm for a period exceeding one year.

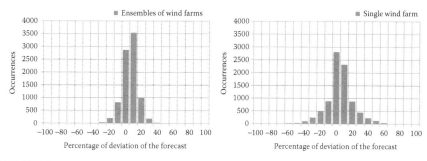

FIGURE 2.9
Distribution of the forecasting errors for (a) a set of wind farms and (b) a single wind farm.

ensembles are first transformed in a more tractable problem, or several point forecasts are provided and then converted into a probabilistic forecast.

2.3.2.1 Uncertainty Forecast in Wind Power

In order to provide an example of a regional probabilistic wind power forecast, the deterministic forecasts showed above will use the KDE technique, among other variables, as information about the prediction horizon. The values of the bandwidth were not optimized and were obtained by experience. The Gaussian kernel (the most used kernel) was chosen for all variables.

For each time step of the resulting probabilistic forecast, the estimated density was compared with the parametric beta distribution for different levels of the forecast wind power. The parameters of the beta PDF were obtained for each time step ahead using the method of moments, which is related with the mean and variance of the determined density for each time step [93,94]. The comparison is shown in Figure 2.10 and indicates that the beta distribution fits well with the density obtained by KDE, for different values of forecast power and different time horizons.

Although a parametric distribution, beta distribution could be advantageous in order to model the uncertainty existing in bounded variables with a maximum and minimum value, as in the case of the output of renewable energy resources. The inversion of the parametric density is easier to obtain, whereas for nonparametric density functions, it may be necessary to use specific programs. The beta PDF can be represented using four parameters (α, β, maximum value, and minimum value), which can be an important simplification when using these forecasts in other optimization problems such as scheduling.

The use of KDE in this case was useful, allowing use of multivariate forecasting problems having more than one explanatory variable. First, we obtain the estimation of the kernel density (which may be controlled by bandwidth parameters), after which the density is fitted to a beta distribution.

The deterministic and probabilistic forecasts for the total horizon (168 hours ahead) are presented in Figures 2.11 and 2.12.

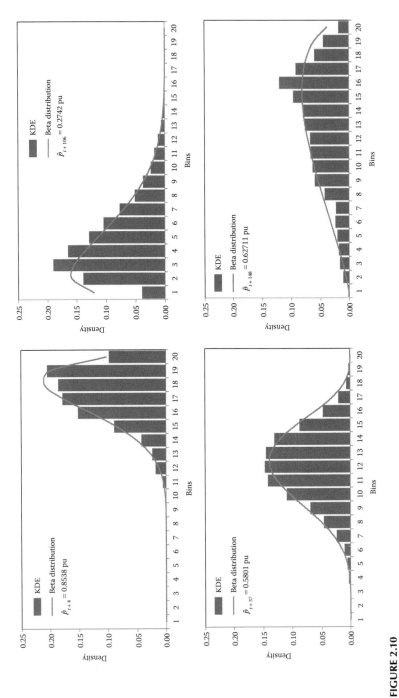

FIGURE 2.10
(See color insert.) A comparison between KDE and beta distribution obtained.

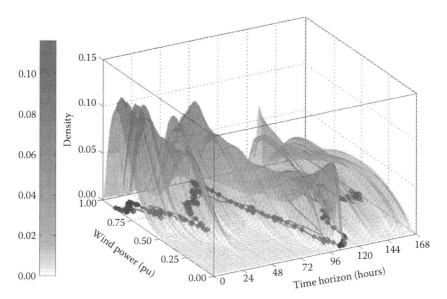

FIGURE 2.11
(See color insert.) Representation of the obtained probability density function of each step ahead of the forecast (for 168 hours ahead). The black dotted line represents the observed wind power and the red line the deterministic forecast.

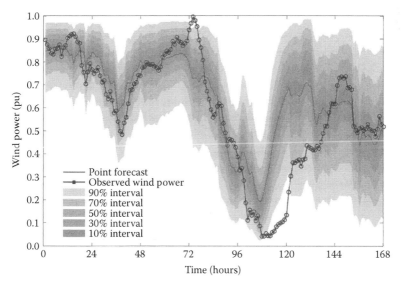

FIGURE 2.12
(See color insert.) Example of a point and a probabilistic wind power forecast for the next 168 hours for prediction intervals of 10%, 30%, 50%, 70%, and 90%.

It is interesting to verify that, as the time horizon increases, the uncertainty is modeled by larger distributions, due to the larger errors. The same conclusion can be extracted by analyzing Figure 2.12.

There are also other ways to represent the uncertainty [92], such as by a set of quantiles or forecasting intervals, as shown in Figure 2.12.

For instance, the confidence interval of 50% is defined by two forecast quantiles with nominal proportions of 25% and 75%, respectively.

2.3.2.2 Wind Power Forecast Scenarios

To deal with uncertainties in a scheduling problem, various stochastic approaches have been developed, the majority of the publications being focused on the impact of wind power forecast uncertainty. In [19,20], wind power scenarios are used to capture the temporal correlation between forecast errors that is not provided by probabilistic forecasts produced independently of the time horizon. In these cases, the wind power scenarios were calculated based on the methodology presented in [95], where the author describes the prediction errors for each look-ahead time through a covariance matrix. Another similar approach based on the covariance matrix is presented in [93].

Following the stages for the scenario generation presented in [95], along with knowledge of the probabilistic distribution forecasts (Figure 2.12) for each step ahead of the time horizon, it is possible to obtain wind power forecasting scenarios, as presented in Figure 2.13.

The 14 forecast scenarios represented in Figure 2.13 combine information about the uncertainty forecast with interdependence between the historical errors.

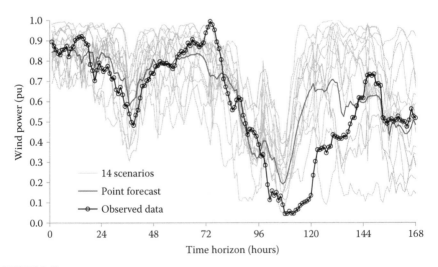

FIGURE 2.13
(See color insert.) Wind power forecast scenarios calculated based on the methodology presented in the literature. (Adapted from P. Pinson et al., *Wind Energy*, 2008.)

2.3.3 Solar Power Forecast

Most of the error in solar power forecasting is related to the prediction of the primary resource, the irradiance arriving to the energy conversion systems. Moreover, other variables such as the PV system temperature (which can be estimated from ambient temperature) and wind can be used but have less impact on solar power forecasting. For this reason, most of the developed works in this research area are centered around forecasting irradiance. Clouds are the main factor affecting the variability of the direct component of the irradiance in the short term, so studies on cloud forecasting are also found in the literature. The majority of the methods in the literature do not predict solar irradiance, because it is the most difficult piece of information to forecast.

Solar power forecasting methods can be divided into physical and statistical methods. Also, combined approaches can be found in the literature.

2.3.3.1 Physical Models

Physical models for irradiance forecasting consider how irradiance is related to other atmospheric variables with different approaches, depending on the computational power and input data available.

Several NWPs (and reference models) are specifically evaluated in [96] for intraday forecasts of solar irradiance. The region considered is continental United States and the models are model and then converted back to irradiance North American Mesoscale Forecast System (NAM), ECMWF, and GFS. In general, ECMWF performs better than the other two; MOS-corrected output results in GFS having a better overall result, whereas ECMWF is more accurate in cloudy conditions. Furthermore, there are studies where phase errors in mesoscale models are analyzed [97], the conclusion being that when higher errors occur in these models, they are not due to the higher resolution but due to inaccurate space–time locations of phenomena.

2.3.3.2 Statistical and Machine Learning Models

Statistical and machine learning methods perform forecasts by determining the parameters of a model [a formal structure such as autoregressive moving average (ARMA), neural network, SVR, and regression tree] that minimize some statistical measure of error when known measured behaviors are used. A review of artificial intelligence techniques applied to solar applications is presented in [98]. The different works are now discussed, separated by the main type of input used.

2.3.3.2.1 Ground Measurements

When ground measures of solar irradiance and other meteorological variables are available, many techniques can be applied (mainly divided into time series and neural networks) for single points or correlated networks of points.

In [99], a 1-hour-ahead irradiance (global, direct, and diffuse) is predicted from other variables: sunshine duration, temperature, and relative humidity. Two methods are compared: adaptive α-model and feed-forward neural network, the latter yielding better results. A method to forecast global irradiance has been discussed in [100], where cloud cover is used as input (estimated from direct irradiance measures), forecast with an autoregressive integrated moving average (ARIMA) model and then converted back to irradiance (giving irradiance forecast). Two other methods are also tested using ARIMA but with different irradiance amounts (global in one method, direct and diffuse in the other). Chaabene et al. show a method producing 5-minute-irradiance forecasts, based on a Kalman filter and, previously, daily conditions are forecast by an estimator of accumulated daily irradiance and minimum and maximum daily temperature, based on a neuro-fuzzy model [101]. Evaluation of six different methods (ARIMA, Feedforward Neural Network [FFNN], hybrid models, transfer functions, simple regression and Unobserved Component Models [UCM]) normally applied to forecasting Global Horizontal Irradiance (GHI) using time series of ground data is commented in [102]. Computing performance (neural networks are expensive computationally) and forecasts with different time resolutions (5, 15, and 30 minutes) are also considered. ARIMA, hybrid, and transfer function methods give the best results overall. At high resolutions, all methods perform equally well, whereas at low resolutions, ARIMA is the best.

2.3.3.2.1 All-Sky Cameras

All-sky cameras (also *sky camera* or *sky imager*) obtain colored images of the sky above a plant, normally with a wide lens. Its application to solar forecasting is currently a very active area of research. Methods using this kind of data are specialized in intra-hour forecasts, and give good results for DNI (direct normal irradiance). Currently, methods detect clouds, estimate cloud movement, and extrapolate future position assuming linear movement of clouds. Models of formation and dissolution of clouds (or other phenomena such as several layers of clouds and convection) have not been currently evaluated and are not found in the literature, which may not, however, be important for the time frame considered (less than 1 hour, sometimes only some minutes).

DNI is forecast in the method presented in [17] using a sky camera, and giving 1-minute samples with a forecast horizon going from 3 to 15 minutes. Cloud velocity is extracted, and a set of cloud fractions in the direction of cloud velocity is obtained. From the array of cloud fractions, assuming linear movement and a simple relation, future DNI values are obtained. Chow et al. [18] present a technique to measure and forecast clouds and irradiance in the order of minutes ahead, using a sky imager. Irradiance is computed using a clear-sky model, which is reduced to 40% when a cloud is detected as passing in front of the sun. Clouds are detected and converted to a binary mask and then cloud motion estimation is performed with a cross-correlation method.

2.3.3.2.2 Satellite Images

Satellite images coming from stationary meteorological satellites allow the estimation of irradiance over wide areas, and existing forecasting methods based on these images give good results for up to 6 hours. Methods are based on the determination of parameters of clouds (cloud type, cloud mask, cloud index, etc.), then its motion is estimated and then assuming persistence of this motion, future position of clouds is obtained. Later, the application of models such as Heliosat-2 [103] permits to obtain a forecast of irradiance.

Studies discussed in [104,105] are centered on the forecasting of cloud motion from satellite images. Cloud motion vectors (CMV) are first determined by cross-correlation (block matching), all pixels are filled (in places where there is no cloud speed or where cloud speed cannot be estimated, an interpolation scheme is used), and then a rigid motion is assumed using backward trajectory based on a semilagrangian scheme. For validation, cloud contours are extracted. Evans focuses on the calculation of cloud motion vectors using an iterative technique that allows determination of smooth vector fields [106].

2.3.3.3 Combined Approaches

In [107], NWP forecasts [from ECMWF and Germany's Deutscher Wetterdienst (DWD)] are combined with satellite forecasts (obtained from CMV extracted from Meteosat Second Generation images), linearly weighting models with a different regression for each forecast horizon, and then eliminating resulting bias. For day ahead, the two NWPs are combined; for intraday, satellite forecasts are added to the mixture. The mixture improves the performance and results are even better in regional forecasts. It advances lines of research: integration of power measures in hour-ahead forecasts, consideration of additional NWPs (for day ahead), application of methods for the integration of different forecasts, and usage of infrared images to determine CMV (satellite forecasts).

2.3.3.4 Uncertainty in PV Production Forecast

Techniques developed in the wind power forecasting area have been applied to solar power forecasting, but uncertainty forecasts could be further explored. For instance, Bacher et al. model the uncertainty using quantile regression [108].

As we have seen, in PV power forecasting, the most important variable used to predict solar power is global irradiance, which is conditioned mainly by the evolution of clouds. Although not taken into account in most of the reviewed forecasting models, we will use ambient temperature forecast in our model, which is relatively stable and well predicted by NWP models.

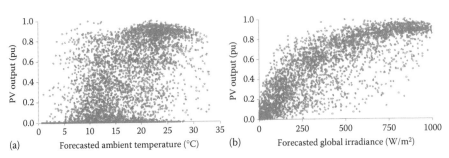

FIGURE 2.14
Relationship measured between (a) the PV output and forecast temperature and (b) the PV output and global irradiance.

Figure 2.14 presents the relationship between the PV production of an ensemble of plants (almost a year) and global irradiance and temperature forecast resulting from an NWP provided at 00 UTC for the next 24 hours.

Meteorological forecasts shown are averages weighted by the nominal power of each PV plant. Apart from these variables, solar time and estimated radiation by the clear sky model (ignoring the presence of clouds in the atmosphere) were also used as inputs. By relating the production of PV power plants and the ambient temperature, it is clear that for higher temperatures, there is a reduction in the maximum value of power achieved by the power plants (lower efficiency).

In order to provide an example of deterministic and probabilistic forecasts applied to the production of an ensemble of PV power plants, we will use the KDE technique again; the resulting histogram is fit to a beta distribution. The forecasts are tested for cloudy, partially cloudy, and cloudless days for the next 24 hours (Figure 2.15).

Although WRF is used for forecasts on both clear and cloudy days, the solar irradiance computed using WRF still has considerable error, leading to high uncertainties in PV power forecasts, especially on days affected by the presence of clouds. The forecast is much accurate in summer, for days with clear sky conditions.

2.3.4 Mini Hydropower Forecast

The mini-hydropower plants are low-rated power plants localized in rivers with relatively reduced water flows and low storage capacity, compared with the reservoirs of large hydropower plants. If they are of *run-of-river type*, the flow regulation capacity is neglected. In a very simple approach, it can be considered that energy production in this kind of power plants depends mainly on the water level stored in the reservoirs in the case of facilities with a reservoir, or from flow in the case of run-of-river power plant; in both cases, the approaches depend from rainfall. However, the relationship between rainfall and inflow is highly nonlinear [109–111].

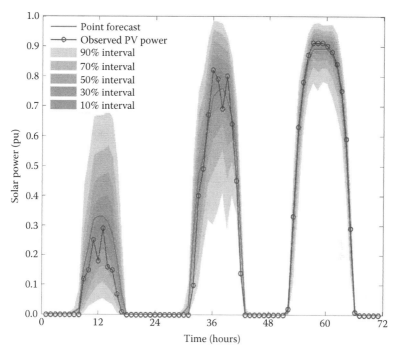

FIGURE 2.15

(See color insert.) Example of a probabilistic PV power forecast. Each forecast is provided at 00 UTC for the next 24 hours.

Contrary to wind production, where the wind speed and the power produced are connected, in the hydropower plants, rainfall may not correspond to the power production [65]. In this sense, several authors [109,111–116] have dedicated their studies to develop inflow forecast methods, taking into account the scope of predictions and physical features that characterize the watershed area, inflow characteristics, operation strategies, and forecast horizons. In [117], water inflows are considered to depend not only on rainfall but also on environment temperature and, in more specific cases, snow melting, air humidity, large watersheds, and localized storms.

Hydro forecast models do not have the same maturity level as wind power models, namely, probabilistic forecasts; the majority of the published works are based on statistical models.

Statistic models look for strong relations between the inflow or power historical data variables; the information is measured in real time, using recursive techniques. Statistical models present an advantage, in that they do not need physical information. However, in the parameter estimation process, it is necessary to determine a large set of historical data and information in real time. The artificial neural network (ANN), in spite of not being a pure statistical method, has intrinsically the capacity to reach nonlinear mathematical relations, contained in the watershed models. Beyond that, it shows relative ease of implementation, being one of the most common tools used in inflow and power production forecasting in hydropower stations.

In [113,115,116], the reservoir inflow forecasting for a mini-hydropower plant based on rainfall forecasting is proposed. Two temporal horizons were considered: very short term, using historical data of rainfall and inflow measured along the watershed, and short term, based on rainfall forecasting and, consequently, inflow to reservoir. In both cases, the values were subsequently processed using ANN of multilayer perceptron (MLP) [116].

In [109], a study on daily river inflow forecasting, without considering power production, is presented. In this study, two forecast methods, namely, self-organizing neural networks and autoregressive moving average (ARMA), are compared based on time series.

Many other topics based on ANN were studied, such as echo state network, self-organizing nonlinear autoregressive model with exogenous (SONARX) input, radial-based functions (RBF), and artificial neuro fuzzy inference system (ANFIS) [114].

In [118], the author used ANN with several learning algorithms such as back-propagation, conjugate gradient, cascade correlation, and Levenberg-Marquardt for short-term forecasts of the diary stream flow. In [119], inflow forecasts with basic gradient descendent method, resilient back-propagation, scaled conjugate gradient, and Levenberg-Marquardt are presented; the results are compared with the ARIMA method. Similar studies are presented in [120], where daily forecasts of inflows using the ANN-type MLP and RBF are presented. Using past values of rainfall and runoff as network inputs to forecast the inflow with MLP presented slightly better results. According to [118] and [121], on average, 90% of applications of hydrological studies use MLP with back-propagation.

2.3.4.1 Short-Term Forecast of Regional SHP Production

The forecasting model described here takes advantage of the short-term forecasting model presented in [65], named H4C. H4C is used to provide a deterministic forecast of the aggregated power production of several SHPs for the next few days with hourly resolution, using historical values of the SHPs production as input variables, as well as forecasts of precipitation resulting from WRF. The idea is to use the KDE technique instead of ANN to obtain a probabilistic forecast, as we did above with wind and PV power.

Figure 2.16 presents the temporal evolution of the total power production in a certain region of a set of SHPs, and the average precipitation (forecast for the next 24 hours ahead provided by WRF) weighted by the rated power of each mini hydropower plant and forecast precipitation for areas near the power plants.

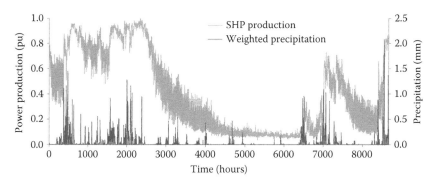

FIGURE 2.16
(See color insert.) Temporal evolution of the power production of the aggregated SHPs and the average precipitation.

Analyzing Figure 2.16, it is clear that a direct relation between the production and precipitation (due to phase-shift errors in precipitation forecast, different watershed characteristics and, maybe, operation strategies) cannot be obtained. In order to establish a mathematical relation between those variables, a new variable called *hydrological power potential* (HPP) is defined. To estimate it, first we convert the hourly values (of power and precipitation) to daily moving averaged values, as in Equation 2.12 [65].

$$H_{d,h} = B\left(H_{d,h-1} + A \cdot R_{d,h}\right) \tag{2.12}$$

where:
$H_{d,h}$ is the calculated HPP for the hour h and the day d in pu
$H_{d,h-1}$ is the calculated HPP for the hour $h - 1$ and the day d, in pu
A is a parameter related to the incremental response of the electric power generation to the precipitation, in pu/mm
B is a dimensionless parameter that is related to the speed of decrease of such generation in dry days. This parameter has values lower than 1
$R_{d,h}$ is the forecast precipitation for the hour h and the day d, in mm

Parameters A and B are obtained by the least squares error method, taking as error the difference between the daily average recorded power production value and the corresponding HPP value. The result is presented in Figure 2.17.

Next, as in [65], a sigmoid curve (Equation 2.13) will be used to adjust the maximum and minimum power of the SHPs.

$$P_{est\_d,h} = P_{min} + \frac{P_{max} - P_{min}}{\left(1 + e^{\left\{-8\left[(H_{h,d} - h_c)/h_s\right]\right\}}\right)} \tag{2.13}$$

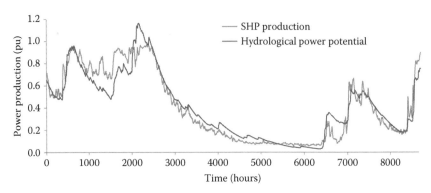

FIGURE 2.17
(See color insert.) Temporal evolution of the daily moving averaged power production of the aggregated SHPs and the corresponding calculated HPP.

where:
   $P_{est\_d,h}$ is the forecast power production for hour h and day d, in pu
   P_{max} is the maximum power production, in pu
   P_{min} is the minimum power production, in pu

h_c and h_s are the parameters to adjust the sigmoid function of 2.13 and are calculated by Equation 2.14.

$$\begin{cases} h_c = \dfrac{h_0 + h_m}{2} \\ h_s = h_m - h_0 \end{cases} \qquad (2.14)$$

where:
   h_0 is the value at which HPP reaches the minimum value of power production
   h_m is the value at which HPP reaches the maximum value of power production

Figure 2.18 shows the relation between the HPP with the real averaged power production and with the adjusted sigmoid, $P_{est\_d,h}$.
   When the measured data of hydropower production is available (forecasting system connected to the monitoring system) for hour h, $P_{record\_d,h}$, it could be used to adjust the value of HPP calculating a new value of H_h, called $H_{new\_d,h}$, which corresponds to the new adjusted hydrological power potential value for instant h (data assimilation—Equation 2.15) [65].

$$H_{new\_d,h} = h_c - \frac{h_s}{8} \ln\left[\frac{1}{P_{record\_d,h} - P_{min}} \cdot (P_{max} - P_{min}) - 1 \right] \qquad (2.15)$$

This adjustment is made with the intention of increasing the accuracy of the forecasting models and reducing the forecasting errors introduced by NWP.

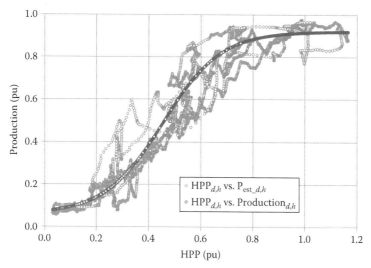

FIGURE 2.18
(See color insert.) Relation between HPP with the parameterized sigmoid and the measured values (average).

After this process, an ANN can be used to convert this information to hourly power production. Among the HPP, hourly information can be used as input in order to model the production pattern (if exists).

2.3.4.2 Uncertainty in SHP Production Forecast

Instead of using an ANN to convert the HPP to power production, we will use the KDE technique in order to provide not only a single-value forecast but also some quantification of the uncertainty existing in predictions. Here, information of the time horizon is also used to model the increase of uncertainty with the horizon length and the degradation of the quality of the precipitation forecasts. The achieved result is presented in Figure 2.19.

As in case of wind power, after obtaining a predictive PDF for each step ahead of the forecast, the uncertainty can be represented in many forms. For each step ahead, the uncertainty is represented by a beta distribution.

2.3.5 Load Forecast

During the last few years, a huge set of forecasting models based on statistical and artificial intelligence were used to provide short-term load predictions [122], such as similar-day approach, regression methods, time series, neural networks, expert systems, fuzzy logic, and support vector machines. Also, adaptive forecasting models are referred in [123].

The analyzed data of consumption, in this chapter, corresponds to a national aggregate demand (consumption centers geographically dispersed).

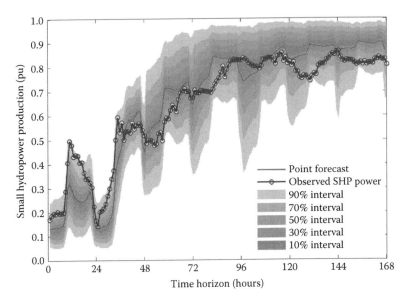

FIGURE 2.19
(See color insert.) Prediction intervals of a probabilistic forecast of hydro production obtained from an ensemble of small hydropower plants provided at 00 UTC for the next 168 hours.

As in RESs, this aggregation has a smoothing effect that reduces variability, making this a more predictable variable. Typically, time factors such as hour of the day (to distinguish peak and off-peak periods) and day of the week (weekday, weekend, or public holiday) are used as explanatory variables in load forecasting models in the short term (timeframe from 1 hour to 1 week). In this sense, the variation of the load along the time is dependent of the consumer's behavior which varies every hour and depends on the day (if it is a workday or non-workday). This pattern is explicit in Figures 2.20 and 2.21.

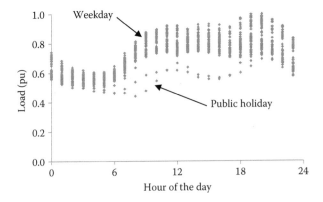

FIGURE 2.20
(See color insert.) Relationship between electric load consumption (pu) and time of the day in hours for a weekday (Wednesday).

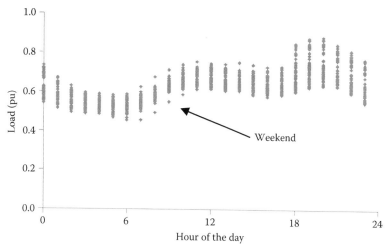

FIGURE 2.21

Relationship between electric load consumption (pu) and time of the day in hours in weekend (Sunday).

From Figure 2.20, it is possible to conclude that information about public holidays (nonworkday) is also useful information, once the social behavior on these days is similar to a Sunday, as shown in Figure 2.21. The magnitude of load is higher in weekdays than in weekends or holidays, so it is clear that the use of information on hours of the day and whether the day considered is a workday or not may be useful for the accuracy of the forecasting model. Also, the uncertainty seems to be higher during peak periods than during off-peak periods.

Also, weather variables, such as ambient temperature, affect the demand of electricity. Figure 2.22 shows a slight increase of consumption in the summer period that is explained by the use of cooling (due to higher ambient temperatures).

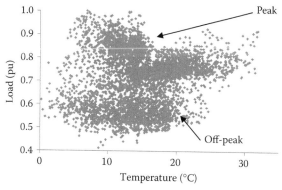

FIGURE 2.22

Relationship between temperature (next 24 hours forecast using WRF) and electric load (weekdays).

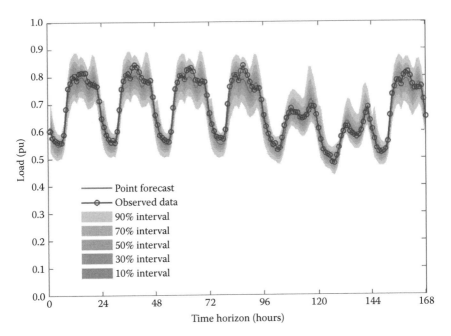

FIGURE 2.23
Example of a probabilistic load forecast for the next 7 days.

The increase in consumption during winter periods is much more perceptive. In this case, the situation may be attributed to use of electricity in heating due to lower temperatures.

When recently observed information is available (from SCADA), delayed values of the variable used to predict data to be integrated in forecasting models in order to obtain better performance in predictions (consumption in the previous hour, consumption in the same hour of the same day of the last week, etc.) may be advantageous.

Forecasts of ambient temperature for the next 168 hours (WRF), hour of the day, day of the week, holiday's, and time horizon information were used as explanatory variables of a deterministic forecasting model, in order to obtain point forecasts of electric load consumption. This variable is posteriorly used as input of KDE, and then a beta distribution was used to fit the density obtained, providing the results that are presented in Figure 2.23.

The uncertainty of this type of variable seems to be low and stable along the entire forecast horizon.

2.3.6 Net Load Forecast—Dependence and Uncertainty Aggregation

In a generation power system with renewable sources power plants, the renewable productions due to this *clean* resource have some privileges when the generation mix scheduling is done. Although its variability and forecasting challenges are considered as priorities, the conventional (thermal and large hydro) units are committed based on economic and technical aspects. While the conventional production is controllable and easy to manage, the remaining variables of the scheduling process, such as the load and renewable productions, have to be forecast. Associated with these forecasts there are always errors that can call into question the results of unit commitment (UC).

In this sense, as renewable generation are nondispatchables sources of energy, they can be seen as negative loads [124], and by this way, the definition of net load is created. Fundamentally, net load represents the amount of load that must be fed by conventional generation units. Thereby, in the UC and economic dispatch formulation, the load demand of balance equation is replaced by the net load.

Figure 2.24 shows an example of net load forecast with uncertainty in a power system with several renewable power productions. One of the added values is the uncertainty aggregation of several renewable and load demand forecasts in one variable. The process starts with the meteorological forecasts. Depending on the production technology, the treatment of spot forecasts is done independently and converted into power production and load demand (as shown above for each technology and demand). At the end, the aggregated production is subtracted from the load demand forecast, resulting in a net load point forecast. At this point, several initial variables were converted into only one variable.

As mentioned above, all the values that result from forecasts have an associated uncertainty that can be modeled by several probabilistic techniques. Generally, as wind production has much more weightage than the remaining renewable production such as solar and small hydro, the common approaches for the net load forecasts with uncertainty are done considering only wind power production. Some approaches consider the load demand and wind production random functions that are described by point forecasts and zero-mean normally distributed errors (due to the central limit theorem). Therefore, considering wind production and the load demand as independent random variables (uncorrelated) modeled as normal distributions $N(\mu_W, \sigma_W)$ and $N(\mu_L, \sigma_L)$, respectively, the net load forecast results from the subtraction of the means and the deviations are given by $\sigma_{NL}^2 = \sigma_L^2 + \sigma_W^2$ [124–126].

If the uncertainties are characterized by different types of PDFs, assuming that all renewable sources and load are statistically independent, to evaluate the probability distribution of the sum (or subtraction) of two or more independent random variables, a convolution must be applied to their individual distributions [127–129].

A more complex methodology can be found in [130], where the authors use the copula theory in order to model the stochastic dependence between random variables. The net load distribution is calculated taking into account the dependence between wind power and system load.

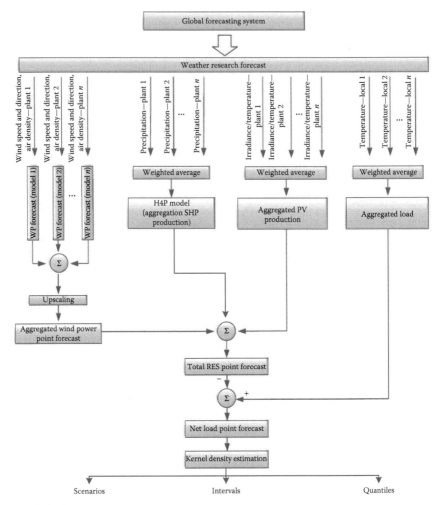

FIGURE 2.24
Methodology used to achieve net load forecast with uncertainty quantification.

A good forecast of the *net load* could give trustable information to the grid operator in order to improve the system operation costs. It is notorious that the quality of the forecasts is essential to the success of the scheduling.

As we have seen in Figure 2.24, load demand is a much more predictable variable (low forecasting errors) when compared with the predictability of RESs. Furthermore, the uncertainty keeps relatively stable along the 168 hours ahead contrary to, for instance, the wind power. The uncertainty existing in RES forecast is reflected in net load forecast. Figure 2.25 presents a 168-hour-ahead net load forecast, following the scheme in Figure 2.24.

The time horizon information was also used as explanatory variable of this target variable in order to model the increasing of the uncertainty with

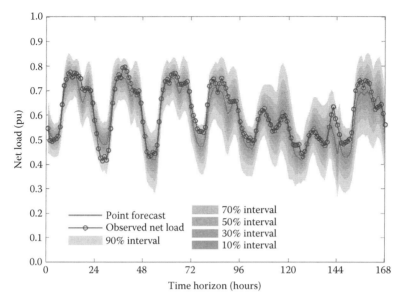

FIGURE 2.25
(See color insert.) Representation of the net load uncertainty by a set of intervals forecast, based on methodology explained in Figure 2.24.

the increasing of the time horizon, observed in historical data. Following a similar approach as the one presented in the subsection of wind power forecast, a set of scenarios based on covariance matrix could be produced in order to provide net load scenarios to feed stochastic scheduling tools.

2.4 Forecast Service and Usage

In the framework of the European project SiNGULAR two platforms were developed: one providing updated forecasts of RES generation and net load and another that is fed by probabilistic forecasts for UC in the electrical system of São Miguel Island in the Azores.

2.4.1 Interface for Forecasting Service

Within the scope of the SiNGULAR project, a forecasting web service platform containing information for five different insular systems was developed. The forecasting system is available online at singular.smartwatt.net. This system is operational, producing forecasts for the next 7 days for each variable in each target location with a time step of an hour. When historical information of consumption or production is available, the uncertainty

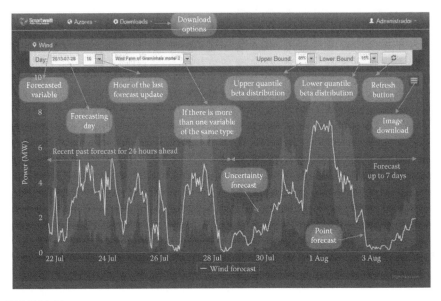

FIGURE 2.26
(See color insert.) Options and information provided by the forecasting platform developed for the wind farm of Graminhais in São Miguel Islands in the Azores.

is also modeled. Figure 2.26 shows the aspect of the developed forecasting platform. Furthermore, it presents some of the options and information provided by this service. It is possible to download the past forecast data and the forecast for the next few days with time discretization of an hour. The forecasts are refreshed four times a day.

This type of information could be used to feed the own system operator (SO) models. Also, within the scope of this project, a methodology based on risk analysis for scheduling optimization, designed for insular electricity networks, in this case, São Miguel Island, was developed and presented below.

2.4.2 Usage of Uncertainty Forecast in Islands' Scheduling

The integration of large amounts of RESs in typical insular systems, especially during peak load, can decrease the fuel consumed by thermal generators reducing the system operation costs. However, there are some technical challenges that need to be overcome for an efficient management of this type of insular grids due to the highly stochastic nature of several RES output. For instance, in São Miguel's power system (Azores), during off-peak periods, the system is already saturated with RES, so it is necessary, most of the times, to limit the wind power output in order to maintain all committed thermal generators operating above their technical minimum values.

The optimization of committed thermal units, by minimizing the number and size of online thermal units, can maximize the integration of RES and minimize its curtailment. On the other side, an excessive reduction of the maximal thermal capacity committed could lead to situations wherein the spinning reserve is not sufficient to meet high values of net load. Therefore, due to the uncertainty (that exists) in load and RES generation forecasts, in general, it is not possible to find a completely robust scheduling solution where technical minimum and maximum values of thermal units fit all range of the probable net load.

To mitigate this kind of problem, a short-term scheduling approach, especially designed for insular electricity networks, based on risk analysis was developed. The generation scheduling optimization was designed to minimize the sum of the estimated costs based on risk cost analysis. The probabilistic estimation of costs is not based on scenarios, but is based on the expected risk estimation approach that uses directly the predictive PDFs. In this approach, a reserve level is not predefined, but reliability and irregular operation probability minimization lead to solutions with enough dynamic reserve levels.

The model developed consists of evaluating the adequacy of each possible combination of thermal generators online in a UC problem for each hour of a probabilistic net load forecast, avoiding the need for the development of a large number of scenarios, modeling explicitly the impact of the forecast uncertainty and considering the possibility of single thermal unit failure. The start-up costs are taken into account using a dynamic programing technique. In the objective function of this problem, the fuel consumption curves of thermal units (in normal operation mode), the probabilities of the thermal generators operating inside/outside of their appropriate range of power (i.e., risk of load shed and wind curtailment necessity based on probabilistic forecasts), and the probability of normal operation after the occurrence of wind curtailment were also considered. A more detailed description of this methodology applied to São Miguel's power system can found in [131].

2.4.2.1 Short Description of the Methodology

The proposed methodology is divided into two stages. The first one consists of a preprocessing stage, which is done only once (offline), and the second one, which is processed online whenever updated forecasts are available.

The preprocessing stage consists of the following steps:

1. *Determination of fuel consumption curves for each thermal unit.* In the case of São Miguel Island, they were estimated using the information of specific fuel consumption of each unit.
2. *Definition of all possible combinations of thermal generators online (GENSET) and associated technical constraints* (maximum/minimum output). As there are eight units in the analyzed case study, it is possible to

define 255 possible combinations of units that could be in operation in order to meet a specific net load value.

3. *Resolution of an economic dispatch problem.* For each GENSET defined in (2), the fuel consumption curves of each generator obtained in (1) can be used to optimize the operating point by the Lagrange method:

   a. Precomputation of the *combined* optimal fuel consumption and the production curves of each generator of each GENSET as a function of the net load. The idea of this offline process is to precompute a database of the *combined* optimal system fuel consumption curves and production curves of each generator belonging to a specific GENSET as a function of the net load when different generators are combined in operation. For this, an economic dispatch problem is solved using the fuel consumption curves calculated in (1), optimizing the operating point using the Lagrange method, but respecting the maximum and minimum constraints of each generator of the GENSET.

      At this point, the preprocessing part of the methodology is complete. The online processing comprises the following steps:

4. *Get updated probabilistic forecasts of RES and load.* The forecasts used to feed the methodology presented are from the platform present above in Figure 2.26. The renewable generation system is composed by two geothermal power plants, seven small hydropower plants and one wind farm (Table 2.1).

   After obtaining the density using KDE for each hour of each variable, an approximation to a beta distribution is made, as exemplified in point 2.3.2.1.

5. *Obtaining the net load uncertainty forecast.* In this work, it was assumed that forecast renewable sources are statistically independent random variables and their sum can be derived by convolution. The PDF of the net load is computed by the convolution of the load PDF forecast and the symmetric of the RES PDF forecast. After convolving these random variables, the density obtained is, again, approximated to a beta distribution. At this point, it is also necessary to calculate the net load, not considering the wind generation, in order to emulate the net load uncertainty behavior when wind power is curtailed.

6. *Evaluation of the adequacy of each GENSET.* Evaluation of the adequacy of each GENSET. In the UC problem, for each hour ahead of the net load forecast, each GENSET is evaluated individually independently of the scheduling solution in the previous hours. A predictive PDF and the cumulative distribution function of net load for a specific hour are represented in Figure 2.27. In order to show all the possible areas where the system may operate, the technical limits of a specific GENSET (combination of generators in operation) are represented.

TABLE 2.1

Explanatory Variables Used in Forecasting Models of Wind Power, Hydro Power, Geothermal Power, and Load for São Miguel Island

Forecast Variable	Explanatory Variables—Model
Wind power	Forecasts from WRF (next 7 days): • Wind speed (m/s) • Wind direction (°) • Air density (kg/m³) The WRF variables are used directly into the KDE technique.
Hydropower	Forecasts from WRF (next 7 days): • Precipitation (mm) First using H4P model and then precipitation forecasts are estimated from the HPP. After that, the HPP is used as input in KDE.
Geothermal power	This power production is a relatively constant source of energy. The output is defined by set points that rarely suffer large changes and is being used as the base of the load diagram. In this case, the main source of errors is the unexpected outages of some units. Using the historical information, the set point is estimated by a moving average of the final hours of production. Obviously, in the operation mode, it is assumed that all geothermal units are available and the only uncertainty is related with small oscillations around a predefined set point. The uncertainty is modeled using KDE.
Load	Forecasts from WRF (next 7 days): • Temperature (°C) • Hour of the day • Day of the week (including public holiday information) • Week of the year The variables are used directly into the KDE technique.

As it can be observed Figure 2.27, there are three different possible areas of operation: normal operation area, load shed risk area, and wind curtailment risk area. For each hour, the best solution is the one that maximizes the probability (area) of normal operation and minimizes the load shed and wind curtailment probabilities. The latter two areas penalizes the evaluation cost function. The penalization per unit of shed energy of load is higher than the wind curtailed. Ideally, to avoid load- or RES-shed measures, or even prevent, the thermal generation units operate below their technical minimum, for each hour should be selected as a GENSET solution that respects the restrictions $min_{GENSET} \leq min_{L\_net,h}$ and $max_{GENSET} \geq max_{L\_net,h}$.

However, in the case of São Miguel Island, due to the discrete solution set, this condition is impossible to be achieved most of the time. When there is a probability of the thermal units operating above their technical minimum (the same as wind curtailment), the impact limit of all the available wind power is analyzed. In this case, net load forecasts are calculated only considering geothermal and hydro

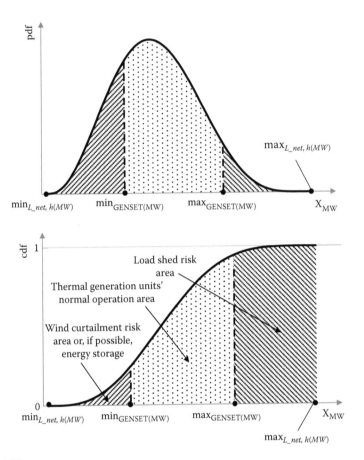

FIGURE 2.27
Representation of the net load uncertainty associated to a specific hour and the risk associated to a specific committed set of generators (GENSET). The variables $min_{L\_net, h}$ and $max_{L\_net, h}$ represent the minimum and maximum value of the predictive PDF and CDF (cumulative distribution function) of the net load forecast for a specific hour, respectively. The variables min_{GENSET} and max_{GENSET} represent the technical minimum and maximum power output of the GENSET in evaluation.

production. If even after this extreme measure, there is still a chance of wind curtailment, then the risk-based cost function is penalized proportionally to that probability.

In this analysis, both the start-up costs and the possibility of contingency of a single thermal unit were ignored.

7. *Inclusion of thermal generation unit's contingency N − 1.* Considering that there is a probability of failure of any thermal unit, the cost of neighborhood GENSETs to the evaluated one is included in the objective cost function. This parcel is affected by the probability that this situation happens, penalizing the objective of the risk-based cost function.

8. *Inclusion of the start-up costs of thermal generation.* After evaluating the adequacy of each GENSET singly, the costs resulting for each hour ahead are combined with the start-up costs of thermal generation units using dynamic programming. The best solution for each hour will depend on the solution of the previous hour and sometimes the GENSET with lower risk-based cost (at 7) is not the selected one due to the consideration of start-up costs of offline units that may endear the solution.

2.4.2.2 Achieved Results

Figure 2.28 shows the results of net load forecasts for the last week of October 2014. The dashed line represents the observed net load after the wind curtailment by the SO. The dotted line represents theoretical net load if the system operator takes advantage of all the available wind resource (avoid wind curtailment). The red line represents the net load point forecast. The blue area represents the confidence interval of 90% defined by two forecast quantiles with nominal proportions equal to 5% and 95%.

Analyzing Figure 2.28, we find that the wind power is curtailed basically in off-peak hours where the net load is lower. This limitation is done in order to try to keep the thermal generators operating above their technical minimum value.

Figure 2.29 shows the scheduling solution provided by SO for the same week and the solution provided by the methodology developed, which uses probabilistic forecasts as inputs.

The maximum capacity of the dispatched thermal generation units at each instant by SO is sufficient to meet the net load. However, in some instances,

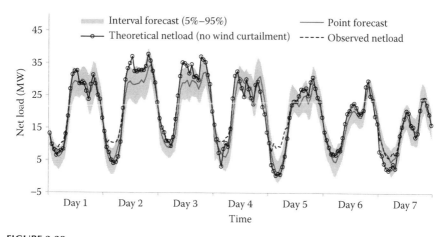

FIGURE 2.28

Comparison between forecast net load for electric system of S. Miguel and the observed net load. It is also Represented also is the theoretical net load taking advantage of all available wind resources.

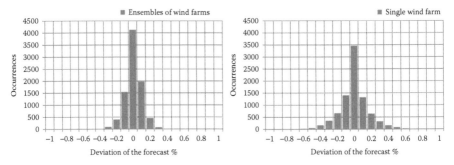

FIGURE 2.29
(See color insert.) Scheduling solutions provided by: (a) the methodology developed and (b) system operator (SO).

the number of committed units is too much higher than necessary. This excess of committed units leads to the following two problems:

1. Sometimes the net load is so low that the generators are operating above their minimum, especially during off-peak hours.
2. Even after the wind power curtailment (in order to raise net load), generators still operate under undesirable conditions.

Analyzing the results, we conclude that with the model developed, it is possible to minimize the wasted wind energy and reduce the energy produced by thermal units under undesirable conditions. However, for some hours when forecasts have higher errors, there could be a necessity of load shed. Analyzing the performances of the probabilistic net load forecast using some indicators such as reliability and sharpness, we conclude that there is the necessity to improve the probabilistic forecasts once the success of this approach, and the generality of the approaches, is strongly dependent of forecasting quality. Otherwise, the results can be compromised.

The reliability and sharpness (Figure 2.30) of the probabilistic net load forecasts were calculated for a period of 5900 hours (only periods where all

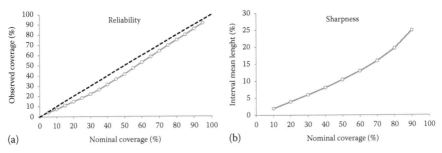

FIGURE 2.30
(See color insert.) Reliability and sharpness of probabilistic net load forecast.

geothermal units available were considered for these calculations). The forecasting method used in this approach tends to systematically underestimate the uncertainty since nominal quantiles proportions are greater than the observed ones (the deviations achieve the 8%). This situation is not beneficial for our scheduling application, because uncertainty is underestimated, leading to situations in which the resulting reserve is not sufficient to meet the net demand.

The values of sharpness are relatively low, with a nominal coverage of 90% corresponding to around 25% of the net load rated. To rate net load, the highest value registered in the data was considered. These low values of sharpness lead to *narrow* uncertainty intervals resulting in the underestimation of the uncertainty, as seen in the reliability index.

References

1. S. K. Sheikh and M. G. Unde, Short-term load forecasting using ANN technique, *International Journal of Engineering Sciences and Emerging Technologies*, vol. 1, no. 2, pp. 97–107, February 2012.
2. X. Wang, P. Guo and X. Huang, A review of wind power forecasting models, *Energy Procedia*, vol. 12, pp. 770–778, 2011.
3. X. Zhao, S. Wang and T. Li, Review of evaluation criteria and main methods of wind power forecasting, *Energy Procedia*, vol. 12, pp. 761–769, 2011.
4. C. Monteiro, R. Bessa, V. Miranda, A. Botterud, J. Wang and G. Conzelmann, *Wind Power Forecasting: State-of-the-Art 2009*, Report ANL/DIS-10-1, Argonne National Laboratory, Chicago, IL, 2009.
5. R. Ulbricht, U. Fischer, W. Lehner and H. Donker, First steps towards a systematical optimized strategy for solar energy supply forecasting, in *Proceedings of the 2013 ECML/PKDD International Workshop on Data Analytics for Renewable Energy Integration (DARE)*, Prague, Czech Republic, September 23, 2013.
6. T. Reynolds, Y. Glina, S. Troxel and M. McPartland, *Wind Information Requirements for NextGen Applications Phase 1: 4D-Trajectory Based Operations (4D-TBO)*, Project Report, Lincoln Laboratory, Massachusetts Institute of Technology, Lexington, MA, February 20, 2013.
7. G. Giebel, P. Sørensen and H. Holttinen, *Forecast Error of Aggregated Wind Power*, Report for the TradeWind project, Risø, Roskilde, Denmark, 2007.
8. U. Focken, M. Langea, K. Mönnicha, H.-P. Waldla, H. G. Beyerb and A. Luigb, Short-term prediction of the aggregated power output of wind farms— A statistical analysis of the reduction of the prediction error by spatial smoothing effects, *Journal of Wind Engineering and Industrial Aerodynamics*, vol. 90, pp. 231–246, 2002.
9. T. Adams and F. Cadieux, Wind power in Ontario: Quantifying the benefits of geographic diversity, in *Proceedings of the 2nd Climate Change Technology Conference*, McMaster Universiy, Hamilton, Ontario, Canada, May 12–15, 2009.

10. V. C. Leaney, D. Sharpe and D. Infield, The applicability of spatial analysis techniques to monitoring wind farms, in *Proceedings of the European Wind Energy Conference*, Nice, France, pp. 216–219, March 1–5, 1999.

11. E. Lorenz, T. Scheidsteger, J. Hurka, D. Heinemann and C. Kurz, Regional PV power prediction for improved grid integration, *Progress in Photovoltaics: Research and Applications*, vol. 19, pp. 757–771, 2011.

12. A. Jain and B. Satish, Cluster based short term load forecasting using artificial neural, in *Proceedings of the 2009 IEEE PES Power Systems Conference Exposition*, Washington, DC, March 15–18, 2009.

13. G. Kariniotakis, P. Pinson and N. Siebert, The state of the art in short-term prediction of wind power—From an offshore perspective, in *Proceedings of the 2004 SeaTechWeek*, pp. 1–13, Brest, France, 2004.

14. W. Sokołowska, J. Opalka, T. Hossa and W. Abramowicz, The quality of weather information for forecasting of intermittent renewable generation, in *Proceedings of the 28th EnviroInfo 2014 Conference*, University of Oldenburg, Germany, September 2014.

15. S. G. Benjamin, J. M. Brown, K. J. Brundage, B. E. Schwartz, T. G. Smirnova, T. L. Smith, L. L. Morone, *RUC-2 – The Rapid Update Cycle Version 2*, NWS Technical Procedures Bulletin, Environmental Modeling Center, National Centers for Environmental Prediction, Camp Springs, MD, 1998.

16. A. Hammer, D. Heinemann, E. Lorenz and B. Luckehe, Short-term forecasting of solar radiation: A statistical approach using satellite data, *Solar Energy*, vol. 67, pp. 139–150, 1999.

17. R. Marquez and C. F. Coimbra, Intra-hour DNI forecasting based on cloud tracking image analysis, *Solar Energy*, vol. 91, pp. 327–336, May 2013.

18. C. W. Chow, B. Urquhart, M. Lave, A. Dominguez, J. Kleissl, J. Shields and B. Washom, Intra-hour forecasting with a total sky imager at the UC San Diego solar energy testbed, *Solar Energy*, vol. 85, pp. 2881–2893, November 2011.

19. J. Wang, A. Botterud, R. Bessa, H. Keko, L. Carvalho, D. Issicaba, J. Sumaili and V. Miranda, Wind power forecasting uncertainty and unit commitment, *Applied Energy*, vol. 88, pp. 4014–4023, November 2011.

20. A. Botterud, Z. Zhou, J. Wang, J. Valenzuela, J. Sumaili, R. Bessa, H. Keko and V. Miranda, Unit commitment and operating reserves with probabilistic wind power forecasts, in *Proceedings of the 2011 IEEE Trondheim PowerTech*, pp. 1–7, Trondheim, Norway, 2011.

21. E. Kalnay, *Atmospheric Modeling, Data Assimilation and Predictability*, University Press, Cambridge, 2003.

22. J. Charney, R. Fjortoft and J. Von Neumann, Numerical integration of the barotropic vorticity equation, *Tellus*, vol. 2, pp. 237–254, 1950.

23. D. Shea, S. Worley, I. Stern and T. J. Hoar, *An Introduction to Atmospheric and Oceanographic Data*, NCAR Technical Notes, Mesoscale and Microscale Meteorology Division, National Center for Atmospheric Research, Boulder, CO, 1994.

24. G. Barr and R. J. Chorley, *Atmosphere, Weather, and Climate*, Routledge, London, 2003.

25. W. M. Washington and C. L. Parkinson, *An Introduction to Three-Dimensional Climate Modeling*, University Science Books, Oxford, 1986.

26. M. Kanamitsu, Description of the NMC global data assimilation and forecast system, *Weather and Forecasting*, vol. 4, pp. 335–342, 1989.

27. A. Simmons, D. Burridge, M. Jarraud, C. Girard and W. Wergen, The ECMWF medium-range prediction models: Development of the numerical formulations and the impact of increased resolution, *Meteorology and Atmospheric Physics*, vol. 40, pp. 28–60, 1989.

28. http://www.emc.ncep.noaa.gov/GEFS/.php.

29. F. Molteni, R. Buizza, T. Palmer and T. Petroliagis, The ECMWF ensemble prediction system: Methodology and validation, *Quarterly Journal of the Royal Meteorological Society*, vol. 122, pp. 73–119, 1996.

30. http://www.emc.ncep.noaa.gov/gc_wmb/Documentation//TPBoct05/T382. TPB.FINAL.htm.

31. C. A. Reynolds, P. J. Webster and E. Kalnay, Random error growth in NMC's global forecasts, *Monthly Weather Review*, vol. 122, pp. 1281–1305, 1994.

32. Y. Zhu, G. Iyengar, Z. Toth, M. S. Tracton and T. Marchok, Objective evaluation of the NCEP global ensemble forecasting system, *Preprints of the 15th AMS Conference on Weather Analysis and Forecasting*, vol.15, pp. 79–82, Norfolk, VA, 1996.

33. E. N. Lorenz, Effects of analysis and model errors on routine weather forecasts, in *Proceedings of the ECMWF Seminars on Ten Years of Medium-Range Weather Forecasting*, Reading, September 4–8, 1989.

34. Z. Toth and E. Kalnay, Ensemble forecasting at NMC: The generation of perturbations, *American Meteorological Society*, vol. 74, pp. 2317–2330, 1993.

35. Z. Toth, O. Talagrand, G. Candille and Y. Zhu, Probability and ensemble forecasts, *Environmental Forecast Verification: A Practitioner's Guide in Atmospheric Science*, I. Jolliffe and D. Stephenson (eds.), Wiley, Chichester, 2002.

36. G. S. Dietachmayer and K. Droegemeier, Application of continuous dynamic grid adaptation techniques to meteorological modeling, *Monthly Weather Review*, vol. 120, pp. 1707–1722, 1992.

37. Y.-L. Lin, *Mesoscale Dynamics*, Cambridge University Press, Cambridge, 2010.

38. W. C. Skamarock, J. B. Klemp, J. Dudhia, D. O. Gill, D. M. Barker, X. Y. Huang, W. Wang and J. G. Powers, *A Description of the Advanced Research WRF Version 3*, NCAR—Mesoscale and Microscale Meteorology Division, National Center for Atmospheric Research, Boulder, CO, 2008.

39. J. Dudhia, *The Weather Research and Forecast Model Version 2.0: Physics Update*, Preprints of the 6th WRF/15th MM5 Users' Workshop, National Center for Atmospheric Research, Boulder, CO, 2005.

40. J. Dudhia, *The Weather Research and Forecasting Model: 2007 Annual Update*, Technical Notes, National Center for Atmospheric Research, Mesoscale and Microscale, Meteorology Division, Boulder, CO, 2007.

41. J. C.-F. Lo, Z.-L. Yang and R. A. Pielke, Assessment of three dynamical climate downscaling methods using the Weather Research and Forecasting (WRF) model, *Journal of Geophysical Research*, vol. 113, issue D9, pp. 1–16, 2008.

42. W. C. Skamarock, J. B. Klemp, J. Dudhia, D. O. Gill, D. M. Barker, W. Wang and J. G. Powers, *A Description of the Advanced Research WRF Version 2*, NCAR Tech Notes, Mesoscale and Microscale Meteorology Division, Boulder, CO, 2005.

43. WMO, *Guide to Climatological Practices*, WMO No. 100, 3rd edition, WMO, Geneva, Switzerland, 2010.

44. T. Black, The new NMC mesoscale eta model: Description and forecast examples, American Meteorological Society Weather Forecasting Journal, vol. 9, issue 2, pp. 265–278, 1994.

45. http://laps.noaa.gov/.
46. J. A. McGinley, S. Albers and P. Stamus, Validation of a composite convective index as defined by a real-time local analysis system, *Weather and Forecasting*, vol. 6, pp. 337–356, 1991.
47. S. Albers, The LAPS wind analysis, *Weather and Forecasting*, vol. 10, pp. 342–352, 1995.
48. S. Albers, J. McGinley, D. Birkenheuer and J. Smart, The Local Analysis and Prediction System (LAPS): Analyses of clouds, precipitation, and temperature, *Weather and Forecasting*, vol. 11, pp. 273–287, 1996.
49. D. Birkenheuer, The effect of using digital satellite imagery in the LAPS Moisture Analysis, *Weather and Forecasting*, vol. 14, pp. 782–788, 1999.
50. T. R. Loveland, B. C. Reed, J. F. Brown, D. O. Ohlen, Z. Zhu, L. Yang and J. W. Merchant, Development of a global land cover characteristics database and IGBP DISCover from 1 km AVHRR data, *International Journal of Remote Sensing*, vol. 21, pp. 1303–1330, 2000.
51. K. Verdin and S. Greenlee, Development of continental scale digital elevation models and extraction of hydrographic features, in *Proceedings of the 3rd International Conference/Workshop on Integrating GIS and Environmental Modelling*, Santa Fe, New Mexico, 1996.
52. J. Olson, *Global Ecosystem Framework-Definitions: USGS EROS Data Center Internal Report*, Sioux Falls, SD, 1994.
53. P. Sellers, Y. Mintz, Y. Sud and A. Dalcher, A simple biosphere model (SiB) for use within general circulation models, *Journal of Atmospheric Science*, vol. 43, pp. 505–531, 1986.
54. C. A. Hiemstra, G. E. Liston, R. A. Pielke, D. L. Birkenheuer and S. C. Albers, Comparing Local Analysis and Prediction System (LAPS) assimilations with independent observations, *Weather and Forecasting*, vol. 21, pp. 1024–1040, 2006.
55. P. Katsafados, A. Papadopoulos, K. Lagouvardos, V. Kotroni, E. Mavromatidis and I. Pytharoulis, *Statistical Evaluation of the Local Analysis and Prediction System Over Greece*, 2nd LAPS User Workshop, David Skaggs Research Center, Boulder, CO, October, 2012.
56. T.-W. Yu and V. Gerald, Evaluation of NCEP operational model forecasts of surface wind and pressure fields over the oceans, *Preprints Proceedings of 20th Conference on Weather Analysis and Forecasting/16th Conference on Numerical Weather Prediction*, Seattle, WA, 2004.
57. O.-Y. Kima, C. Lu, J. A. McGinley, S. C. Albers and J.-H. Oh, Experiments of LAPS wind and temperature analysis with background error statistics obtained using ensemble methods, *Atmospheric Research*, vol. 122, pp. 250–269, 2012.
58. A. Capecchi, D. Tiesi, M. Conte, F. Miglietta, B. Pasi, A. Gozzini, J. Sairouni and J. Mercader, Recent and future developments in Europe with LAPS system coupled with regional NWP models, 2nd LAPS User Workshop, David Skaggs Research Center, Boulder, CO, October, 2012.
59. R. Celci, H. Cazale and J. Vassal, Prevision de la houle. La methode des densites spectroangulaires, *Bulletin d'information du comité central oceanographie et d'etude des cotes*, vol. 9, pp. 416–534, 1957.
60. The Wamdi Group, The WAM model—A third generation ocean wave prediction mode, *Journal of Physical Oceanography*, vol. 18, pp. 1775–1810, 1988.

61. N. Booij, R. C. Ris and L. H. Holthuijsen, A third-generation wave model for coastal regions—Model description and validation, *Journal of Geophysical Research*, vol. 104, pp. 7649–7666, 1999.

62. H. L. Tolman, User manual and system documentation of WAVEWATCH III version 3.14, National Weather Service – National Centers for Environmental Prediction, Camp Springs, MD, May 2009.

63. G. Giebel, R. Brownsword, G. Kariniotakis, M. Denhard and C. Draxl, *The State-of-the-Art in Short-Term Prediction of Wind Power: A Literature Overview*, 2nd edition, 2011.

64. W. Glassley, J. Kleissl, C. P. van Dam, H. Shiu, J. Huang, G. Braun and R. Holland, Current state of the art in solar forecasting, Final Report, *California Renewable Energy Forecasting, Resource Data and Mapping*, 2010.

65. C. Monteiro, I. J. Ramirez-Rosado and L. A. Fernandez-Jimenez, Short-term forecasting model for electric power production of small-hydro power plants, *Renewable Energy*, vol. 50, pp. 387–394, 2013.

66. G. Reikard, P. Pinson and J.-R. Bidlot, Forecasting ocean wave energy: The ECMWF wave model and time series methods, *Ocean Engineering*, vol. 38, no. 10, pp. 1089–1099, July 2011.

67. P. Pinson, G. Reikard and J.-R. Bidlot, Probabilistic forecasting of the wave energy flux, *Applied Energy*, vol. 93, pp. 364–370, May 2012.

68. J. Juban, L. Fugon and G. Kariniotakis, Probabilistic short-term wind power forecasting based on kernel density estimators, in *Probabilistic Wind Power Forecasting—European Wind Energy Conference*, Milan, Italy, May 2007.

69. R. Bessa, J. Sumaili, V. Miranda, A. Botterud, J. Wang and E. Constantinescu, Time-adaptive kernel density forecast: A new method, in *Proceedings of the 17th Power Systems Computation Conference*, Stockholm, Sweden, August 2011.

70. J. Jeon and J. W. Taylor, Using conditional kernel density estimation for wind power density forecasting, *Journal of the American Statistical Association*, vol. 107, pp. 66–79, 2012.

71. J. B. Bremnes, Probabilistic wind power forecasts using local quantile regression, *Wind Energy*, vol. 7, pp. 47–54, 2004.

72. H. A. Nielsen, H. Madsen and T. S. Nielsen, Using quantile regression to extend an existing wind power forecasting system with probabilistic forecasts, *Wind Energy*, vol. 9, pp. 95–108, 2006.

73. G. Fu, F. Y. Shih and H. Wang, A kernel-based parametric method for conditional density estimation, *Pattern Recognition*, vol. 44, pp. 284–294, 2011.

74. S. Demir and Ö. Toktamis, On the adaptive Nadaraya-Watson kernel regression estimators, *Hacettepe Journal of Mathematics and Statistics*, vol. 39, pp. 429–437, 2010.

75. G. Dudek, Short-term load forecasting based on kernel conditional density estimation, *Przegląd Elektrotechniczny*, vol. 86, pp. 164–167, 2010.

76. P. Pinson, H. A. Nielsen, J. K. Møller and H. Madsen, Non-parametric probabilistic forecasts of wind power: Required properties and evaluation, *Wind Energy*, vol. 10, pp. 497–516, 2007.

77. P. Pinson and G. Kariniotakis, Conditional prediction intervals of wind power generation, *IEEE Transactions on Power Systems*, vol. 25, pp. 1845–1856, November 2010.

78. A. Togelou, G. Sideratos and N. D. Hatziargyriou, Wind power forecasting in the absence of historical data, *IEEE Transactions on Sustainable Energy*, vol. 3, pp. 416–421, 2012.

79. R. D. Prasad and M. S. R. C. Bansal, Some of the design and methodology considerations in wind resource assessment, *IET Renewable Power Genereration*, vol. 3, pp. 53–64, 2009.

80. L. Ma, S. Luan, C. Jiang and Y. Z. H. Liu, A review on the forecasting of wind speed and generated power, *Renewable and Sustainable Energy Reviews*, vol. 13, pp. 915–920, 2009.

81. R. G. Kavassery and K. Seetharaman, Day-ahead wind speed forecasting using f-ARIMA models, *Renewable Energy*, vol. 34, pp. 1388–1393, 2009.

82. J. P. S. Catalão, H. M. I. Pousinho and V. M. F. Mendes, An artificial neural network approach for short-term wind power forecasting in Portugal, *Engineering Intelligent Systems for Electrical Engineering and Communications*, vol. 17, pp. 5–11, 2009.

83. I. J. R. Rosado, L. A. F. Jimenez, C. Monteiro, J. Sousa and R. Bessa, Comparison of two new short-term wind-power forecasting systems, *Renewable Energy*, vol. 34, pp. 1848–1854, 2009.

84. J. P. S. Catalão, H. M. I. Pousinho and V. M. F. Mendes, Short-term wind power forecasting in Portugal by neural network and wavelet transform, *Renewable Energy*, vol. 36, pp. 1245–1251, 2011.

85. K. Bhaskar and S. N. Singh, AWNN-assisted wind power forecasting using feed-forward neural network, *IEEE Transactions on Sustainable Energy*, vol. 3, pp. 306–315, 2012.

86. H. M. I. Pousinho, V. M. F. Mendes and J. P. S. Catalão, Application of adaptive neuro-fuzzy inference for wind power short-term forecasting, *IEEJ Transactions on Electrical and Electronic Engineering*, vol. 6, pp. 571–576, 2011.

87. R. Jursa and K. Rohrig, Short-term wind power forecasting using evolutionary algorithms for the automated specification of artificial intelligence models, *International Journal of Forecasting*, vol. 24, pp. 694–709, 2008.

88. R. J. Bessa, V. Miranda and J. Gama, Entropy and correntropy against minimum square error in offline and online three-day ahead wind power forecasting, *IEEE Transactions on Power Systems*, vol. 24, pp. 1657–1666, 2009.

89. H. M. I. Pousinho, V. M. F. Mendes and J. P. S. Catalão, A hybrid PSO-ANFIS approach for short-term wind power prediction in Portugal, *Energy Conversion and Management*, vol. 52, pp. 397–402, 2011.

90. J. Contreras, F. J. N. R. Espinola and A. J. Conejo, ARIMA models to predict next-day electricity prices, *IEEE Transactions on Power Systems*, vol. 18, pp. 1014–1020, 2003.

91. M. Lange, U. Focken and D. Heinemann, Previento—Regional wind power prediction with risk control, in *Proceedings of the World Wind Energy Conference*, Berlin, Germany, 2002.

92. J. Juban, L. Fugon and G. Kariniotakis, Uncertainty estimation of wind power forecasts: Comparison of probabilistic modelling approaches, in *Proceedings of the European Wind Energy Conference and Exhibition*, Brussels, Belgium, April 2008.

93. Á. J. Duque, I. Sánchez, E. Castronuovo and J. Usaola, Simulation scenarios and prediction intervals in wind power forecasting with the Beta distribution, in 11th Spanish-Portuguese Conference on Electrical Engineering (CHLIE), vol. 1, no. 4, pp. 1–5, Zaragoza, Spain, 2009.

94. H. Bludszuweit, J. Dominguez-Navarro and A. Llombart, Statistical analysis of wind power forecast error, *IEEE Transactions on Power Systems*, vol. 23, pp. 983–991, 2008.
95. P. Pinson, H. Madsen, H. A. Nielsen, G. Papaefthymiou and B. Klöckl, From probabilistic forecasts to statistical scenarios of short-term wind power production, *Wind Energy*, vol. 12, pp. 51–68, 2008.
96. P. Mathiesen and J. Kleissl, Evaluation of numerical weather prediction for intra-day solar forecasting in the continental United States, *Solar Energy*, vol. 85, pp. 967–977, May 2011.
97. C. Keil and G. C. Craig, A displacement-based error measure applied in a regional ensemble forecasting system, *Monthly Weather Review*, vol. 135, pp. 3248–3259, September 2007.
98. A. Mellit and S. A. Kalogirou, Artificial intelligence techniques for photovoltaic applications: A review, *Progress in Energy and Combustion Science*, vol. 34, pp. 574–632, October 2008.
99. A. Mellit, H. Eleuch, M. Benghanem, C. Elaoun and A. M. Pavan, An adaptive model for predicting of global, direct and diffuse hourly solar irradiance, *Energy Conversion and Management*, vol. 51, pp. 771–782, April 2010.
100. D. Yang, P. Jirutitijaroen and W. M. Walsh, Hourly solar irradiance time series forecasting using cloud cover index, *Solar Energy*, vol. 86, pp. 3531–3543, December 2012.
101. M. Chaabene and M. B. Ammar, Neuro-fuzzy dynamic model with Kalman filter to forecast irradiance and temperature for solar energy systems, *Renewable Energy*, vol. 33, pp. 1435–1443, July 2008.
102. G. Reikard, Predicting solar radiation at high resolutions: A comparison of time series forecasts, *Solar Energy*, vol. 83, pp. 342–349, March 2009.
103. C. Rigollier, M. Lefèvre and L. Wald, The method Heliosat-2 for deriving shortwave solar radiation from satellite images, *Solar Energy*, vol. 77, pp. 159–169, 2004.
104. T. M. Hamill and T. Nehrkorn, A short term cloud forecast scheme using cross correlations, *Weather and Forecasting*, vol. 8, pp. 401–411, June 1993.
105. T. Nehrkorn, T. M. Hamill and L. W. Knowlton, Nowcasting methods for satellite imagery, Technical Report, Atmospheric and Environmental Research inc, Cambridge, MD, 1993.
106. A. Evans, Cloud motion analysis using multichannel, *IEEE Geoscience and Remote Sensing Letters*, vol. 3, pp. 392–396, July 2006.
107. E. Lorenz, J. Kühnert and D. Heinemann, Short term forecasting of solar irradiance by combining satellite data and numerical weather predictions, in *Proceedings of the 27th European Photovoltaic Solar Energy Conference and Exhibition*, 2012.
108. P. Bacher, H. Madsen and H. A. Nielsen, Online short-term solar power forecasting, *Solar Energy*, vol. 83, pp. 1772–1783, October 2009.
109. M. Valenga and T. Ludermir, Self-organizing modeling in forecasting daily river flows, in *Proceedings of 5th Brazilian Symposium on Neural Networks*, pp. 210–214, Belo Horizonte, Brazil, 1998.
110. C. Liu and W. Liu, Study on forecasting discharge of hydropower station with backwater effect based on improved EBP neural network, *Power Engineering Society General Meeting*, vol. 3, pp. 1451–1454, 2003.
111. C. M. Zealand, D. H. Burn and S. P. Simonovic, Short term streamflow forecasting using artificial neural networks, *Journal of Hydrology*, vol. 214, pp. 32–48, January 1999.

112. M. Valenca and T. Ludermi, Monthly stream flow forecasting using an neural fuzzy network model, *Proceedings of the 6th Brazilian Symposium on Neural Networks*, pp. 117–119, Rio de Janeiro, Brazil, 2000.

113. T. Stokelj, D. Paravan and R. Golob, Short and mid term hydro power plant reservoir inflow forecasting, in *Proceedings of PowerCon 2000. International Conference on Power System Technology*, vol. 2, pp. 1107–1112, Perth, Australia, 2000.

114. R. Sacchi, M. Ozturk, J. Principe, A. A. F. Carneiro and I. N. da Silva, Water inflow forecasting using the eco state network: A Brasilian case study, in *Proceedings of the International Joint Conference on Neural Networks*, pp. 2403–2408, Orlando, FL, 2007.

115. D. Paravan, T. Stokelj and R. Golob, Selecting input variables for HPP reservoir water inflow forecasting using mutual information, in *Proceedings of the 2001 IEEE Porto Power Tech Proceedings*, vol. 2, pp. 1–6, Porto, Portugal, 2001.

116. D. Paravan, T. Stokelj and R. Golob, Improvements to the water management of a run-of-river HPP reservoir: Methodology and case study, *Control Engineering Practice*, vol. 12, pp. 377–385, April 2004.

117. R. Golob, T. Štokelj and D. Grgič, Neural-network-based water inflow forecasting, *Control Engineering Practice*, vol. 6, pp. 593–600, May 1998.

118. O. Kişi, Streamflow forecasting using different artificial neural network algorithms, *Journal of Hydrologic Engineering*, vol. 12, pp. 532–539, September 2007.

119. W. Huang, B. Xu and A. Chan-Hilton, Forecasting flows in Apalachicola River using neural networks, *Hydrological Processes*, vol. 18, pp. 2545–2564, September 2004.

120. Y. Dibike and D. Solomatine, River flow forecasting using artificial neural networks, *Physics and Chemistry of the Earth, Part B: Hydrology, Oceans and Atmosphere 2001*, vol. 26, pp. 1–7, 2001.

121. P. Coulibaly, F. Anctil and B. Bobee, Daily reservoir inflow forecasting using artificial neural networks with stopped training approach, *Journal of Hydrology*, vol. 230, pp. 244–257, May 2000.

122. E. A. Feinberg and D. Genethliou, Load forecasting, in *Applied Mathematics for Reestructured Electric Power Systems—Optimization, Control, and Computational Intelligence*, J. H. Chow, F. F. Wu, and J. Momoh (eds.), Springer, 2005, pp. 269–282.

123. A. K. Singh, I. Ibraheem, S. Khatoon, M. Muazzam and D. K. Chaturvedi, Load forecasting techniques and methodologies: A review, in *Proceedings of the 2nd International Conference on Power, Control and Embedded Systems, Allahabad, Uttar Pradesh*, 2012.

124. D. Pozo and J. Contreras, A chance-constrained unit commitment with n-k security criterion and significant wind generation, *IEEE Transactions on Power Systems*, vol. 28, pp. 2842–2851, August 2013.

125. A. Kalantari, J. F. Restrepo and F. D. Galiana, Security-constrained unit commitment with uncertain wind generation : The loadability set approach, *IEEE Transactions on Power Systems*, vol. 28, pp. 1787–1796, May 2013.

126. G. Liu and K. Tomsovic, Quantifying spinning reserve in systems with significant wind power penetration, *IEEE Transactions on Power Systems*, vol. 27, pp. 2385–2393, 2012.

127. R. C. Williamson, *Probabilistic Arithmetic*, PhD thesis, University of Queensland, Brisbane, Australia, 1989.

128. M. A. Matos and R. J. Bessa, Setting the operating reserve using probabilistic wind power forecasts, *IEEE Transactions on Power Systems*, vol. 26, May 2011.

129. M. Matos and R. Bessa, Operating reserve adequacy evaluation using uncertainties of wind power forecast, in *Proceedings of the IEEE Bucharest PowerTech*, Bucharest, Romania, pp. 1–8, June 2009.

130. G. Papaefthymiou and D. Kurowicka, Using copulas for modeling stochastic dependence in power system uncertainty analysis, *IEEE Transactions on Power Systems*, vol. 24, pp. 40–49, February 2009.

131. S. A. Smartwatt, Mathematical formulation of the different forecasting models and data analysis, D2.1 Report of SINGULAR—Smart and Sustainable Insular Electricity Grids under Large-Scale Renewable Integration, Project Consortium [ENERGY.2012.7.1.1], EU, November 2013.

3

Probabilistic Harmonic Power Flow Calculations with Uncertain and Correlated Data

Gianfranco Chicco, Andrea Mazza, Angela Russo, Valeria Cocina, and Filippo Spertino

CONTENTS

3.1 Introduction

Early research on probabilistic harmonic power flow (PHPF) was conducted more than 40 years ago, when mathematical analysis was used on instantaneous values of current from individual harmonic components (Arrillaga and Watson 2003; Ribeiro 2005).

When evaluating voltage and current harmonics caused by nonlinear loads, the input data of the power system model are affected by unavoidable uncertainties. Time variations of linear load demands, network configurations, and operating modes of nonlinear loads are the main factors causing these uncertainties (Anders 1990). The variations have a random character and, therefore, are expressed in a probabilistic way (Fuchs and Masoum 2009).

The problems associated with direct application of the fast Fourier transform to compute harmonic levels of non-steady-state distorted waveforms, and various ways to describe recorded data in statistical terms applied to a set of recorded data, have been described by Baghzouz et al. (1998). Harmonics are typically represented by statistical measures and histograms, with the challenging point that the probability density functions (PDFs) are often multimodal and as such are not representable using classical probability distributions (Morrison et al. 2009). Furthermore, the summation of random harmonic phasors, the characteristics of harmonic voltages caused by dispersed nonlinear loads, and some applications of these probabilities have been described by Baghzouz et al. (2002).

The probabilistic network analysis of transmission and distribution systems in the presence of harmonics can be carried out using either *direct* or *integrated* methods (Caramia et al. 2009b). In direct methods, the probability functions of harmonic voltage are calculated for assigned probability functions of harmonic currents injected by nonlinear loads (Zhang and Xu 2008). Conversely, in integrated methods, the probability functions of harmonic voltage and current harmonics are calculated together by properly taking into account the interactions between voltage distortions and harmonic currents (Fuchs and Masoum 2009).

The direct method can be applied by considering either the deterministic or the probabilistic framework. As explained by Caramia et al. (2009b), in the *deterministic* field, the direct method for network analysis consists of two steps. The first step comprises the evaluation of the waveform of the current absorbed by nonlinear loads, assuming that the voltage at the converter terminal is an ideal pure sinusoid. Then, Fourier analysis is carried out and the current harmonics are obtained. This process is repeated for all converters present in the system. The second step comprises the evaluation of the voltage harmonics at system bus bars, in which the following network harmonic relations are applied, for $i = 1, \ldots, N$:

$$\overline{V}_i^{(h)} = \sum_{j=1}^{N} \dot{Z}_{ij}^{(h)} \overline{I}_j^{(h)}$$ (3.1)

where:
$\dot{Z}_{ij}^{(h)}$ is the (i, j)th term of the harmonic impedance matrix

$\overline{I}_j^{(h)}$ is the current harmonic of order h injected into the network at bus j

$\overline{V}_i^{(h)}$ is the voltage harmonic of order h at bus i, respectively

By writing the N equations (Equation 3.1) with the partitioning of the real and imaginary parts for each equation, $2N$ scalar equations are obtained.

In the *probabilistic* framework, the input data and output variables of the direct methods are random variables and therefore have to be characterized by PDFs, either joint or marginal. In order to run the probabilistic direct methods, first, the harmonic current PDFs injected into the network bus bars are evaluated for fixed PDFs of the input data, employing proper harmonic current models. Then, the evaluation of the voltage harmonic PDFs at the system buses is carried out by assuming the harmonic current PDFs, calculated first, as input data and using the relations of Equation 3.1 as the harmonic voltage model.

The complete model for the PHPF, which represents the natural extension to the probabilistic field of the harmonic power flow, includes the equations characterizing the operation of nonlinear loads. The harmonic power flow model formulated by Caramia et al. (2009b) is expressed by representing the power balance equations (active, reactive, and apparent) at the linear and nonlinear load bus bars, the active power balance and voltage regulation equations at the generator bus bars, the harmonic current balance equations at the linear load bus bars, the harmonic and fundamental current balance equations at the nonlinear load bus bars, and the commutation angle equations at the nonlinear load bus bars.

To obtain the PDFs of the random state vector starting from the knowledge of the PDFs of the random input vector, several probabilistic techniques have been applied to the nonlinear system:

- Nonlinear Monte Carlo simulation, which applies the Monte Carlo procedure starting from the knowledge of the PDFs of the input variables (Goeke and Wellssow 1996; Au and Milanovic 2006).
- Linear Monte Carlo simulation, based on the Monte Carlo procedure, that relies on a linearized form of the nonlinear system (Carpinelli et al. 2001).
- Convolution process approach, based on the convolution process, that can be solved by numerical methods (Caramia et al. 2009b).
- Approximate distribution approaches, in which the true PDFs of the problem variables are approximated with PDFs whose analytical expressions are determined once a finite number of moments of the true PDFs are known; some approaches based on Johnson, Pearson, Gram–Charlier, and Edgeworth approximation were presented and tested by Esposito et al. (2001) and by Esposito and Varilone (2002).

The computational efforts and result accuracy depend on the probabilistic technique involved. In particular, the nonlinear Monte Carlo simulation guarantees high accuracy of the results but requires very high computational effort. The simplified probabilistic techniques based on linearized models (linear Monte Carlo simulation, convolution approach, Pearson's approach) guarantee a significant reduction of computational effort but do not provide an acceptable accuracy of the results in the case of both significantly high input data PDF variance and multimodal distributions.

Another approach for probabilistic network analysis of power systems considering harmonics, discussed in Caramia et al. (2009b), is the probabilistic iterative harmonic analysis, based on a procedure in which the converter equations, as well as the network equations, are solved in a successive manner (Carbone et al. 1994). In solving the converter equations, the current harmonics injected into the network are estimated by assuming the voltage supply at each converter bus bar to be fixed. When the network model is solved, the bus bar voltage harmonics are evaluated by assuming that the current harmonics injected by converters are fixed. This analysis can be applied to both balanced and unbalanced power systems. It is characterized by high result accuracy but requires high computational effort, mainly in the case of unbalanced power systems.

In Russo et al. (2014), a new probabilistic method for harmonic analysis of multiconverter power systems that applies the point estimate method (PEM) is described. The PEM enables the researcher to obtain the first moments of the output random variables of interest through the solution of only a few deterministic harmonic power flows compared with the enormous number of repetitions required by the classical Monte Carlo simulation procedure. Moreover, this method guarantees substantial reductions of computational efforts and high accuracy in the output random variables characterization.

Some studies on harmonic load-flow problems have been based on the possibility theory, as an alternative tool to the probability analysis (Romero et al. 2011, 2012). Possibility distributions (or fuzzy numbers) instead of probabilities are the inputs used to describe the uncertainty in the magnitude and composition of the loads. The use of this theory could be better suited to describe and quantify the real nature of the uncertainty involved in harmonic studies. Further contributions addressed the probabilistic treatment of the time-dependent behavior of interharmonics (Ribeiro 2005; Testa and Langella 2005).

Some contributions that appeared in the relevant literature deal with probabilistic harmonic analysis of distribution systems in the presence of distributed generation (DG). The problem of the calculation of the waveform distortions in distribution systems with wind farms has been considered by Caramia et al. 2006a. The authors proposed a PHPF, which properly includes among the input random variables not only those linked to the linear and nonlinear load presence but also those linked to wind speeds.

In Ruiz-Rodriguez et al. (2014), the statistical characterization of grid-connected photovoltaic (PV) harmonic currents at different fundamental-frequency current output intervals based on historical time-series data has been presented. The authors illustrate the method for determining the relationship between PV harmonic current emission and background harmonic voltages, and the method for obtaining the distribution functions of the PV harmonic currents. Instead of using the Gram–Charlier, Edgeworth, or Cornish–Fisher approximation series, to reconstruct harmonic current emission PDFs from moments or cumulants, the authors use the Legendre series, which perform better in terms of both convexity and accuracy in the reconstruction.

In this chapter, a computational tool to determine the probabilistic outputs of the network variables (voltages, currents, losses, etc.) is described. This computational tool is incorporated in the interval power flow analysis implemented in the overall distribution system analysis tool DERMAT (Mazza et al. 2014). The general structure of the modules implemented in the tool, including uncertainty modeling and incorporating the harmonic models of the network components, generations, and loads, is represented in Figure 3.1. The details of the iterative procedures implemented are reported in Section 3.4.2.

The next sections address the various aspects concerning the definition of the data, the execution of the computational procedures, and the output of the results. Section 3.2 deals with the modeling of uncertainty for the loads and generations connected to the network. Section 3.3 discusses the harmonic modeling for loads and generation. Section 3.4 illustrates the implementation of the PHPF. Section 3.5 introduces network performance indicators used to

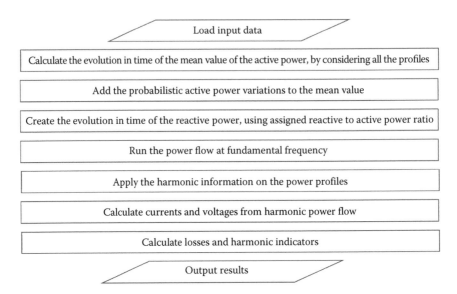

FIGURE 3.1
Structure of the modules for the probabilistic harmonic power flow incorporated in DERMAT.

characterize the outputs. Section 3.6 shows the results of a case study application. The last section concludes the preceding discussion.

3.2 Modeling the Uncertainty of Load and Generation

3.2.1 Probabilistic Modeling of the Network Load

The probabilistic characterization of different types of electrical loads has been addressed in a number of contributions. Jardini et al. (2000) take into account the partitioning of the loads into different categories. Details on residential load aggregations are discussed in Heunis and Herman (2002) and Carpaneto and Chicco (2008), showing the most effective probability distributions that can be used in the aggregate load representations, with details on the beta, gamma, and log-normal probability distributions. Recent modeling of the power demand provided by electric vehicles is addressed in Lojowska et al. (2012), Ye et al. (2014), and Li and Zhang (2014).

For the application in the DERMAT tool, the data input for each load (or load aggregation) is given by indicating the evolution in time of the mean value and of the standard deviation of the load power. If the time steps with which the load patterns are different with each other, the procedure illustrated in Chicco et al. (2014b) is run to reformulate all the load patterns with the same time step used for the probabilistic power flow calculations.

3.2.2 Probabilistic Modeling of the Generation

The probabilistic model of the generation mainly depends on the uncertainty of the meteorological variables affecting the power production. After constructing the PDFs of the random variables characterizing the uncertainty, an effective method uses the statistical method of *cumulants* to synthesize the properties of the random variables. The calculation of the cumulants is an alternative with respect to the calculation of the probabilistic moments of a probability distribution. From the cumulants, the PDF of the random variable is reconstructed using the Gram–Charlier series or the Cornish–Fisher expansion. The power output is then obtained from specific models linking the environmental variables to the characteristics and efficiencies of the chain from the generation source to the electrical network interface (Ruiz-Rodriguez et al. 2011). The evolution in time of the power output is calculated using a specific external procedure and is then used as input in the DERMAT calculation tool.

3.2.3 Modeling the Correlations

Correlations appear in the probabilistic model mainly because of the effect of the environmental variables. For example, the correlations among the solar irradiance patterns have been calculated for five PV plants (indicated as GI, MA, GA1, GA2, and RU) located at sites with variable distances with each other, as shown in Figure 3.2 (Chicco et al. 2014c). The measurement of

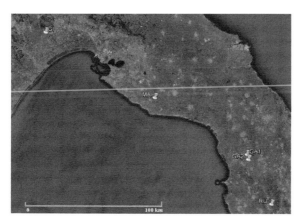

FIGURE 3.2
Location of the measured PV sites.

TABLE 3.1

Correlation Matrix of the Solar Irradiance
Patterns at the Five Sites

	GI	MA	GA1	GA2	RU
GI	1	0.90	0.88	0.87	0.87
MA	0.90	1	0.90	0.90	0.89
GA1	0.88	0.90	1	0.94	0.91
GA2	0.87	0.90	0.94	1	0.91
RU	0.87	0.89	0.91	0.91	1

the solar irradiance patterns refers to the actual tilt angle of the PV panels (orientated toward south), that is, 25° for GA1 and 30° for the other sites. The solar irradiance measurements have been carried out for each minute for a period of two years (2012 and 2013). A correlation matrix is shown in Table 3.1. It can be noted that the correlations are remarkably high even at distances of over 100 km among some sites, due in particular to the characteristics of the region (Chicco et al. 2014a), in which the number of sunny days during the year is significantly high. This provides the rationale for introducing the correlations in the probabilistic representation of PV generators. Similar considerations hold for wind systems (Alexiadis et al. 1999; Feijoo et al. 1999, 2011; Bechrakis and Sparis 2004).

The classical way of considering correlations takes into account normal probability distributions (Feijoo et al. 2011). In this case, taking N correlated random variables $Z_1, \ldots, Z_N$ included in the vector $\mathbf{z}$, these variables are described through their mean values

$$\mu_Z = \left\{ \mu_{Z_1}, \ldots \mu_{Z_N} \right\} \tag{3.2}$$

and correlation coefficient matrix

$$\mathbf{R}_Z = \begin{vmatrix} 1 & \rho_{Z_1 Z_2} & \cdots & \rho_{Z_1 Z_{N-1}} & \rho_{Z_1 Z_N} \\ \rho_{Z_2 Z_1} & 1 & \cdots & \rho_{Z_2 Z_{N-1}} & \rho_{Z_2 Z_N} \\ \vdots & \vdots & \vdots & \vdots & \vdots \\ \rho_{Z_{N-1} Z_1} & \cdots & \cdots & 1 & \rho_{Z_{N-1} Z_N} \\ \rho_{Z_N Z_1} & \rho_{Z_N Z_2} & \cdots & \rho_{Z_N Z_{N-1}} & 1 \end{vmatrix} \tag{3.3}$$

The Cholesky factorization is used to express the correlation coefficient matrix as the product

$$\mathbf{R}_Z = \mathbf{L}_Z \cdot \mathbf{L}_Z^T \tag{3.4}$$

where:

the superscript T denotes matrix transposition

The generation of the correlated variables starts with the extraction of N independent standard normal random variables, included into a vector $\mathbf{g}$. The correlated normal variables are found by calculating the product $\mathbf{z}' = \mathbf{L}_Z \mathbf{g}$.

If the probability distributions of the variables of interest are not normal, the above procedure has to be modified to make it possible to handle any type of probability distribution (Mazza et al. 2014). At the beginning of the analysis, the N correlated random variables $X_1, \ldots, X_N$ (whose probability distributions may be different) are included in the vector $\mathbf{x}$ and are assumed to have mean values $\mu_X = \{\mu_{X_i}, \ldots, \mu_{X_N}\}$ and correlation coefficient matrix

$$
\mathbf{R}_X = \begin{vmatrix}
1 & \rho_{X_1 X_2} & \cdots & \rho_{X_1 X_{N-1}} & \rho_{X_1 X_N} \\
\rho_{X_2 X_1} & 1 & \cdots & \rho_{X_2 X_{N-1}} & \rho_{X_2 X_N} \\
\vdots & \vdots & \vdots & \vdots & \vdots \\
\rho_{X_{N-1} X_1} & \cdots & \cdots & 1 & \rho_{X_{N-1} X_N} \\
\rho_{X_N X_1} & \rho_{X_N X_2} & \cdots & \rho_{X_N X_{N-1}} & 1
\end{vmatrix} \tag{3.5}
$$

For the purpose of handling correlations, the central part of the procedure remains the one described above using the variables with normal distribution and the Cholesky transformation. To generalize the procedure, two important steps are added:

1. At the beginning of the procedure, a mechanism has to be introduced to pass from the correlation coefficient matrix $\mathbf{R}_X$ referring to the original probability distributions to the correlation matrix $\mathbf{R}_Z$ referring to normal random variables (Qin et al. 2013).

2. At the end of the procedure, each random variable Z_i', belonging to the vector $\mathbf{z}'$ referring to a normal probability distribution, has to be transformed into a random variable X_i' referring to the initial probability distribution of the same variable (Morales et al. 2010). This step is made by considering that the values of the cumulative distribution function (CDF) from the two probability distributions, that is, the CDF $F_Z(\cdot)$ for the standard normal distribution and the CDF $F_X(\cdot)$ for the initial probability distribution, have to be equal. The random variable X_i' is then found by first calculating the CDF value $F_Z(Z_i')$ from the instance Z_i', thus obtaining the random variable X_i' from the inverse transformation $X_i' = F_X^{-1}[F_Z(Z_i')]$. The random variable X_i' is expressed in standard form, so that it could be necessary to apply the appropriate scale factor to obtain its final value.

3.3 Harmonic Modeling

3.3.1 Harmonic Models for Loads

3.3.1.1 Harmonic Current Spectra

The magnitude and phase of the harmonic source can be calculated, for example, from the typical harmonic current spectrum of the device (Gomez-Expósito et al. 2009). A commonly used procedure to establish the current source model is as follows:

1. The harmonic-producing load is treated as a constant power load at the fundamental frequency, and the fundamental frequency power flow of the system is solved.

2. The current injected from the load to the system is then calculated and is denoted as $I^1 \angle \vartheta^1$.

3. The magnitude and phase angle of the harmonic current source representing the load are determined as follows:

$$I^h = I^1 \frac{I^h_{\text{spectrum}}}{I^1_{\text{spectrum}}}, \quad \vartheta^h = \vartheta^h_{\text{spectrum}} + h\left(\vartheta^1 - \vartheta^1_{\text{spectrum}}\right) \qquad (3.6)$$

where:
   the subscript *spectrum* stands for the typical harmonic current spectrum of the load

The meaning of the magnitude formula is to scale up the typical harmonic current spectrum to match the fundamental frequency power flow result (I^1). The phase formula shifts the spectrum waveform to match the phase angle ϑ^1. Since the hth harmonic has h times higher frequency, its phase angle is shifted by h times of the fundamental frequency shift.

The typical harmonic current spectrum can be obtained (Gomez-Expósito et al. 2009) by the following:

- Measurements
- Analytical models of nonlinear loads
- Time-domain simulations

Examples of harmonic current spectra are shown below. Figure 3.3 plots a spectrum of residential and commercial loads (Robinson 2003). Figure 3.4 shows a spectrum for industrial loads, based on the data taken from Preda et al. (2012). In residential applications, the load profiles are constructed (Jiang et al. 2012) using a comprehensive approach based on the probability of the

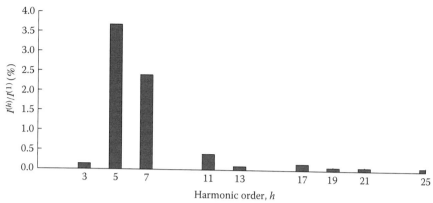

FIGURE 3.3
Residential and commercial loads: harmonic current spectrum. (Based on Robinson, D. 2003. *Harmonic management in MV distribution systems.* PhD Thesis, University of Wollongong, Australia.)

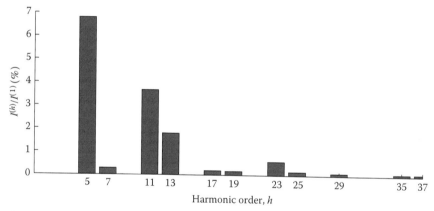

FIGURE 3.4
Harmonic current spectrum for industrial loads. (Based on Preda, T. N., Uhlen, K. and Nordgård, D. E. 2012. Instantaneous harmonics compensation using shunt active filters in a Norwegian distribution power system with large amount of distributed generation. *Proceedings of the 3rd IEEE International Symposium on Power Electronics for Distributed Generation Systems,* Aalborg, Denmark, June 25–28, 2012.)

time of use of the various loads, the size of the household and the pattern of occupancy, and the probabilistic model of the equipment switch-on. Jiang et al. (2012) conclude that a bottom-up-based approach is needed to carry out the harmonic assessment for residential loads. The harmonic current spectra are indicated with reference to the 95th percentile of the individual demand distortion and total demand distortion (IEEE 1993) for a group of houses supplied by a secondary feeder. Jiang et al. (2014) present an assessment of the harmonics obtained from electric vehicles, using the harmonic model formulated in Equation 3.6.

Furthermore, a great deal of research work (not reviewed here) is available on the harmonic current spectra of individual appliances.

3.3.1.2 Harmonic Summation for Load Aggregations

In the basic representation of an aggregation of multiple loads for harmonic studies, the loads can be considered as independent harmonic sources, that is, assuming that the corresponding phasors are independent of each other. Furthermore, in the typical case, each harmonic order is assumed to be independent of the other harmonic orders. Finally, the harmonic phasors are considered to be independent of time. On the above hypotheses, the analysis of an aggregation of loads is carried out by calculating the phasor sum of random harmonic components.

Let us consider N harmonic loads connected to the point of common coupling (PCC), each of which is described by a phasor representing the load current at each harmonic order. Let us assume a common angle reference located at the ascending zero of the first (fundamental) harmonic of the supply voltage at the PCC. The quantity of interest is the total harmonic current referring to the aggregation of harmonic loads. For each harmonic order, the maximum current of the load aggregation (worst case) is obtained by calculating the algebraic sum of the current magnitude of each phasor. For practical purposes, this worst case leads to an excessive overestimation of the harmonic load, because the phase angles of the harmonic currents may have a considerable variability at some harmonic orders, and in many cases, this variability increases with the increase of the harmonic order, even reaching the full 360° range. When the range of variation of the phase angles increases, there is greater possibility of obtaining compensation among the harmonic phasors, significantly reducing the magnitude of the phasor sum below the maximum current corresponding to the worst case. A specific assessment of the sum of the phasors referring to a group of aggregate loads is then useful to determine the expected values (and more generally the probability distributions) of the harmonic currents in practical cases.

The calculation of the sum of random phasors has been addressed using different approaches. A basic distinction among these approaches (Pierrat 1991; Carbone et al. 1994; Cavallini et al. 1998; Langella and Testa 2009) leads to identifying mainly analytical approaches and semiempirical approaches.

3.3.1.2.1 Analytical Approaches

Analytical approaches are aimed at calculating the probability distribution of the magnitude of the total load current at each harmonic order by formulating analytical expressions taking into account the phase angle variations. Sherman (1972) considered a large number N of statistically independent random phasors with fixed or variable magnitudes and phase angles uniformly variable in the range $[0,2\pi]$, and expressed the probability that the phasor sum exceeds a certain level with a formula involving the Bessel function of the first kind of order zero. This result was extended by Rowe (1974) assuming that the magnitude of each phasor has a uniform variation from zero to a maximum and the phase angles have a uniform variation in the range [0,2π], with magnitude and angle begin statistically independent, obtaining for N a sufficiently large Rayleigh PDF for the magnitude of the phasor sum.

Pierrat (1991) showed an interesting interpretation of the case in which the magnitude of the N initial phasors is distributed in the range from zero to a maximum and the variation of the phase angles is limited in a given range $\Delta\varphi$ (chosen from 0 to 2π) is and symmetrical with respect to the horizontal axis. The determination of the PDF of the magnitude of the phasor sum is interpreted as a random-walk problem on the Cartesian axes. The resulting PDF depends on $\Delta\varphi$ and varies from the Rayleigh type for $\Delta\varphi = 2\pi$ to the normal type for $\Delta\varphi = 0$. The corresponding representation of the phasor sum on the Cartesian plane is a circle for $\Delta\varphi = 2\pi$ and tends to a segment for $\Delta\varphi = 0$, following an elliptic law. Elliptical distributions have also been used in a dedicated case study on the traction system (Morrison and Clark 1984).

Symmetries are very helpful in finding out analytical solutions. Lehtonen (1993) presented a particular solution for the summation of the harmonic phasors by considering the special case of a joint PDF symmetrical with respect to the bisector of the first quadrant, for which the standard deviations related to the horizontal and vertical axes are equal. More generally, axis rotation was addressed (Morrison and Clark 1984; Langella et al. 1997) by computing a suitable rotation angle of the initial Cartesian reference in such a way to obtain a null correlation among the random variables representing the phasor sum in the new Cartesian reference. Such a rotation angle is generally determined in function of the variances and of the correlation coefficient of the initial phasor sum. A simpler expression for the rotation angle, only depending on the mean values of the initial phasor sum components, is obtained in the case of symmetrical joint PDF, where one of the rotated axes coincides with the axis of symmetry.

3.3.1.2.2 Semiempirical Approaches

The semiempirical approaches lead to direct formulations of the magnitude of the total load current for each harmonic order. Some semiempirical approaches (Lemoine 1976; Lagostena and Porrino 1986; Crucq and Robert 1989) resorted to the following expression of the magnitude Z_h of the phasor sum with respect to the magnitudes a_{ih} of the $i = 1, \ldots, N$ phasors contributing to the sum at harmonic order h, depending on two parameters (k_h and β_h):

$$Z_h \approx k_h \sqrt[\beta_h]{\sum_{i=1}^{N} |a_{ih}|^{\beta_h}} \tag{3.7}$$

The values $k_h = 1$ and $1 \le \beta_h \le 2$ are used in the formulations of Lemoine (1976) and Lagostena and Porrino (1986). The arithmetic sum of the phasor magnitudes (for $\beta_h = 1$) has a null exceeding probability. Conversely, the Euclidean norm ($\beta_h = 2$) could have a high exceeding probability. Crucq and Robert (1989) showed that β_h highly depends on the phase angle variation, introducing the formulation (3.7) with $k_h \leq 1$ and $1 \leq \beta_h \leq 2$, with k_h and β_h depending on the magnitude and phase ranges of the phasors but not on the number N of phasors. They proposed a set of values of k_h and β_h, whose validity was confirmed from Monte Carlo simulations. The effectiveness of these values was confirmed in the application AC/DC converters presented in Cavallini et al. (1994). An expression formulated as in Equation 3.7 is also found in the publication IEC 61000-3-6 (2008) based on the analysis on rectifier bridges, used when the phase angle of multiple harmonic sources is not known. This formulation corresponds to assuming $k_h = 1$ for any harmonic order h, and β_h variable in function of the harmonic order h, that is, $\beta_h = 1$ for $h < 5$, $\beta_h = 1.4$ for $5 \leq h \leq 10$, and $\beta_h = 2$ for $h > 10$. A more recent proposal (Xiao and Yang 2012) is to use $\beta_h = 1.1$ for any harmonic order h, that will result as a conservative value for $h > 5$. For variable-speed induction motor drives, in order to avoid underestimation of the total current, Cuk et al. (2013) suggest the use of values of β_h between 1.1 and 1.5. Specific values of β_h based on empirical studies are reported in Wikston (2009) for multiple arc furnaces at different percentiles of the harmonic sum.

A further formulation was presented by Cavallini and Montanari (1995) by considering the *random walk* problem with the PDF of the square magnitude of the phasor sum belonging to the family of the χ^2 distributions. Under the hypotheses of having some equal probabilistic moments in the actual and approximated distributions, and assuming uncorrelated Cartesian components of the phasor sum, the PDF of the phasor sum is expressed as a function of mean value and standard deviation of the Cartesian components, in a formulation involving the Gamma function. This approximation has provided acceptable results to represent the magnitude of the sum of harmonic currents for loads with similar characteristics. This formulation was extended by Cavallini et al. (1998) to address correlated random variables in Cartesian coordinates, by introducing the generalized Gamma distribution.

3.3.1.2.3 Other Approaches

Further methods for summing up harmonic currents based on Monte Carlo simulations have been typically used to obtain the PDF of the harmonic sum in the absence of other formulations (Wang et al. 1994; Cavallini et al. 1995; Ye et al. 1999; Gorgette et al. 2000). In addition, the discrete convolution of the joint PDF was exploited in Emanuel and Kaprielian (1986), Kaprielian et al. (1994), and Chicco et al. (2001). Diversity factors for load aggregations are defined in Mansoor et al. (1995) and Au and Milanovic (2006).

3.3.1.2.4 Simplifying Assumptions for Harmonic Summation

In addition to the hypotheses listed in Section 3.3.1.1, the probabilistic approach to harmonic summation generally assumes that the number of random phasors is high enough to invoke the central limit theorem and express the result as a normal probability distribution (Baghzouz and Tan 1987; Kazibwe et al. 1989; Lehtonen 1993; Dwyer et al. 1995; Ye et al. 1999; Baghzouz 2005). Some studies (Sherman 1972; Kazibwe et al. 1989; Wang et al. 1994) aimed at determining the minimum number of aggregated phasors making it possible to use the normal approximation, finding out that this number could vary from 3 to 10 if the PDF of the harmonic currents is the same.

Another assumption is that the polar (or orthogonal) components of each phasor are statistically independent of each other (Baghzouz and Tan 1987). This assumption is needed to split the complex random phasors into their polar (or orthogonal) components and to use the marginal PDFs instead of the joint PDF. This assumption is not needed when the joint PDF is used.

The concept of summation of the harmonic phasors implicitly assumes that all the loads are connected to the same point, that is, the same node of the network. This assumption has been often used also to deal with cases in which the electrical distance among the loads is relatively low, as in Ye et al. (1999), for urban railway systems, or in Haiduck and Baghzouz (2002), for a cluster of gaming machines. In the latter case, in function of the number of machines in play mode, most of the total harmonic amplitudes decrease to a minimum value then increase again. This effect has been associated to the cancelation of harmonics taking place because of the phase angle dispersion at specific combinations of machines for each harmonic component.

3.3.2 Harmonic Models for Generators

The generators can be as follows:

1. Directly connected through synchronous generators. The synchronous-based DG units can be modeled as a source behind an impedance (Ravikumar Pandi et al. 2013).
2. Directly connected through asynchronous generators. The model of an asynchronous generator can be included (Papathanassiou and Papadopoulos 2006).
3. Connected through a power-electronic interface [e.g., pulse width modulation (PWM) converters system or six-pulse thyristor converters]. In the harmonic studies, inverter-based DG units can be modeled as sources injecting harmonic currents.

In particular, for generators directly connected through synchronous generators (Figure 3.5), for the purpose of determining the network harmonic admittances, the synchronous generators can be modeled as a series combination of resistance and inductive reactance, that is,

$$\overline{Y}_g^h = \frac{1}{R_g \sqrt{h} + j X_d'' h} \tag{3.8}$$

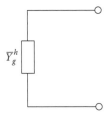

FIGURE 3.5
Synchronous generator harmonic model.

where:
  R_g is the resistance derived from the machine power losses
  X_d'' is the generator's subtransient reactance (Arrillaga and Watson 2003).
      A frequency-dependent multiplying factor can be added to the reactance terms to account for skin effect

Furthermore, some cases of wind and PV generation are addressed in the next subsections.

3.3.2.1 Wind Generation Systems

3.3.2.1.1 Harmonic Current Spectra and Total Harmonic Distortion for Wind Systems

For wind systems, different structural solutions for grid connection correspond to various types of harmonic spectra. An example is reported in Figure 3.6, showing the spectrum for a wind generator connected by converters (Mendonca et al. 2012). Furthermore, the 3D plot shown in Figure 3.7

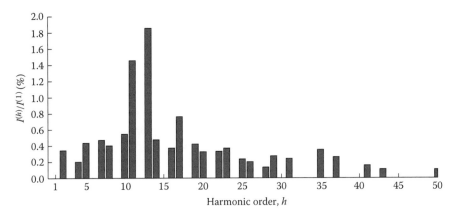

FIGURE 3.6
Wind generator connected by converters: harmonic current spectrum. (Based on Mendonca, G. A., Pereira, H. A. and Silva, S. R. 2012. Wind farm and system modelling evaluation in harmonic propagation studies, *Proceedings of the International Conference on Renewable Energies and Power Quality*, Santiago de Compostela, Spain, March 28–30, 2012.)

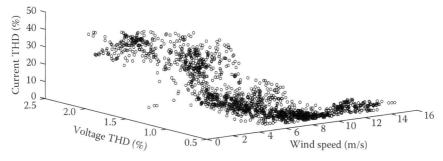

FIGURE 3.7
Voltage and current THD values for a wind generator in operation at different wind speeds.

indicates that the voltage and current total harmonic distortion (THD) values are relatively small for high wind speeds and increase considerably when the wind speed decreases (Chicco et al. 2006).

3.3.2.1.2 Harmonic Summation for Grid Connected Wind Systems

Concepts similar to those indicated in Section 3.3.1.2 can be applied for determining the sum of random phasors of the harmonics currents produced by aggregations of generators.

Concerning the summation of harmonic currents, IEC 61400-21 (2008) deals with grid-connected wind turbines and adopts the same formula as IEC 61000-3-6 (2008). A specific adjustment to the values adopted in the IEC publication is proposed in Medeiros et al. (2010), in which the coefficient β_h for harmonic order h is calculated through an iterative procedure based on the Newton method. In that reference, simulations on a number of test cases are shown to indicate that the proposed adjustment leads to better results than the one provided using the values of β_h taken from IEC 61400-21.

3.3.2.2 PV Generation Systems

3.3.2.2.1 Harmonic Current Spectra and THD for PV Systems

The harmonic behavior of PV systems depends on the level of power supplied by the PV generators and on the characteristics of the inverters. When the power supplied is close to the rated power, the harmonic distortion of the current is generally low, but increases significantly when the power supplied decreases (Chicco et al. 2009a). About the effect of the inverters, in most cases, the harmonics produced as a result of the inverter control contain components at harmonic orders higher than the maximum harmonic order considered by the standards (i.e., the 50th harmonic). These components are not monitored by the power quality (PQ) analyzers and do not contribute to the conventional harmonic distortion indicators.

Indicative examples of individual harmonic distortion and THD values based on real measurements on PV systems are reported in Figure 3.8,

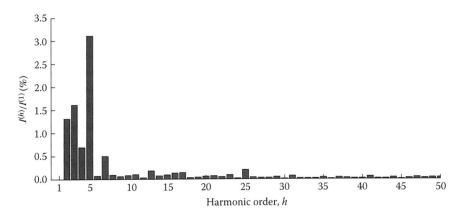

FIGURE 3.8
Harmonic current spectrum for a photovoltaic generator, based on experimental measurements.

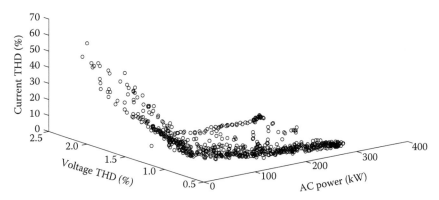

FIGURE 3.9
Voltage and current THD values for a PV system in operation with different AC output power values.

showing the spectrum of the harmonic currents (normalized with respect to the amplitude of the current at the first harmonic), and in the 3D plot shown in Figure 3.9, indicating that the voltage and current THD values are relatively small for high AC output power (corresponding to high values of the solar irradiance for typical ambient temperatures) and increase considerably when the AC output power decreases. The current THD value has been calculated with reference to the actual current at fundamental frequency. An alternative is to use as reference the current rating (i.e., an invariant value), as in Spertino and Graditi (2014).

3.3.2.2.2 *Harmonic Summation for Multiple Inverters Operation in PV Plants*

The main aspects concerning grid-connected PV inverter operation refer to their impact on the distribution system voltage profile and losses, and to the voltage and current waveform distortion. Concerning voltage profile and losses, the specific contribution of PV systems is often relatively limited, since the rated power of the PV systems are much smaller than the short-circuit power of the distribution systems at the PCC.

Taking into account the waveform distortion, the relevant issues are harmonics and interharmonics depending on the interaction between PV systems and the grid at the PCC. In the presence of multiple inverters connected to the grid, it becomes relevant to characterize the sum of the harmonic and interharmonic components.

For this purpose, the summation of harmonic phasors has been addressed in the literature by taking into account the cancelation effect that may arise in distribution networks at different harmonic orders, depending on the phase angles of the harmonic phasors.

The harmonic behavior of a group of PV inverters connected to the PCC is analyzed below using a simplified but meaningful approach based on the so-called harmonic summation ratio (Chicco et al. 2009a, 2009b).

Let us suppose that a generic number N of PV inverters are connected to the PCC. Let $i^{(1)}(t)$ be the current waveform of one of the PV inverters and $i^{(N)}(t)$ is the current waveform due to the contribution of all inverters seen from the PCC. Then, the current waveforms are subject to the Fourier transformation to obtain the currents in the frequency domain, for all the harmonic orders $h = 1,\ldots,H$ (e.g., with $H = 50$). The resulting currents are $I_h^{(1)}$ for one of the PV inverters and $I_h^{(N)}$ due to the contribution of all inverters seen from the PCC.

Let us further consider the rated power $P_{rINV,AC}^{(i)}$ of the inverters $i = 1,\ldots,N$, where in particular $P_{rINV,AC}^{(1)}$ is the rated power of the individual inverter monitored.

The effect of the presence of N inverters connected to the PCC is represented by calculating, for each harmonic order $h = 1,\ldots,H$, the *harmonic summation ratio*

$$\zeta_h^{(N)} = \frac{I_h^{(N)}}{I_h^{(1)}} \frac{P_{rINV,AC}^{(1)}}{\sum_{i=1}^{N} P_{rINV,AC}^{(i)}} \tag{3.9}$$

In Equation 3.9, the harmonic summation ratio should belong to the range $0 \le \zeta_h^{(N)} \le 1$, for $h = 1,\ldots,H$. Conceptually, when the harmonic summation ratio is equal to unity, the harmonic contributions sum up arithmetically; that is, the corresponding current phasors of the various PV inverters at the harmonic order considered are in phase with each other, so that:

$$I_h^{(N)} = \sum_{i=1}^{N} I_h^{(i)} \tag{3.10}$$

Conversely, when the harmonic summation ratio is equal to zero, there is a cancelation effect among the whole set of phasors at the corresponding harmonic order. For equal PV inverters, assuming current phasors with the same amplitude, this happens if the current phasors are regularly distributed in the complex plane (with regular phase shifts of $2\pi/N$). However, in practical calculations, it could also happen that one or more values of $\zeta_h^{(N)}$ are higher than unity. This can be due to the effects of measurement uncertainty, or to the fact that the characteristics of the monitored PV inverter do not correspond to the expected ones.

In addition to these limit conditions, it is possible to identify another condition, according to which the current phasors sum up in an Euclidean way. In this case, the magnitude of the total current is equal to the square root of the sum of squares of the magnitudes of the individual currents, that is,

$$I_h^{(N)} = \sqrt{\sum_{i=1}^{N}\left[I_h^{(i)}\right]^2} \tag{3.11}$$

If all PV inverters are equal, the harmonic summation ratio is as follows:

$$\zeta_h^{(N)} = \frac{1}{\sqrt{N}} \tag{3.12}$$

3.4 Implementation of the PHPF

3.4.1 General Aspects

3.4.1.1 Harmonic Analysis

The harmonic analysis tool incorporated in DERMAT is implemented for assessing the voltage and current harmonic distortion at all harmonic orders (until the predefined maximum order). This tool is able to build the model of the distribution network at the harmonic frequencies and to properly characterize the harmonic sources, including the distributed generators, at each time step of analysis (without loss of generality, the analysis indicated here refers to one hour) in the given time interval (e.g., one day).

The harmonic analysis tool illustrated in this chapter is based on the application of the current injection method (Arrillaga and Watson 2003). The linear portion of loads and generators are represented as *node impedances* (constant for all the harmonics), whereas the nonlinear portions are modeled as *current generators*, having different frequency for different harmonic considered.

When the harmonics generated by the nonlinear components are reasonably independent of the voltage distortion level in the AC system, the derivation of the harmonic sources can be decoupled from the analysis of harmonic penetration and a direct solution is possible. Since most nonlinearities manifest themselves as harmonic current sources, (IEEE 2001, 2004), this is normally called the *current injection method*. In such cases, the expected voltage levels or the results of a fundamental frequency load flow are used to derive the current waveforms, and with them the harmonic content of the nonlinear components.

In the harmonic power flow solved with the current injection method, conceptually, the representation of the system is divided into two main parts, namely, the construction of the bus impedance matrix $\mathbf{Z}_{bus}$ and the handling of the harmonic sources.

3.4.1.1.1 Construction of the Bus Impedance Matrix $\mathbf{Z}_{bus}$

The network is generally represented using the bus impedance matrix $\mathbf{Z}_{bus} \in C^{N,N}$, containing the branch impedances and the constant impedance loads. For the harmonic power flow, the matrix $\mathbf{Z}_{bus}$ also contains the model of the linear portion of the load defined by its assigned power but converted into an admittance load in order to be used by the current injection method.

More specifically, the elements considered as inputs in the construction of the matrix $\mathbf{Z}_{bus}$ are as follows:

1. Branch impedances (represented using the Π model)
2. Transformer impedances (represented using the Π model)
3. Nodal admittance loads (e.g., capacitor banks)
4. Admittance modeling the *linear* portion of the load, reported in impedance terms in the matrix $\mathbf{Z}_{linear}$

The entries in the bus impedance matrix change with the harmonic order. In particular, the resistance representing the linear portion of the load can be assumed to remain constant for all the harmonics (if there is no specific information on the need for changing its value at different harmonics). The reactances of the network elements and the model of the linear portion of the load defined by its assigned power but converted into an impedance load have different values depending on the harmonic order.

By indicating with H, the maximum harmonic order considered in the analysis, the calculation of H bus impedance matrices (one for each harmonic order) is needed. At each harmonic order, the corresponding matrix $\mathbf{Z}_{bus}$ can be calculated as the inverse of the bus admittance matrix $\mathbf{Y}_{bus}$, or by direct construction using a set of rules, detailed in Grainger and Stevenson (1994). The latter procedure is used in this chapter.

3.4.1.1.2 Handling the Harmonic Sources

Both load and generation can produce harmonics, mainly because of the widespread use of electronic interfaces at the grid connection.

The representation of the harmonic content of loads and generation is possible only by *measuring* a certain number of devices of similar size and by obtaining from them the harmonic behavior for that particular device class.

For this reason, we have described the behavior of *five* classes of loads and generation, in particular, residential loads, industrial loads, tertiary sector loads, PV generation, and wind generation.

Introducing a current generator injecting in the network an assigned current at each harmonic order represents the nonlinear part of the load.

First of all, the current at fundamental frequency referring to the nonlinear portion of the load or generation power is computed. Because the harmonic information refers to the different load/generation types with different power profiles, the fundamental current for the nonlinear portion of the load is still computed by considering nodes, profiles, and time steps. After that, using the information on the amplitude and the angles of different harmonics, the *nodal* harmonic currents are obtained. For all of them, the *amplitude ratio* and the *angle* are indicated.

The amplitude ratio represents the ratio between the *amplitude* of the current at the harmonic h and the *fundamental current* obtained by considering the nonlinear portion. The fundamental current referring to the nonlinear portion of the load or generation (by considering only one node for sake of simplicity) is computed as follows:

$$I_{\text{nonlinear}}^{(1)} = \frac{\sqrt{P_{\text{nonlinear}}^2 + Q_{\text{nonlinear}}^2}}{V} \tag{3.13}$$

Then, the amplitude ratio for $h = 1,\ldots,H$ is defined as follows:

$$A^{(h)} = \frac{I_{\text{nonlinear}}^{(h)}}{I_{\text{nonlinear}}^{(1)}} \tag{3.14}$$

The *angle* provided as input represents the angle given by the typical spectrum, whose values come from the measurement of (aggregate or single) loads and generations. However, it is necessary to determine the *absolute* angle of the current.

The computation of the real angle needs the knowledge of the following elements:

1. The first harmonic current angle with respect to the first harmonic voltage; with the term *first harmonic* we are indicating the *first harmonic* depending on the *nonlinear* portion of loads and generators (because the *linear portions* was already considered as entries in the building of $\mathbf{Z}_{\text{bus}}$), that is,

$$\theta_{\text{nonlinear}}^{(1)} = \arctan\left(\frac{Q_{\text{nonlinear}}}{P_{\text{nonlinear}}}\right) \tag{3.15}$$

2. The angle of the first harmonic of voltage $\theta_V^{(1)}$ (coming from the classical power flow)
3. The spectrum angle $\theta_{\text{spectrum}}^{(h)}$ of the harmonic h considered
4. The spectrum angle $\theta_{\text{spectrum}}^{(1)}$ of the first harmonic

From all these elements, we obtain the angle of the hth harmonic of the current:

$$\theta^{(h)} = \theta_{\text{spectrum}}^{(h)} + h\left\{\left[\theta_V^{(1)} + \theta_{\text{nonlinear}}^{(1)}\right] - \theta_{\text{spectrum}}^{(1)}\right\} \tag{3.16}$$

3.4.1.2 Harmonic Power Flow Solutions

Thanks to knowledge of both the impedance matrix and the currents at all the harmonic orders, according with the current injection method, it is possible to compute the harmonic nodal voltage values and the branch voltage values.

The outputs of the harmonic analysis tool are as follows:

- *Results for the system nodes*: harmonic voltage amplitude and angle, for all harmonic orders through the maximum order, at each time step between 0 and 24 hours
- *Results for the system branches*: harmonic current amplitude and angle, for all harmonic orders through the maximum order, at each hour between 0 and 24 hours; losses at harmonic frequencies at each hour between 0 and 24 hours

These outputs are further elaborated in order to calculate the harmonic branch currents, harmonic losses, and the harmonic indicators such as system indicators (e.g., THD) when the power profiles of loads or generators are characterized in a probabilistic way, so that operational conditions closer to the reality can be determined.

3.4.1.3 PHPF Calculation

The probabilistic calculations are carried out using an external loop implemented to apply the Monte Carlo method. For this purpose, the load and generation data are introduced on the basis of their probabilistic model, and the harmonic data of the generations and loads refer to the random values of the corresponding power extracted for the Monte Carlo repetition under analysis. In addition, the possibility to include correlations among different loads or generators is considered; for example, in case of renewable-based local generation, the correlation may be given by similar weather conditions in relatively small geographical areas.

3.4.2 Structure of the Computational Procedure

The computations are performed by applying the Monte Carlo analysis to the various time steps, then calculating the statistics of all the output variables (i.e., mean value, standard deviation, and more generally the whole PDF represented using relevant histograms).

The computational procedure is organized in different blocks:

1. *Load input data.* In this part of the program, the input data needed for the successive computation are loaded, in particular:
   a. *Network branch data.* For each branch, additional information regarding the modeling of the components at harmonic frequencies

should be specified (e.g., the values of line resistance at the harmonic frequencies).

b. *Transformer data.* For each transformer, additional information regarding the modeling at harmonic frequencies should be specified (e.g., the values of transformer resistance at the harmonic frequencies).

c. *Generator impedance data.* Data regarding the harmonic model of generators (excluding the ones connected by converters) have to be provided.

d. *Normalized active power profiles.* They represent the *typical shape* of different types of loads and generation. They are always positive and their values lie in the range [0,1]. In the probabilistic framework, they represent the evolution in time of the *mean value* of the characteristic shape and are used as references for generating random active power values according with the probabilistic description of the loads.

e. *The tanφ profiles.* They represent the value of the reactive power to active power ratio for each type of load and generation considered.

f. *Load and generator data.* They are characterized by the *reference power,* and from the *probabilistic data.* For each bus, the harmonic injections (generators connected through power electronics interfaces and nonlinear loads) have to be specified. In particular, the following data have to be provided:

  i. Portion of active power required by nonlinear loads connected at this node (expressed in normalized magnitude with respect to the total load active power).

  ii. Portion of reactive power required by nonlinear loads (expressed in normalized magnitude with respect to the total reactive power of the load).

  iii. Portion of active power generated by the generator connected through power electronic interfaces (expressed in normalized magnitude with respect to the total generated active power P_{gen}).

  iv. Portion of reactive power generated by the generator connected through power electronic interfaces (expressed in normalized magnitude with respect to the total generated reactive power Q_{gen}).

g. *Harmonic data.* The harmonic data provide information about the harmonic behavior of loads and generation. In this work, each harmonic behavior is linked with a normalized active power profile. For each harmonic source, the harmonic current spectrum (amplitude and angle of all the harmonic components) has to be provided. The amplitudes are expressed in per unit of the component at fundamental frequency. In particular, to exemplify, the following harmonic sources have been considered:

    i. Industrial nonlinear loads.
   ii. Residential nonlinear loads.
  iii. Tertiary nonlinear loads.
   iv. PV generators.
    v. Wind generators connected through power electronic interfaces.

2. *Creation of the mean value of the power, for all the profiles and all the time steps.* In this block, the *mean value of the active power* is determined using the *reference power* (characterizing each load/generator in each node) and the *normalized active power profiles*.

3. *Application of the* probabilistic *behavior to the active power.* At this point, the probabilistic behavior is added to the active power for every profile and node. In fact, each probabilistic description of loads or generators is associated to the corresponding normalized active power profile and the corresponding node. This permits to add the uncertainty exactly at the *correct node* and at the *correct profile*, for the entire time interval considered.

In particular, two different probabilistic entities are handled:

   a. *Correlated entities.* In this case, the probabilistic behavior depends not only on the *probabilistic description* of the entity but also on the behavior of the other entities correlated to it.

   b. *Uncorrelated entities.* In this case, the behavior of the load or generator is completely independent of the other entities installed in the network.

Then, using the information about $\tan\varphi$, the *mean value of the reactive power* is computed (the value of the reactive power in a given time step is directly linked to the value of active power at the same time step).

4. *Application of the harmonic behavior to the active and the reactive power.* If there is no information about the harmonic content of loads or generators, their behavior is always represented as linear.

Both the linear and the nonlinear portions of active and reactive powers are computed: by starting from the active and reactive power deriving from point (3), and by applying the information of the percentage of linear and nonlinear portions, a representation of the fundamental power as *linear* and *nonlinear portions* is obtained.

5. *Run the fundamental power flow.* The fundamental power flow is requested for the computation of all the quantities of the network by considering the first harmonic of active and reactive power (i.e., both linear and nonlinear parts).

6. *Run the harmonic power flow.* The harmonic power flow has the following inputs:

   a. Branch and transformer impedance.

   b. Generator impedance.

   c. Linear active power.

   d. Nonlinear active power.

   e. Linear reactive power.

   f. Nonlinear reactive power.

   g. Pattern spectra (both amplitude and angles).

Starting from the general scheme of the modules reported in Figure 3.1, the flowchart illustrating the structure of the harmonic power flow calculations is shown in Figure 3.10, with a detailed description reported in Figure 3.11. These figures contain the definition of the data in matrix form.

3.5 Calculation of the Network Performance Indicators

3.5.1 Overview on the Performance Indicators

PQ indices represent, for compactness and practicality, the quickest and most useful way to describe the characteristics of PQ disturbances (Caramia et al. 2009a). They are convenient for condensing complex time- and frequency-domain waveform phenomena into a single number.

An important classification of PQ indices is based on the portion of the electrical system to which the index refers:

- *Site indices*: these refer to a single customer PCC.
- *System indices*: these refer to a segment of the distribution system or, more generally, to the entire electrical system.

While the site index treatment requires the collection of different observations in time, for system indices, observations at different sites, usually a representative set of the system under study, must also be collected.

Weighting factors can be introduced to take into account the sites not monitored and the difference in importance between different monitored sites. For example, weighting factors may be based on the number of substations or customers or on the rated power of each site.

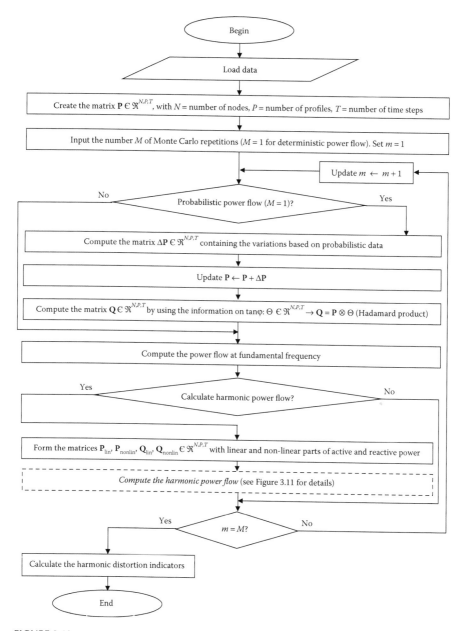

FIGURE 3.10
Flowchart of the harmonic power flow incorporated in DERMAT, operating on the basis of probabilistic load and generation data.

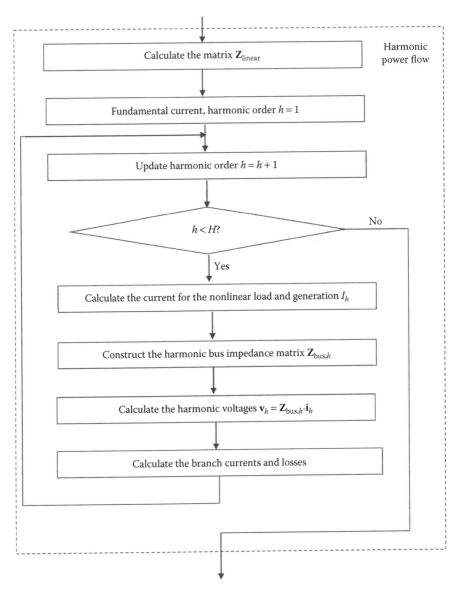

FIGURE 3.11
Details on the harmonic power flow calculations.

3.5.2 Losses Indicators

The presence of harmonic voltages and currents is the cause of additional losses in the network (Fuchs et al. 1986; IEEE Task Force 1995; Caramia et al. 1998). Therefore, when evaluating the energy losses, the harmonic losses have to be considered.

In the deterministic framework, for a given time with $t = 1,\ldots,T$ time steps of duration τ each, the energy losses in the network are computed from the energy losses at branch b (with $b = 1,\ldots,B$) during each time interval, that is,

$$W_{\text{losses}} = \tau \sum_{t=1}^{T} \sum_{b=1}^{B} \sum_{h=1}^{H} R_b^{(h)} [I_{bt}^{(h)}]^2 \qquad (3.17)$$

where:

R_b^h is the resistance of branch b evaluated at harmonic h

I_{bt}^h is the root-mean-square (RMS) value of the hth harmonic current, flowing in branch b, during the time interval t

More sophisticated models can be adopted in order to perform a more accurate evaluation of losses. For instance, accurate modeling of transformers can be included to account for losses due to harmonic currents and voltages occurring in windings because of the skin effect and the proximity (Fuchs and Masoum 2009).

In the probabilistic framework, the expected value of energy losses can be calculated (Caramia et al. 1998).

3.5.3 Waveform Distortion Aspect and Indicators

3.5.3.1 Site Indices

Waveform distortions (voltage or current) can be characterized by several indices and the most frequently used are as follows (Caramia et al. 2009a):

- Individual harmonics.
- THD factor.
- Individual interharmonics.
- Total interharmonic distortion factor.

The individual voltage or current harmonic A_h is the ratio between the RMS value of harmonic component of order h, X_h, and the RMS value of the fundamental component, X_1, of the voltage or current waveform).

$$A_h = \frac{X_h}{X_1} \qquad (3.18)$$

THD is defined as the RMS of the harmonic content divided by the RMS value of the fundamental component, usually multiplied by 100:

$$\text{THD} = \frac{\sqrt{\sum_{h=2}^{H_{\max}} X_h^2}}{X_1} 100 \qquad (3.19)$$

where:

H_{max} is the order of the highest harmonic taken into account

The spectral components to be included in the calculation of harmonic indices can be obtained by performing a Fourier analysis of the current or voltage waveforms and using adequate data aggregation techniques. In particular, in IEC 61000-4-7 2002 and IEC 61000-4-30 2003, the procedure for the measurements of waveform distortion indices refers to the discrete Fourier transform (DFT) of the waveforms over a rectangular window with a length equal to 10 cycles for systems, where the power system frequency is 50 Hz (i.e., 200 ms). The window width determines the frequency resolution for the spectral analysis, which results in 5 Hz.

Then, the spectral components obtained by applying the DFT to the actual signal inside the window are aggregated in different configurations, called *groups* or *subgroups* (Caramia et al. 2009a), depending on what kind of measurement is to be performed (harmonics, interharmonics, or both).

3.5.3.2 Temporal Aggregation

A time-aggregation technique frequently used consists of the combination of several values of a given parameter, each parameter determined over an identical time interval, to provide a value for a longer time interval. The measures carried out on 200 ms time intervals are further aggregated, as described in Table 3.2, being Q_i the value of the harmonic parameter in the ith time interval (Caramia et al. 2009a).

Further indicators used to quantify the waveform distortion (Arrillaga and Watson 2003 and Caramia et al. 2009a) are the C-message index, the current-telephone influence factor (IT product), the voltage-telephone influence factor (VT product), the K-factor, the peak factor, and the true power factor.

3.5.3.3 PQ Indices and DG

As it is well known and extensively addressed in the relevant literature, DG interacts with PQ levels of the distribution network. In fact, DG can

TABLE 3.2

Aggregation of the Measured Values at Different Time Intervals

Interval Type	Mean Type	Formulation
Very short interval (3 seconds)	Mean over 15 intervals of 0.2 seconds	$Q_{vsh} = \sqrt{\dfrac{1}{15}\displaystyle\sum_{i=1}^{15} Q_i^2}$
Short interval (10 minutes)	Mean over 200 intervals of 3 seconds	$Q_{10min} = \sqrt{\dfrac{1}{200}\displaystyle\sum_{i=1}^{200} Q_{3s,i}^2}$
Long interval (2 hours)	Mean over 12 intervals of 10 minutes	$Q_{2h} = \sqrt{\dfrac{1}{12}\displaystyle\sum_{i=1}^{12} Q_{10min,i}^2}$

introduce several disturbances causing a reduction in PQ levels (Bracale et al. 2011).

The impact of DG on PQ also depends on the connection that can be direct or with a power electronic interface. In the first case, the generators directly connected to the distribution system can influence the background waveform distortions. In fact, generators modify the harmonic impedances and contribute to modify the voltage harmonic profiles at the distribution system bus bars. In addition, the simultaneous presence of shunt capacitors installed to improve the induction generator power factor can generate resonance conditions. In the second case, the power electronic interface can inject harmonic currents, which can impact on voltage distortion. However, it should be noted that DG can also improve the PQ of the distribution system, mainly as a consequence of the increase of short-circuit power and of advanced controls of PWM converters that reduce the low-order harmonics.

Generally speaking, the assessments of the variation in PQ levels due to the presence of DGs can be made through the following assessments:

- Assessment prior to the installation of DG
- Assessment after the installation of DG

Before installing distributed generators, the distribution network operator has to take a decision regarding the size and the location of DG units. At this stage, an analysis of the impact of DG units on PQ level can support the decision of the distribution network operator. For instance, depending on the DG properties, preliminary knowledge of PQ improvements or deteriorations can suggest the management strategies for the distribution system.

After installing distributed generators, the assessment of PQ levels is aimed at verifying the effectiveness of existing limits on disturbances. Then, when PQ levels reach unacceptable values, appropriate corrective actions can be taken.

The two approaches described above require completely different tools and quantities for PQ assessment because the scopes of their use vary widely.

In particular, when referring to PQ assessment prior to the installation of DG, suitable indicators can be introduced. The indicators, which are compact and practical, have been demonstrated to be the best way to assess PQ levels, as it is proven by their large use in standards, recommendations, and guidelines.

In the previous subsections, these indicators have been defined in a general way. These indicators can be easily customized for the different PQ aspects and indices as indicated below.

3.5.3.4 Power Quality Variation Indices

In Bracale et al. (2011), Caramia et al. (2005), and Caramia et al. (2006b), some indices that quantify the variations on the quality service in consequence of installations of DG have been proposed. These indices, for each site and for the system, offer proper metrics for evaluating the modification of disturbances in the presence of DG and can be calculated by a probabilistic approach for taking into account the time-varying nature of the PQ disturbances.

Considering a generic PQ index X, it is possible to introduce the PQ variation of index X due to the DG installation (called $X_{\Delta\%}$) as follows:

$$X_{\Delta\%} = \frac{X_{OLD} - X_{NEW}}{X_{OLD}} \cdot 100 \qquad (3.20)$$

where:

X_{NEW} is the value of a single or global index X, after the new installation of DG
X_{OLD} is the value of a single or global index X, before the new installation of DG

The generic index in Equation 3.20 can be defined for a large category of indices, each of which allows the quantification of the increase or decrease of the PQ level of the electrical service due to the installation of new generation units.

An application of the PQ variation index, with reference to harmonic distortion, can be found in Ortega et al. (2013), where the problem of the assessment of the impact of PV systems is addressed.

3.5.3.5 Normalized Power Quality Variation Indices

In the presence of installations of DG, it could also be useful to evaluate the variation of the network performance in terms of the PQ level for installed unit power of DG. This indication can be obtained by means of the normalized PQ variation index defined as follows (Bracale et al. 2011):

$$NPQ_{\Delta\%} = \frac{X_{\Delta\%}}{P_{DG}} = 100 \left(\frac{X_{OLD} - X_{NEW}}{X_{OLD} P_{DG}} \right) \qquad (3.21)$$

where:

P_{DG} is the total installed power of the new distributed generators

This indicator can be used to assess where a new generator can be installed to obtain the maximum reduction of disturbances in the distribution network or to compare the impact on PQ of different sizes, structures, and/or connections of DG plants.

3.5.3.6 System Indicators

The system indices reported in the literature are as follows (Caramia et al. 2009a):

- The system total harmonic distortion 95th percentile, $STHD_{95}$.
- The system average total harmonic distortion, SATHD.
- The system average excessive total harmonic distortion ratio index, $SAETHDRI_{THD^*}$.

To define these indices, let us consider a distribution system with N bus bars and M monitoring sites.

The $STHD_{95}$ index is defined as the 95th percentile value of a weighted distribution; this weighted distribution is obtained by collecting the 95th percentile values of the M individual index distributions, with each index distribution obtained from the measurements recorded at a monitoring site. The weights L_s (for $s = 1,\ldots,M$) can be linked to either the connected powers or the number of customers served from the area represented by the monitored data. The system index $STHD_{95}$ allows us to summarize the measurements both temporally and spatially by handling measurements at M sites of the system in a defined time, assumed to be significant for the characterization of the system service condition. Due to the introduced weights, this system index allows the assignment of different levels of importance to the various sections of the entire distribution system.

The system average THD (SATHD), is based on the mean value of the distributions rather than the 95th percentile value. The SATHD index gives average indications on the system voltage quality. It can be computed as follows:

$$SATHD = \frac{\sum_{s=1}^{M} L_s \mu(THD_s)}{\sum_{s=1}^{M} L_s} \tag{3.22}$$

where:
$\mu(THD_s)$ is the mean value of THD values at monitoring site s

The system average excessive THD ratio index, $SAETHDRI_{THD^*}$, quantifies the number of measurements that exhibit a THD value exceeding the THD* threshold. This index is computed by counting, for each monitoring site in the system, the measurements that exceed the THD* value and normalizing this number to the total number of the measurements conducted at the site under analysis. It can be computed as follows:

$$SAETHDRI_{THD^*} = \frac{\sum_{s=1}^{M} L_s \left(N_{THD^*s} / N_{Tot,s}\right)}{\sum_{s=1}^{M} L_s} \tag{3.23}$$

where:
N_{THD^*s} is the number of steady-state measurements at monitoring site s that exhibit a THD value greater than the specified threshold THD* and $N_{Tot,s}$, the total number of measurements at monitoring site s

3.6 Case Study Application

3.6.1 System Data

The system used to present the results is the 16-node test system taken from Civanlar et al. (1988), with base voltage 23 kV and base power 1 MVA.

This system has $N = 16$ nodes, $S = 3$ supply nodes and $B = 16$ total branches. In order to obtain a radial system configuration, the number of redundant branches to open is $\Lambda = B - N + S = 3$, and the total number of radial configurations that can be obtained by opening three branches is 190 (Andrei and Chicco 2008). The system is represented in Figure 3.12, where the dashed lines correspond to one of the solutions with three open redundant branches. In practice, opening the redundant branches results in partitioning the network into three independent subnetworks, each of which is supplied from the corresponding slack node.

On these bases, the example shown here adopts partially different types of inputs for the nodes of the three subnetworks, with the aim of illustrating specific aspects of the calculations carried out with the PHPF tool. In particular, the nodes of the three subnetworks contain the following generations and loads:

- Subnetwork 1 (nodes S1, 4, 5, 6, and 7): uncorrelated loads only, no local generations
- Subnetwork 2 (nodes S2, 8, 9, 10, 11, and 12): uncorrelated loads and correlated local generations (at nodes 9 and 10)
- Subnetwork 3 (nodes S3, 13, 14, 15 and 16): correlated loads (at nodes 13, 14, and 15) and uncorrelated local generations (at nodes 15 and 16)

Figure 3.13 shows the hourly patterns of the net power supplied by the slack nodes during a typical day. Subnetworks 2 and 3 present different behavior, because in subnetwork 2, the net power at the slack S2 is always positive, also when DG is considered, whereas in subnetwork 3, the effect of DG leads to negative values of the net power in some periods of the day.

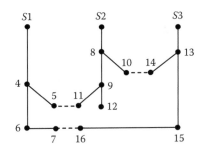

FIGURE 3.12
Layout of the 16-node test system.

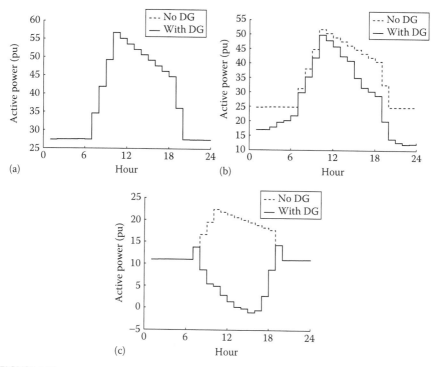

FIGURE 3.13
Mean value of active power injected with and without DG: (a) in node S1, (b) in node S2, and (c) in node S3.

All the types of the local generations are assumed as 100% nonlinear, whereas the loads are assumed as 70% nonlinear both on active and reactive powers.

The load and generation data used are reported in Tables 3.3 and 3.4, respectively.

The correlated loads are characterized by the correlation matrix represented in Figure 3.14, where the correlation coefficients are assumed constant during all the day.

The correlated local generation is characterized by the correlation matrix reported in Figure 3.15.

3.6.2 Results and Discussions

To exemplify the results of the calculations of the PHPF implemented in DERMAT, the example presented in this section has been run on the 16-node network. The results are graphically shown for some network nodes, belonging to the observed set of nodes $N = \{4, 7, 9, 10, 15, 16\}$. Further sets of nodes are defined below in order to show specific results.

TABLE 3.3

Reference Active and Reactive Powers
for Loads

Load Node	P_{ref} (pu)	Q_{ref} (pu)
4	34.00	16.47
5	16.00	7.75
6	12.00	5.81
7	5.00	2.42
8	28.00	13.56
9	10.00	4.84
10	14.00	6.78
11	3.00	1.45
12	5.00	2.42
13	5.00	2.42
14	8.00	3.87
15	4.00	1.94
16	10.00	4.84

TABLE 3.4

Reference Active and Reactive Powers for Local Generators
and Their Types

Generation Node	P_{ref} (pu)	Q_{ref} (pu)	Type
9	−10.00	0	Wind
10	−6.00	0	Wind
15	−10.45	0	PV
16	−12.65	0	PV

Nodes	13	14	15
13	1	0.3	0.5
14	0.3	1	0.4
15	0.5	0.4	1

FIGURE 3.14
Load correlation matrix entries (inside the thick borders).

Nodes	9	10
9	1	0.3
10	0.3	1

FIGURE 3.15
Local generation correlation matrix entries (inside the thick borders).

The variety of the solutions is represented by showing the following results:

1. *Local indicators*
   a. Evolution in time of voltages and currents (plotting the average values and the standard deviation bands around the average values)
   b. Evolution in time of the THD
   c. PDFs of the voltage THD at selected nodes for a given hour
   d. Evolution in time of the PDFs of the total losses
2. *System indicators*
   a. PDFs of the voltage THD at different hours (all nodes)
   b. PDFs of the voltage THD at different nodes (all hours)
   c. PDFs of the voltage THD (all hours and nodes)

Figure 3.16 contains the evolution in time of mean values and standard deviation bands of the currents and voltages for the nodes belonging to the observed set N in the presence of DG. The results obtained at each hour are represented as points at each hour, linearly interpolated with each other for the sake of legibility of the figure.

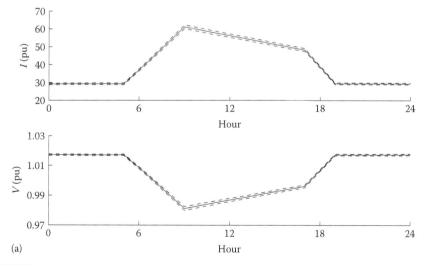

(a)

FIGURE 3.16
Evolution in time of load currents and voltages at selected nodes (with DG): (a) node 4. (*Continued*)

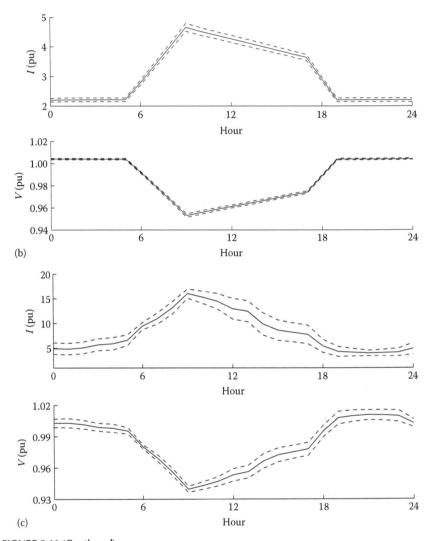

FIGURE 3.16 (*Continued*)
Evolution in time of load currents and voltages at selected nodes (with DG): (b) node 7 and
(c) node 9.   (*Continued*)

Subnetwork 1 is passive. The uncertainty band around the mean values,
shown for both load current and voltage at nodes 4 and 7, is relatively small,
resulting in well-defined voltage and current values.

More interesting is the behavior of voltages and currents at nodes 9 and 10,
belonging to subnetwork 2. These two nodes present correlated generation
and exhibit larger variation bands. In the tests carried out, some differences
have been found in the standard deviation of the results by considering
or not the correlation in generations and loads, as shown in Figure 3.17.

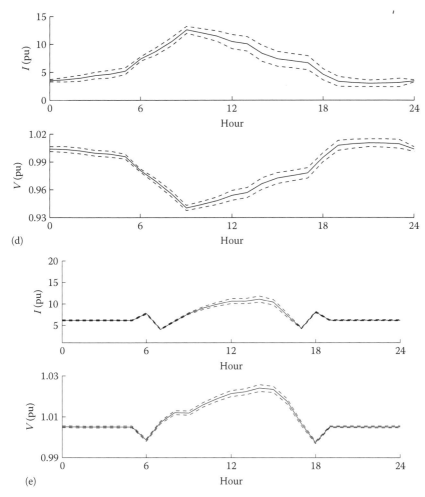

FIGURE 3.16 (*Continued*)
Evolution in time of load currents and voltages at selected nodes (with DG): (d) node 10 and
(e) node 15. (*Continued*)

In particular, considering the correlation, the standard deviation has become
higher. The mean values of the results remained consistently the same. This
indicates that neglecting correlations in the probabilistic power flow leads to
underestimating the width of the current and voltage variation bands.

In Figure 3.18, the evolution in time of currents and voltages without DG of
nodes forming the set $N_{gen} = \{9, 10, 15, 16\}$ have been reported: it is possible to
note how DG affects the shape and the value of both currents and voltages,
as well as the uncertainty of the values, especially at node 16.

Figure 3.19 shows the box plots of the voltage THD at each hour. The representation is based on the *box and whiskers* technique (Tukey 1977), in which the
boxes contain the internal indication of the median value (50th percentile),

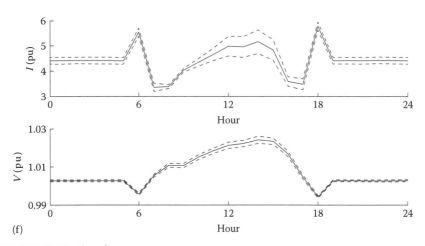

FIGURE 3.16 (*Continued*)
Evolution in time of load currents and voltages at selected nodes (with DG): (f) node 16.

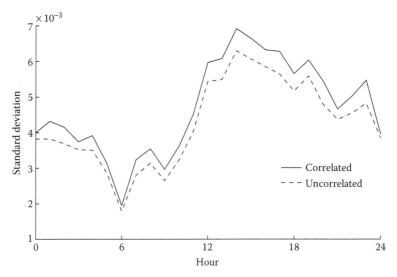

FIGURE 3.17
Standard deviations of the voltage at node 9 for both correlated and uncorrelated cases.

and the upper and lower limits of each box corresponding to the 25th and 75th percentiles, respectively. The distance between the 25th and the 75th percentiles is defined as *interquartile range*. Outside the boxes, the upper limit of the whisker indicates the minimum between 1.5 times the interquartile range above the 75th percentile and the maximum THD value. Likewise, the lower limit of the whisker indicates the maximum between 1.5 times the interquartile range below the 25th percentile and the minimum THD value. Finally, the points located outside the upper and lower limits of the whiskers are marked as outliers.

For the set of nodes N_{gen}, the values with and without DG are shown.

For all the nodes belonging to the set N_{gen}, the same behavior is highlighted. If DG is not considered, the percentage of voltage THD decreases. This fact happens because, as reported previously, the local generation is considered to be 100% nonlinear.

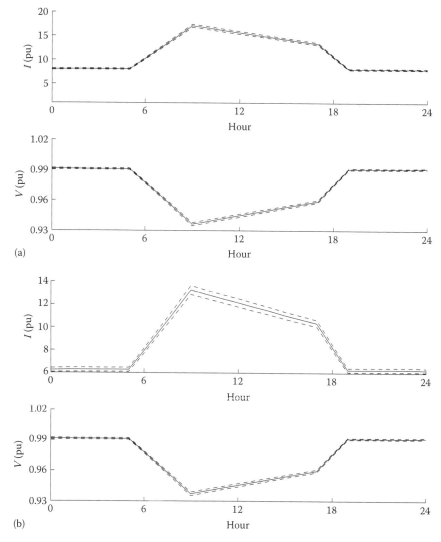

FIGURE 3.18
Evolution in time of load currents and voltages at selected nodes (without DG): (a) node 9; (b) node 10.
(*Continued*)

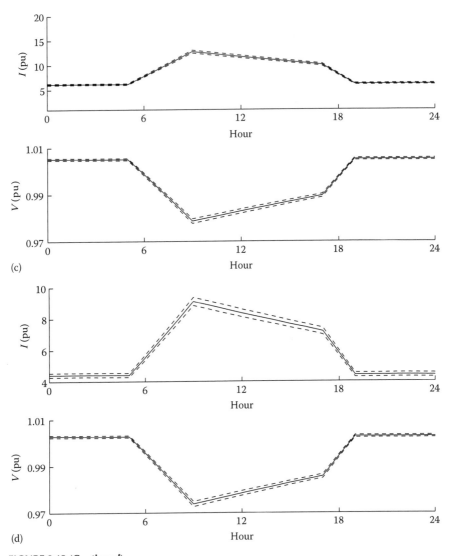

FIGURE 3.18 (Continued)
Evolution in time of load currents and voltages at selected nodes (without DG): (c) node 15;
(d) node 16.

Considering a given hour (10:00 am), the PDFs of the voltage THD at the
nodes belonging to the observed set $N_5 = \{S1,4,5,6,7\}$ are shown in Figure 3.20.
It can be seen that the THD does not always exhibit symmetrical PDF shapes
and its value is higher for nodes located far from the slack node.

In Figure 3.21, the total losses of the network are reported. The presence of
DG in the network permits the reduction of the losses, although this reduction is not so high. This fact depends on the reverse flow of the power in
subnetwork 3 (as previously shown in Figure 3.13).

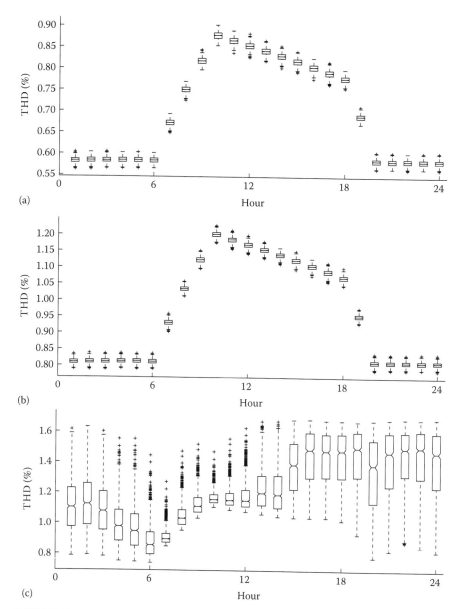

(a)

(b)

(c)

FIGURE 3.19
Evolution in time of the voltage THD: (a) node 4; (b) node 7; (c) node 9 (with DG).   *(Continued)*

3.6.2.1 System Indicators

Figure 3.22 shows the 3D plot of the PDFs of the voltage THD at different hours, calculated by merging the contributions of all nodes, both with and without DG.

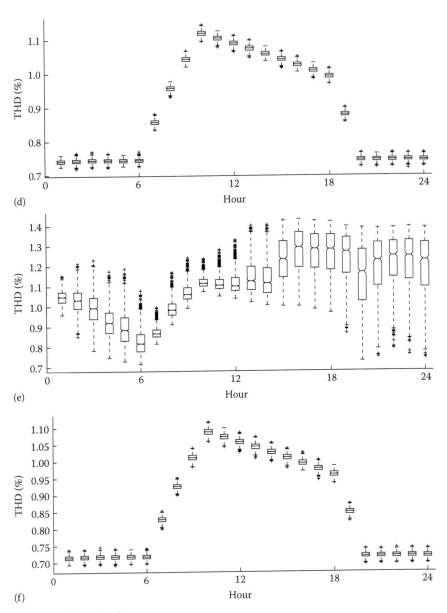

(d)

(e)

(f)

FIGURE 3.19 (*Continued*)
Evolution in time of the voltage THD: (d) node 9 (without DG); (e) node 10 (with DG); (f) node 10 (without DG).

(*Continued*)

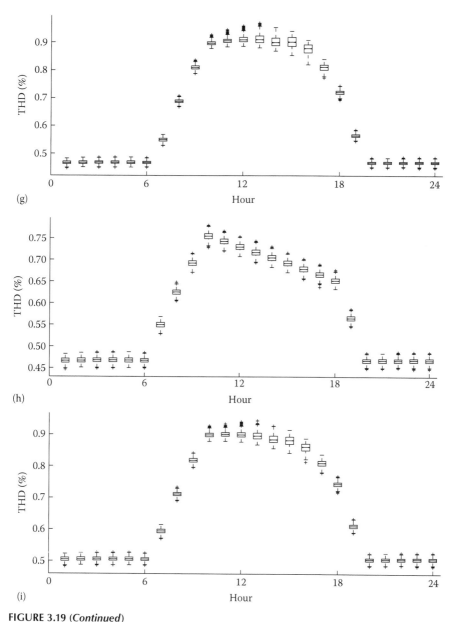

FIGURE 3.19 (Continued)
Evolution in time of the voltage THD: (g) node 15 (with DG); (h) node 15 (without DG); (i) node 16 (with DG).   (Continued)

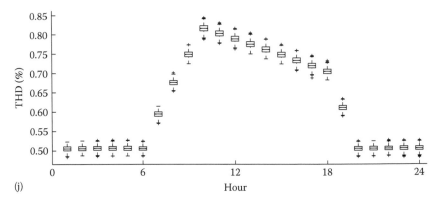

(j)

FIGURE 3.19 (*Continued*)
Evolution in time of the voltage THD: (j) node 16 without DG.

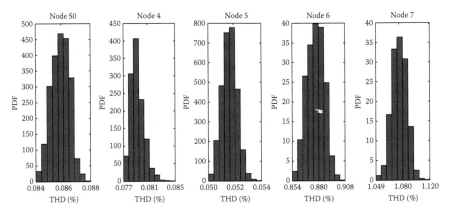

FIGURE 3.20
PDFs of the voltage THD at selected nodes.

We can note the maximum value of THD is higher in the presence of DG, and this increase becomes more evident in the early hours of the morning and in later afternoon, since wind turbines produce more energy in those periods.

Figure 3.23 shows the 3D plot of the PDFs of the voltage THD at different nodes, calculated by merging the contributions at all hours, both with and without DG.

It is possible to observe that the voltages at the three slack nodes $S1$, $S2$, and $S3$ are not distorted. If the voltage THD for the load nodes belonging to the set $N_{Sub1} = \{4,5,6,7\}$ are determined by nonlinear loads, then they are not affected when DG is not considered because no DG is involved in that subnetwork.

For the nodes belonging to the set $N_{Sub2} = \{8,9,10,11,12\}$, the harmonic characteristics of wind generators affect the voltage THD, as well as the current THD. Nodes belonging to the set $N_{Sub3} = \{13,14,15,16\}$ are affected by the harmonic characteristics of the PV generators.

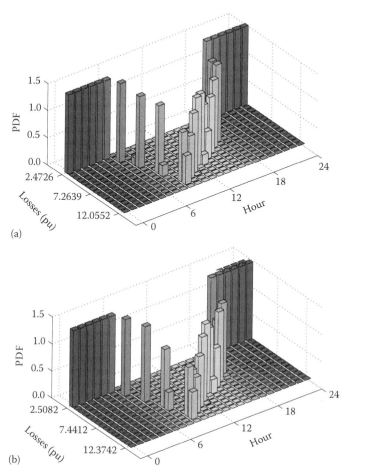

(a)

(b)

FIGURE 3.21

Evolution in time of the PDF of the total losses: (a) with DG and (b) without DG.

It is possible to note that the shape of the PDF of the voltage THD in subnetworks 2 and 3 change in an evident way when DG is considered, with respect to the case without DG. In fact, subnetwork 2 passes from class around 1.6 to class around 1, whereas subnetwork 3 has a THD maximum lower than 1 without DG (while with DG, it was around 1).

The above concepts can be better understood by considering the 2D bar chart of the voltage THD for the set of nodes N_{Sub2} and N_{Sub3}, reported in Figures 3.24 and 3.25, respectively. As explained above, for both subnetworks with the addition of DG, there is an increase of the voltage THD.

Figure 3.26 shows the PDF of the voltage THD determined by taking into account the THD values at all hours and nodes, (1) with DG and (2) without DG. The presence of DG increases the voltage THD, which passes from less

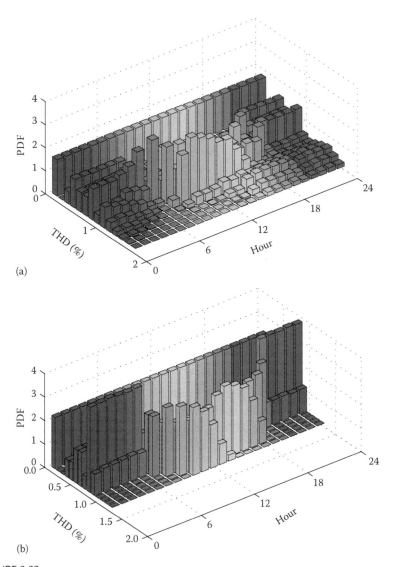

FIGURE 3.22
PDFs of the voltage THD at different hours: (a) with DG and (b) without DG.

than 1.3% to more than 1.6% by adding DG. The first bar in the histogram in each one of the two subplots represents the null THD value at the three slack nodes.

3.6.2.2 System Harmonic Indicators

The system harmonic indicators have been calculated for both the correlated and the uncorrelated cases. The results are reported in Table 3.5. The

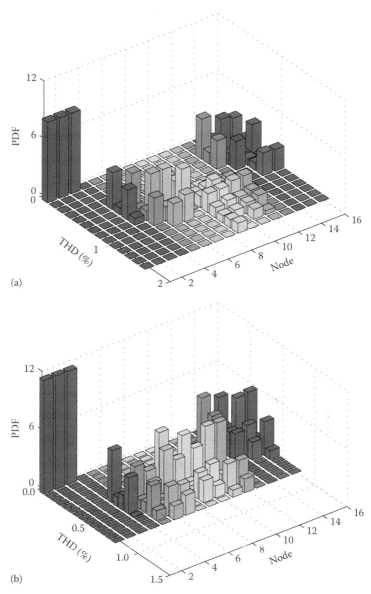

(a)

(b)

FIGURE 3.23
PDFs of the voltage THD at different nodes: (a) with DG and (b) without DG.

values of SATHD and $STHD_{95}$ are very close with each other. Since no value of THD exceeds either the 5% threshold (i.e., the maximum value imposed by IEEE519) or the 8% threshold (the maximum value imposed by EN50160), the indicator $SAETHDRI_{THD^*}$ has a null value.

By analyzing both the correlated and uncorrelated cases, the values of the system harmonic variation indices are reported in Table 3.6. From

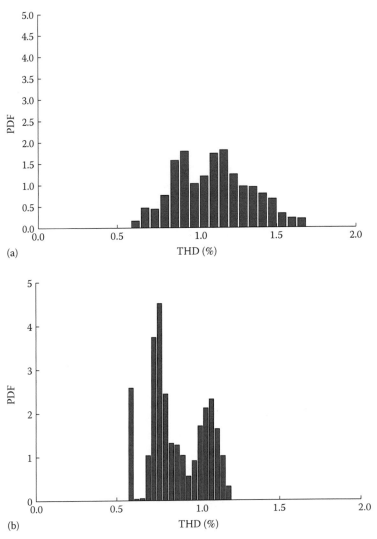

FIGURE 3.24
PDF of the voltage THD (from all values at different hours and nodes of N_{Sub2}): (a) with DG and (b) without DG.

the analysis of Table 3.6, it is evident that for all the considered cases, the harmonic distortion of the voltage becomes worse when the wind and PV generators are involved, leading to negative values of variation indices. The highest variation is registered for the $STHD_{95}\_V$ in the correlated case.

The PQ variation indices for the two subnetworks are reported in Table 3.7.

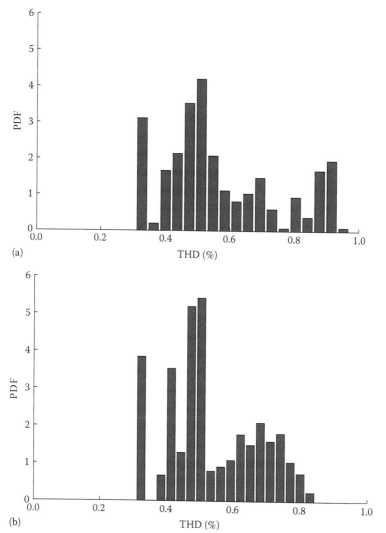

FIGURE 3.25

PDF of the voltage THD (from all values at different hours and nodes of N_{Sub3}): (a) with DG and (b) without DG.

The analysis of results leads to the following considerations:

- In subnetwork 2 with wind generators (installed power 16 MW), the voltage harmonic distortion is worse than in subnetwork 3 having only PV generators (installed power 23.1 MW).

- For subnetwork 2, the SATHD results in 27.15% greater than the value in the case without DG, whereas for subnetwork 3, the SATHD results in 6.6% greater than the value without DG.

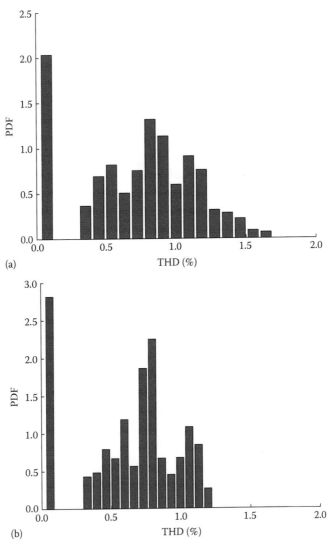

FIGURE 3.26
PDF of the voltage THD from all values at different hours and nodes: (a) with DG and (b) without DG.

TABLE 3.5

System Indicators

	SATHD (%)	STHD$_{95}$ (%)	SAETHDRI$_{THD^*}$ (THD* = 5%) (%)	SAETHDRI$_{THD^*}$ (THD* = 8%) (%)
Uncorrelated case	0.7163	1.356	0	0
Correlated case	0.7155	1.357	0	0

TABLE 3.6

Power Quality Variation Indices for Correlated and Uncorrelated Cases

	SATHD_V (%)	STHD$_{95}$_V (%)	NPQ_V_{SATHD}	NPQ_V_{STHD95}
Uncorrelated	−13.32	−21.33	−0.341	−0.546
Correlated	−13.18	−21.41	−0.337	−0.548

TABLE 3.7

Power Quality Variation Indices

	SATHD_V (%)	STHD$_{95}$_V (%)	NPQ_V_{SATHD}	NPQ_V_{STHD95}
Subnetwork 2	−27.15	−32.55	−1.697	−2.034
Subnetwork 3	−6.57	−18.54	−0.292	−0.803

- As a consequence, the value of NPQ_V_{SATHD} results lower in subnetwork 2 (about −2% for each installed MW of wind generators versus about −0.3% for each installed MW of PV generators).
- Analogous considerations can be made with reference to the NPQ_V_{STHD95} indicator.

3.7 Conclusions

Solving the PHPF is a challenging task, because of the need of accurately modeling the network, loads, and generations by including the uncertainty and their correlations, and to assign the appropriate spectra of the harmonic currents to all the contributions providing harmonic distortion, in order to execute the harmonic power flow with the current injection method.

The overall framework for PHPF calculations has been implemented in the DERMAT tool, specifically developed to perform calculations on distribution systems with distributed energy resources. The structure of the computational modules and the details of the models and formulations have been presented in this chapter. A number of results have been presented by considering a small test network for illustrative purposes. The presence of uncertain and nonlinear loads and local generation sources has been modeled, also including correlations among the power production due to the effect of environmental variables.

The results show the potential of the tool developed in addressing the calculations to obtain the probabilistic representations of the system losses and of the harmonic distortion indicators. In some cases, it has been shown that by modeling the correlations, the standard deviations of the outputs were increasing with respect to the same case considered without modeling the correlations. This result is interesting because it provides a good reason for modeling the correlations to obtain less conservative and more meaningful results. However, the results depend on the level of diffusion of the distributed energy resources in the network and are not generalizable.

The outcomes of the PHPF calculations provide new insights for the definition of key performance indicators expressed in terms of percentiles of the relevant output variables, useful to represent the present need to take into account uncertainty in the representation of the results of distribution system analysis with distributed resources.

Summary

This chapter presents the formulation and application of a comprehensive tool to perform the PHPF calculations for distribution systems characterized by their uncertainty, also with correlated data. The characterization of the uncertainty is provided for the local loads and local generation sources in the framework of a specific tool developed for interval power flow calculations at multiple time steps during a predefined time. The main outputs determined are a number of key performance indicators representing the usage of the distribution networks in the presence of distributed energy resources (harmonic indicators for currents and voltages at the individual nodes and for the whole system and total losses), evaluated in a probabilistic framework. A dedicated case study application is presented on a test system, with the calculation of all the relevant outputs.

Abbreviations

CDF	Cumulative distribution function
DFT	Discrete Fourier transform
DG	Distributed generation
IEC	International Electrotechnical Commission
PCC	Point of common coupling
PDF	Probability density function
PEM	Point estimate method
PHPF	Probabilistic harmonic power flow
PQ	Power quality
PV	Photovoltaic
PWM	Pulse width modulation
RMS	Root mean square
SAETHDRI	System average excessive total harmonic distortion ratio index

SATHD   System average total harmonic distortion
STHD$_{95}$ System total harmonic distortion, 95th percentile
THD   Total harmonic distortion

References

Alexiadis, M. C., Dokopoulos, P. S. and Sahsamanoglou, H. S. 1999. Wind speed and power forecasting based on spatial correlation models. *IEEE Transactions on Energy Conversion* 14: 836–842.

Anders, G. J. 1990. *Probabilistic Concepts in Electric Power Systems*. Wiley, New York.

Andrei, H. and Chicco, G. 2008. Identification of the radial configurations extracted from the weakly meshed structures of electrical distribution systems. *IEEE Transactions on Circuits and Systems I: Regular Papers* 55: 1149–1158.

Arrillaga, J. and Watson, N. R. 2003. *Power System Harmonics* (second edition). Wiley, Chichester.

Au, M. T. and Milanovic, J. V. 2006. Stochastic assessment of harmonic distortion level of medium voltage radial distribution network. *Proceedings of the Probabilistic Methods Applied to Power Systems*, Stockholm, Sweden, June 11–15.

Baghzouz, Y. 2005. An overview on probabilistic aspects of harmonics in power systems. *Proceedings of the IEEE Power and Energy Society General Meeting*, San Francisco, CA, vol. 3, IEEE, Piscataway, NJ, pp. 2394–2396, June 12–16.

Baghzouz, Y., Burch, R. F., Capasso, A. et al. 1998. Time-varying harmonics: Part I—Characterizing measured data. *IEEE Transactions on Power Delivery* 13: 938–944.

Baghzouz, Y., Burch, R. F., Capasso, A. et al. 2002. Time-varying harmonics: Part II—Harmonic summation and propagation. *IEEE Transactions on Power Delivery* 17: 279–285.

Baghzouz, Y. and Tan, O. T. 1987. Probabilistic modeling of power system harmonics. *IEEE Transactions on Industry Applications* 23: 173–180.

Bechrakis, D. A. and Sparis, P. D. 2004. Correlation of wind speed between neighboring measuring stations. *IEEE Transactions on Energy Conversion* 19: 400–406.

Bracale, A., Caramia, P., Carpinelli, G. et al. 2011. Site and system indices for power quality characterization of distribution networks with distributed generation. *IEEE Transactions on Power Delivery* 26: 1304–1316.

Caramia, P., Carpinelli, G., Esposito, T. et al. 2006a. Probabilistic harmonic power flow for assessing waveform distortions in distribution systems with wind embedded generation. *Proceedings of the International Symposium on Power Electronics, Electrical Drives, Automation and Motion*, Taormina, Italy, May 23–26, IEEE, Piscataway, NJ, pp. 818–823.

Caramia, P., Carpinelli, G., Losi, A. et al. 1998. A simplified method for the probabilistic evaluation of the economical damage due to harmonic losses. *Proceedings of the 8th International Conference on Harmonics and Quality of Power*, Athens, Greece, vol. 2, IEEE, Piscataway, NJ, pp. 767–776, October 14–16.

Caramia, P., Carpinelli, G., Russo, A. et al. 2005. New system power quality indices for distribution networks in presence of embedded generation. *Proceedings of the CIGRE Symposium on Power Systems with Dispersed Generation*, Athens, Greece, April 13–16.

Caramia, P., Carpinelli, G., Russo, A. et al. 2006b. Power quality assessment in liberalized market: Probabilistic system indices for distribution networks with embedded generation. *Proceedings of the 9th International Conference on Probabilistic Methods Applied to Power Systems*, Stockholm, Sweden, June 11–15.

Caramia, P., Carpinelli, G. and Verde, P. 2009a. *Power Quality Indices in Liberalized Markets*. Wiley, Chichester.

Caramia, P., Verde, P., Varilone, P. et al. 2009b. Probabilistic modeling for network analysis. In Ribeiro, P. F. (ed.) *Time-Varying Waveform Distortions in Power Systems*. Wiley, Chichester, 95–113.

Carbone, R., Testa, A., Carpinelli, G. et al. 1994. A review of probabilistic methods for the analysis of low frequency power system harmonic distortion. *Proceedings of the 9th International Conference on Electromagnetic Compatibility*, Manchester, September 5–7, IEE, Stevenage, UK, pp. 148–155.

Carpaneto, E. and Chicco, G. 2008. Probabilistic characterisation of the aggregated residential load patterns. *IET Generation, Transmission and Distribution* 2: 373–382.

Carpinelli, G., Esposito, T., Varilone, P. et al. 2001. First-order probabilistic harmonic power flow. *IEE Proceedings on Generation, Transmission and Distribution* 148: 541–548.

Cavallini, A., Cacciari, M., Loggini, M. and Montanari, G. C. 1994. Evaluation of harmonic levels in electrical networks by statistical indexes. *IEEE Transactions on Industry Applications* 30: 1116–1125.

Cavallini, A., Langella, R., Testa, A. and Ruggiero, F. 1998. Gaussian modeling of harmonic vectors in power systems. *Proceedings of the 8th International Conference on Harmonics and Quality of Power*, Athens, Greece, October 14–16, pp. 1010–1017.

Cavallini, A. and Montanari, G. C. 1995. A simplified solution for bi-dimensional random-walks and its application to power quality related problems. *Proceedings of the 30th IAS Annual Meeting*, Orlando, FL, October 8–12, vol. 2, IEEE, Piscataway, NJ, pp. 1125–1132.

Cavallini, A., Montanari, G. C. and Cacciari, M. 1995. Stochastic evaluation of harmonics at network buses. *IEEE Transactions on Power Delivery* 10: 1606–1613.

Chicco, G., Cocina, V., Di Leo, P. et al. 2014a. Weather forecast-based power predictions and experimental results from photovoltaic systems. *Proceedings of the 22nd International Symposium on Power Electronics, Electrical Drives, Automation and Motion*, Ischia, Italy, June 18–20.

Chicco, G., Cocina, V., Mazza, A. et al. 2014b. Data pre-processing and representation for energy calculations in net metering conditions. *Proceedings of the IEEE EnergyCon2014*, Dubrovnik, Croatia, May 13–16, paper 262.

Chicco, G., Cocina, V. and Spertino, F. 2014c. Characterization of solar irradiance profiles for photovoltaic system studies through data rescaling in time and amplitude. *Proceedings of the 49th International Universities' Power Engineering Conference*, Cluj-Napoca, Romania, September 2–5, paper 52.

Chicco, G., Di Leo, P., Scapino, F. and Spertino, F. 2006. Experimental analysis of wind farms connected to the high voltage grid: The viewpoint of power quality. *Proceedings of the IEEE International Symposium on Environment Identities and Mediterranean Area*, Corte-Ajaccio, France, July 10–13, IEEE, Piscataway, NJ, pp. 184–189.

Chicco, G., Napoli, R., Postolache, P. and Toader, C. 2001. Assessing the harmonic content of an aggregation of non-linear loads. *Proceedings of the 4th CNR-CIRED Symposium CEE*, Targoviste, Romania, October 4–5, TipoGal, Galati, Romania, pp. 125–137.

Chicco, G., Schlabbach, J. and Spertino, F. 2009a. Experimental assessment of the waveform distortion in grid-connected photovoltaic installations. *Solar Energy* 83: 1026–1039.

Chicco, G., Schlabbach, J. and Spertino, F. 2009b. Operation of multiple inverters in grid-connected large-size photovoltaic installations. *Proceedings of the 20th International Conference on Electricity Distribution*, Prague, Czech Republic, June 8–11, paper 0245.

Civanlar, S., Grainger, J., Yin, H. and Lee, S. 1988. Distribution feeder reconfiguration for loss reduction. *IEEE Transactions on Power Delivery* 3: 1217–1223.

Crucq, J. M. and Robert, A. 1989. Statistical approach for harmonic measurement and calculation. *CIRED*, Report 2-02, Brighton, UK, pp. 91–96, May 8–12.

Cuk, V., Cobben, J. F. G., Ribeiro, P. F. and Kling, W. L. 2013. Summation of harmonic currents of variable-speed induction motor drives. *Proceedings of the IEEE Power and Energy Society General Meeting*, Vancouver, Canada, July 21–25.

Dwyer, R., Khan, A., McGranaghan, M., Tang, L., McCluskey, R., Sung, R. and Houy, T. 1995. Evaluation of harmonic impacts from compact fluorescent lights on distribution systems, *IEEE Transactions on Power Systems* 10: 1772–1780.

Emanuel, A. E. and Kaprielian S. R. 1986. Contribution to the theory of stochastically periodic harmonics in power systems. *IEEE Transactions on Power Delivery* 1: 285–293.

Esposito, T., Carpinelli, G., Varilone, P. et al. 2001. Probabilistic harmonic power flow for percentile evaluation. *Canadian Conference on Electrical and Computer Engineering* 2: 831–838.

Esposito, T. and Varilone, P. 2002. Some approaches to approximate the probability density functions of harmonics. *Proceedings of the 10th International Conference on Harmonics and Quality of Power*, Rio de Janeiro, Brazil, October 6–9, vol. 1, IEEE, Piscataway, NJ, pp. 365–372.

Feijoo, A. E., Cidrás, J. and Dornelas, J. L. G. 1999. Wind speed simulation in wind farms for steady-state security assessment of electrical power systems. *IEEE Transactions on Energy Conversion* 14: 1582–1588.

Feijoo, A. E., Villanueva, D., Pazos, J. L. and Sobolewski, R. 2011. Simulation of correlated wind speeds: A review. *Renewable and Sustainable Energy Reviews* 15: 2826–2832.

Fuchs, E. F. and Masoum, M. A. S. 2009. *Power Quality in Power Systems and Electrical Machines*. Academic Press, Burlington, MA.

Fuchs, E. F., Roesler, D. J. and Kovacs, K. P. 1986. Aging of electrical appliances due to harmonics of the power system's voltage. *IEEE Transactions on Power Delivery* 3: 301–307.

Goeke, T. and Wellssow, W. H. 1996. A statistical approach to the calculation of harmonics in MV systems caused by dispersed LV customers. *IEEE Transactions on Power Systems* 1: 325–331.

Gomez-Expósito, A., Conejo, A. J. and Cañizares, C. (eds.) 2009. *Electric Energy Systems Analysis and Operation*. CRC Press, Taylor & Francis Group, Boca Raton, FL.

Gorgette, F. A., Lachaume, J. and Grady, W. M. 2000. Statistical summation of the harmonic currents produced by a large number of single phase variable speed air conditioners: A study of three specific designs. *IEEE Transactions on Power Delivery* 15: 953–959.

Grainger, J. and Stevenson, Jr., W. 1994. *Power System Analysis*. McGraw-Hill, New York.

Haiduck, R. and Baghzouz, Y. 2002. Characteristics of harmonic currents generated by a cluster of gaming machines. *Proceedings of the 10th International Conference on Harmonics and Quality of Power*, Rio de Janeiro, Brazil, vol. 1, pp. 398–402.

Heunis, S. W. and Herman, R. 2002. A probabilistic model for residential consumer loads. *IEEE Transactions on Power Systems* 17: 621–625.

IEC 61000-3-6. 2008. *Electromagnetic compatibility (EMC) Part 3: Limits—Section 6: Assessment of emission limits for distorting loads in MV and HV power systems* (second edition). Publication IEC/TR 61000-3-6.

IEC 61400-21. 2008. *Wind turbines—Part 21: Measurement and assessment of power quality characteristics of grid connected wind turbines.*

IEEE. 1993. *IEEE Recommended Practices and Requirements for Harmonic Control in Electrical Power Systems.* IEEE Standard 519-1992, IEEE, Piscataway, NJ, April 1993.

IEEE Task Force. 1985. The effects of power system harmonics on power system equipment and loads. *IEEE Transactions on Power Apparatus and Systems* PAS-104: 2555–2563.

IEEE (Task Force on Harmonics Modeling and Simulation; IEEE PES Harmonic Working Group). 2001. Characteristics and modeling of harmonic sources-power electronics devices. *IEEE Transactions on Power Delivery* 16: 791–800.

IEEE (Task Force on Harmonics Modeling and Simulation; IEEE PES Harmonic Working Group). 2004. Modeling devices with nonlinear voltage-current characteristics for harmonic studies. *IEEE Transactions on Power Delivery* 19: 1802–1811.

Jardini, J. A., Tahan, C. M. V., Gouvea, M. R., Ahn, S. U. and Figueiredo, F. M. 2000. Daily load profiles for residential, commercial and industrial low voltage consumers. *IEEE Transactions on Power Delivery* 15: 375–380.

Jiang, C., Salles, D., Xu, W. and Freitas, W. 2012. Assessing the collective harmonic impact of modern residential loads—Part II: Applications. *IEEE Transactions on Power Delivery* 27: 1947–1955.

Jiang, C., Torquato, R., Salles, D. and Xu, W. 2014. Method to assess the power-quality impact of plug-in electric vehicles. *IEEE Transactions on Power Delivery* 29: 958–965.

Kaprielian, S. R., Emanuel, A. E., Dwyer, R. V. and Mehta, H. 1994. Predicting voltage distortion in a system with multiple random harmonic sources. *IEEE Transactions on Power Delivery* 9: 1632–1638.

Kazibwe, W. E., Ortmeyer, T. H. and Hammam, M. S. A. A. 1989. Summation of probabilistic harmonic vectors. *IEEE Transactions on Power Delivery* 4: 621–628.

Lagostena, L. and Porrino, A. 1986. Prediction of haronic voltage distortion due to different categories of non-linear loads supplied by the electric network. *Proceedings of the CIGRÉ International Conference on Large High Voltage Electric Systems*, Paris, France, August 27–September 4.

Langella, R., Marino, P., Ruggiero, F. and Testa, A. 1997. Summation of random harmonic vectors in presence of statistic dependences. *Proceedings of the 5th PMAPS*, Vancouver, Canada, September 21–25.

Langella, R. and Testa, A. 2009. Summation of random harmonic currents. In Ribeiro, P. F. (ed.) *Time-Varying Waveform Distortion in Power Systems*. Wiley, Chichester.

Lehtonen, M. 1993. A general solution to the harmonics summation problem. *European Transactions on Electric Power* 3: 293–297.

Lemoine, M. 1976. Quelques aspects de la pollution des réseaux par les distorsions harmoniques de la clientele. *RGE* 95: 247–255.

Li, G. and Zhang, X.-P. 2014. Modeling of plug-in hybrid electric vehicle charging demand in probabilistic power flow calculations. *IEEE Transactions on Smart Grid* 3: 492–499.

Lojowska, A., Kurowicka, D., Papaefthymiou, G. and van der Sluis, L. 2012. Stochastic modeling of power demand due to EVs using copula. *IEEE Transactions on Power Systems* 27: 1960–1968.

Mansoor, A., Grady, W. M., Chowdhury, A. H. and Samotyi, M. J. 1995. An investigation of harmonics attenuation and diversity among distributed single-phase power electronic loads. *IEEE Transactions on Power Delivery* 10: 467–473.

Mazza, A., Chicco, G., Bakirtzis, E. et al. 2014. Power flow calculations for small distribution networks under time-dependent and uncertain input data. *Proceedings of the IEEE PES Transmission and Distribution Conference and Exposition*, Chicago, IL, April 14–17.

Medeiros, F., Brasil, D. C., Ribeiro, P. F., Marques, C. A. G. and Duque, C. A. 2010. A new approach for harmonic summation using the methodology of IEC 61400-21. *Proceedings of the 14th International Conference on Harmonics and Quality of Power*, Bergamo, Italy, September 26–29.

Mendonca, G. A., Pereira, H. A. and Silva, S. R. 2012. Wind farm and system modelling evaluation in harmonic propagation studies, *Proceedings of the International Conference on Renewable Energies and Power Quality*, Santiago de Compostela, Spain, March 28–30, 2012.

Morales, J. M., Baringo, L., Conejo, A. J. and Minguez, R. 2010. Probabilistic power flow with correlated wind sources. *IET Generation, Transmission and Distribution* 4: 641–651.

Morrison, R. E., Baghzouz, Y., Ribeiro, P. F. and Duque, C. A. 2009. Probabilistic aspects of time-varying harmonics. In Ribeiro P. F. (ed.) *Time-Varying Waveform Distortion in Power Systems*. Wiley, Chichester.

Morrison, R. E. and Clark, A. D. 1984. Probabilistic representation of harmonic currents in AC traction systems. *IEE Proceedings Part B* 131: 181–189.

Ortega, M. J., Hernández, J. C. and García, O. G. 2013. Measurement and assessment of power quality characteristics for photovoltaic systems: Harmonics, flicker, unbalance, and slow voltage variations. *Electric Power Systems Research* 96: 23–35.

Papathanassiou, S. A. and Papadopoulos, M. P. 2006. Harmonic analysis in a power system with wind generation. *IEEE Transactions on Power Delivery* 21: 2006–2016.

Pierrat, L. 1991. A unified statistical approach to vectorial summation of random harmonic components. *Proceedings of the EPE'91 Conference*, Florence, Italy, vol. 3, EPE Association, Brussels, Belgium, pp. 100–105, September 3–6.

Preda, T. N., Uhlen, K. and Nordgård, D. E. 2012. Instantaneous harmonics compensation using shunt active filters in a Norwegian distribution power system with large amount of distributed generation. *Proceedings of the 3rd IEEE International Symposium on Power Electronics for Distributed Generation Systems*, Aalborg, Denmark, June 25–28, 2012.

Qin, Z., Li, W. and Xiong, X. 2013. Generation system reliability evaluation incorporating correlations of wind speeds with different distributions. *IEEE Transactions on Power Systems* 28(1): 551–558.

Ravikumar Pandi, V., Zeineldin, H. H. and Xiao, W. 2013. Determining optimal location and size of distributed generation resources considering harmonic and protection coordination limits. *IEEE Transactions on Power Systems* 28: 1245–1254.

Ribeiro, P. F. 2005. An overview of probabilistic aspects of harmonics: State of the art and new developments. *IEEE Power Engineering Society General Meeting* 3: 2243–2246.

Robinson, D. 2003. *Harmonic management in MV distribution systems.* PhD Thesis, University of Wollongong, Australia.

Romero, A. A., Zini, H. C. and Rattá, G. 2012. Modelling input parameter interactions in the possibilistic harmonic load flow. *IET Generation, Transmission and Distribution* 6: 528–538.

Romero, A. A., Zini, H. C., Rattá, G. et al. 2011. Harmonic load-flow approach based on the possibility theory. *IET Generation, Transmission and Distribution* 5: 393–404.

Rowe, N. B. 1974. The summation of randomly varying phasors or vectors with particular reference to harmonic levels. *IEE Conference Publication*, IEEE, Stevenage, UK, 110, pp. 177–181.

Ruiz-Rodriguez, F. J., Hernandez, J. C. and Jurado, F. 2011. Probabilistic load flow for radial distribution networks with photovoltaic generators. *Proceedings of the 2011 International Conference on Power Engineering, Energy and Electrical Drives*, Torremolinos, Spain, May 11–13.

Ruiz-Rodriguez, F. J., Hernandez, J. C. and Jurado, F. 2014. Harmonic modelling of PV systems for probabilistic harmonic load flow studies. *International Journal of Circuit Theory and Applications*: 1–25, DOI:10.1002/cta.2021.

Russo, A., Varilone, P. and Caramia, P. 2014. Point estimate schemes for probabilistic harmonic power flow. *Proceedings of the IEEE 16th International Conference on Harmonics and Quality of Power*, Bucharest, Romania, May 25–28, IEEE, Piscataway, NJ, pp. 19–23.

Sherman, W. G. 1972. Summation of harmonics with random phase angles. *Proceedings of the IEE* 119: 1643–1648.

Spertino, F. and Graditi, G. 2014. Power conditioning units in grid-connected photovoltaic systems: A comparison with different technologies and wide range of power ratings. *Solar Energy* 108: 219–229.

Testa, A. and Langella, R. 2005. Interharmonics from a probabilistic perspective. *IEEE Power Engineering Society General Meeting* 3: 2410–2415.

Tukey, J. W. 1977. *Exploratory Data Analysis.* Addison-Wesley, Reading, MA.

Wang, W. J., Pierrat, L. and Wang, L. 1994. Summation of harmonic currents produced by AC/DC static power converters with randomly fluctuating loads. *IEEE Transactions on Power Delivery* 9: 1129–1135.

Wikston, J. 2009. Harmonic summation for multiple arc furnaces. In Ribeiro P. F. (ed.) *Time-Varying Waveform Distortion in Power Systems*. Wiley, Chichester.

Xiao, Y. and Yang, X. 2012. Harmonic summation and assessment based on probability distribution. *IEEE Transactions on Power Delivery* 27: 1030–1032.

Ye, Z., Edward, L., Yuen, K. H. and Pong, M. H. 1999. Probabilistic characterization of current harmonics of electrical traction power supply system by analytic method. *Proceedings of the IECON'99*, San Jose, CA, November 29–December 3, vol. 1, IEEE, Piscataway, NJ, pp. 360–366.

Ye, G., Xiang, Y. and Cobben, J. F. G. 2014. Assessment of the voltage level and losses with photovoltaic and electric vehicle in low voltage network. *Proceedings of the 14th International Conference on Environment and Electrical Engineering*, Krakow, Poland, May 10–12, IEEE, Piscataway, NJ, pp. 431–436.

Zhang, G. and Xu, W. 2008. Estimating harmonic distortion levels for systems with random-varying distributed harmonic-producing loads. *IET Generation, Transmission and Distribution* 2: 847–855.

4

Scheduling Models and Methods for Efficient and Reliable Operations

Emmanouil A. Bakirtzis, Evaggelos G. Kardakos,
Stylianos I. Vagropoulos, Christos K. Simoglou,
and Anastasios G. Bakirtzis

CONTENTS

ABSTRACT   This chapter presents an overview of different methodologies and mathematical optimization models that can contribute toward efficient short-term operation of insular electricity systems with high renewable energy sources (RESs) penetration. Robust algorithms employed for the creation of system load and RES production scenarios that capture the spatial and temporal correlations of the respective stochastic processes, as well as a procedure for the creation of units' availability scenarios using Monte Carlo simulation, are discussed. Advanced unit commitment (UC) and economic dispatch (ED) (UCED) models for the short-term scheduling of the conventional and RES generating units in different short-term timescales (day ahead, intraday, and real time) are described. Indicative test results from the implementation of all models in the power system of the island of Crete, Greece, are illustrated and valuable conclusions are drawn.

4.1 Introduction

The development of sustainable energy systems with reduced fossil fuel emissions, improved energy efficiency, and increased RES penetration is a leading priority of energy road maps in many countries worldwide [1,2]. In this context, many countries have been providing various incentives toward the increase of RES installed capacity, with particular emphasis on generating electricity from wind and, more recently, from solar resources. Large RES plants have already been constructed and operated, whereas the integration of new small and large RES projects continues aggressively.

A large share of the recent RES installed capacity has already taken place in insular electricity grids, since these regions are preferable due to their high RES potential and high conventional generation costs. However, the increasing share of RES in the generation mix of insular power systems presents a big challenge in the efficient short-term operation of the insular networks, mainly due to the limited predictability and the high variability of renewable generation, features that make RES plants nondispatchable, in conjunction with the relevant small size of these networks.

In this context, this chapter provides various methodologies and software tools that are being developed and implemented in the framework of the EU-funded project SiNGULAR for optimal short-term scheduling of insular electricity networks, taking into account the stochastic nature of various system and unit parameters, such as the system load, RES production, and unit availability. Specifically, the algorithms employed for the creation of system load and RES production scenarios that capture the spatial and temporal correlations of the corresponding variables, as well as the procedure followed for the creation of units' availability scenarios using Monte Carlo simulation, are discussed. Appropriate scenario reduction techniques are also applied in order to alleviate computational complexity and burden on the scheduling tools, while preserving the features of the original scenario sets. In addition, the advanced novel UC and ED models that have been developed for the short-term scheduling of the conventional and RES-generating units in different short-term timescales (day ahead, intraday, and real time) are presented. Two distinct modeling approaches differentiated in terms of design and modeling complexity and based on mixed-integer linear programming (MILP) have been developed and are analytically presented. Indicative test results from the implementation of these short-term scheduling models in the insular power system of Crete are illustrated and thoroughly discussed.

4.2 Scenario Generation for the Modeling of the Random System and Unit Parameters

4.2.1 Introduction

The promotion of sustainability in power systems should take into account the inherent characteristic of uncertainty, which poses difficulties in predicting the exact values of many random variables that influence power system operation in different timescales (long term, short term, and real time). For instance, electric load, availability of generation units, and RES production (characterized by high variability and uncertainty) are stochastic variables that have a strong impact on the secure, reliable, and efficient power system operation and management. The great value of predicting these variables led to the development of appropriate forecasting tools that in some cases can be very accurate (e.g., hourly load forecasting for large regions).

Popular forecasting methods include autoregressive moving average (ARMA) models [3] that are used for stationary time series, autoregressive integrated moving average (ARIMA) models for nonstationary processes, seasonal autoregressive integrated moving average (SARIMA) models that capture seasonal patterns of the time series, and autoregressive moving average exogenous (ARMAX) models that include input terms related to exogenous parameters [4]. Another forecasting approach includes probabilistic methods based on probability density functions (PDFs) [5]. Finally, a very popular approach that is able to capture both linear and nonlinear dependencies and has been widely used in forecasting comprises artificial neural networks (ANNs). Relevant bibliography dealing with the design and use of ANNs for load, photovoltaic (PV), and wind generation forecasting can be found in references [6–8].

In power systems with a large share of variable RES production, conventional deterministic scheduling procedures may no longer be adequate, and stochastic approaches have been adopted lately to account for the uncertainty of the various stochastic variables. The effects of stochastic wind and load on the UC and ED problems with high levels of wind power are investigated in [9], where it is shown that stochastic optimization results in better performing schedules than deterministic optimization. A two-stage stochastic programming model for committing reserves in systems with large amounts of wind power is presented in [10], where the proposed model outperforms common deterministic rules for reserve quantification. Finally, the solution of the UC problem under a two-stage stochastic programming formulation is investigated in [11] considering the effects of generation availability and load uncertainty. A detailed literature review on methodologies dealing with the modeling of power system uncertainty on the UC and ED problems is presented in Section 4.3.1.

4.2.2 General Description and State-of-the-Art Literature Review

For the case when uncertainty is modeled through the presence of a scenario set, various methodologies have been proposed in the literature so far for the generation of a representative set of scenarios for random variables. A popular scenario generation technique is the moment matching method that was used in [12] to generate a limited number of scenarios that satisfy specified statistical properties. The basic idea is to minimize some measure of distance between the statistical properties of the generated outcomes and the specified properties. The authors in [13] presented an algorithm that produces a discrete joint distribution consistent with specified values of the first four marginal moments and correlations. The joint distribution is constructed by decomposing the multivariate problem into univariate ones and using an iterative procedure to achieve the correct correlations without changing the marginal moments. An approach that relies on the moment matching technique was proposed in [14]. The approach is based on the idea of integrating simulation and optimization techniques. In particular, simulation is used to generate outcomes associated with the nodes of the scenario tree, which, in turn, provide the input variables for an optimization model that aims at determining the probabilities of the scenarios matching some prescribed targets. An algorithm based on heteroskedastic models and a moment matching approach to construct a scenario tree that is a calibrated representation of the randomness in risky asset returns is presented in [15]. Another widely used scenario generation technique is the path-based method [16]. This method evolves the stochastic process to generate complete paths in a *fan* structure, which is transformed into a scenario tree using *clustering*, also called *bucketing*.

An optimization-based method to generate moment matching scenarios for numerical integration and how it can be used in stochastic programming has been proposed in [17]. The main advantage of this method is its flexibility: it can generate scenarios matching any prescribed set of moments of the underlying distribution rather than match all moments up to a certain order, and the distribution can be defined over an arbitrary set. In the same framework, three approaches for generating scenario trees for financial portfolio problems have been presented in [18]. These are based on simulation, optimization, and hybrid simulation/optimization. Finally, an optimal discretization method that seeks to find an approximation of the initial scenario set that minimizes an error based on the objective function has been described in [19].

Other scenario generation methods are discussed in studies that deal with wind power uncertainty. A method that allows for the generation of statistical scenarios from nonparametric probabilistic forecasts is described in [20], whereas a first-order autoregressive time-series model with increasing noise to approximate the behavior of wind speed forecast errors is presented in [21]. This model allows for the creation of a large number of wind speed scenarios using Monte Carlo simulations, which are then transformed into wind power scenarios with the use of an aggregated power curve model. In addition, simple scenarios around point forecasts are generated in [22] for the optimal scheduling of the generators in a wind-integrated power system considering the demand and wind power production uncertainty. Finally, a new scenario generation methodology that adopts the empirical distributions of a number of forecast bins to model the forecast error of wind power, which are used as inputs to scenario generation, is proposed in [23].

4.2.3 Scenario Generation for System Load and RES Injection

Scenario generation for system load and RES injection is based on a process that combines a scenario generation technique of an original (extended) set of scenarios with a technique to reduce the number of scenarios. An additional methodology for creating spatially cross-correlated scenarios has also been implemented.

The scenario generation technique of the initial (extended) set of scenarios is based on a sampling approach. Specifically, an appropriate ANN-based forecasting model for the random variable under study (e.g., system load, wind, or PV power generation) is first determined and then used for the scenario generation procedure. In the following paragraphs, the basic features of the ANN-based forecasting model are presented and the adopted scenario generation procedure is described in detail.

4.2.3.1 Forecasting with ANNs

In this section, a multi-input/single output, three-layer, feed-forward, back-propagation ANN with hyperbolic tangent sigmoid transfer function in the interval [−1, 1] is discussed for one step-ahead forecasting.

The ANN inputs are (1) past values of the time series of interest, (2) one or more exogenous input parameters correlated to the time series of interest, and (3) appropriate time indices. Time indexing inputs are introduced in the ANN as a pair of variables, $\sin(2\pi k/T)$, $\cos(2\pi k/T)$, where T is the period of the respective time index k [e.g., $T = 24$ for hour-of-the-day (HOD), $T = 7$ for day-of-the-week (DOW), and $T = 365$ for day-of-the-year (DOY)] [6].

TABLE 4.1

Generic Input and Output Structure of the ANN

ANN Inputs	ANN Output
Past values of the stochastic process under study	One-step-ahead forecast value of the stochastic process
Forecast and past values of exogenous inputs	
Time indexing: (e.g., HOD, DOW, and DOY) pair of $\sin(2\pi k/T), \cos(2\pi k/T)$	

Time indexing inputs are important to model seasonality effects related to HOD, DOY, and DOW. A generic description of the ANN inputs and output is given in Table 4.1. The proposed generic framework is very flexible, regarding input selection and parameterization, and is easily adapted to the requirements of the specific application under study.

During the training stage, the ANN is presented with a sequence of historical input-desired output pairs. Its internal parameters (weights) are adjusted, so that the "training error" is minimized.

Once the training stage is completed and the ANN weights have been adjusted, the recall stage begins, where the trained ANN is used for actual forecasting. For one-step-ahead forecasting the ANN runs only once. For two- or more-steps-ahead forecasting, an iterative application of the one-step-ahead forecasting ANN is necessary. With each new iteration, the time indexing inputs are updated accordingly and nonavailable input variables (related to either the time series being forecast or the exogenous parameters) are replaced by the corresponding forecasts obtained so far. With the iterative application of the one-step-ahead forecasting ANN (rolling update of the ANN inputs), the forecast horizon can be extended to any desired number of time steps in the future.

4.2.3.2 Scenario Generation with ANNs

The scenario generation methodology using ANN is illustrated in Figure 4.1 for three time steps ahead. Once the training of the ANN is completed, the time series of the residuals (errors) of the training phase is calculated as the difference between the forecast values (using the trained ANN) and the historical values (real measurements). A statistical analysis of the time series of the residuals investigates whether the time series could be considered a Gaussian white noise signal; that is, it has almost zero autocorrelation coefficients for all time delays other than zero, and the error distribution can be approximated by a normal distribution with zero mean and standard deviation σ [i.e., $N(0,\sigma)$].

As already mentioned, the procedure to generate a set of scenarios for a stochastic process $\mathbf{Y}$ (e.g., electric load, PV, or wind production) comprises an iterative process based on random generation of Gaussian white noises. These error terms follow the approximated distribution of the time series

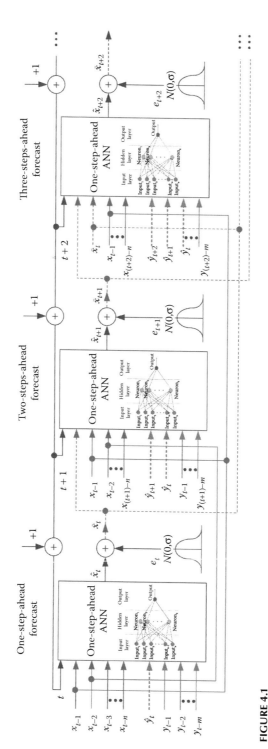

FIGURE 4.1
Scenario generation methodology illustrated for three time steps ahead.

of the residuals derived during the training phase. Based on this logic [24], an appropriate algorithm is proposed to generate a set of scenarios using ANNs, described as follows:

As illustrated in Figure 4.1, time indexing (HOD, DOW, DOY) is specified as an input to the ANN in each stage, exactly as described in the training and the recall phase. For the first step ahead (t), the input vector comprises the following: (1) the previous n recorded (real) values of the time series $\{x_{t-1}, x_{t-n}\}$, (2) the forecast value of any exogenous input $\hat{y}_t$ for the first step ahead, and (3) possible previous m recorded (real) values of the same exogenous inputs $\{y_{t-1}, y_{t-m}\}$. The first-step-ahead ANN output ($\hat{x}_t$) is distorted by the normal distributed error term and yields an input for the two-steps-ahead forecast $t + 1$ ($\tilde{x}_t$), along with the $n - 1$ previous recorded values of the time series $\{x_{t-1}, x_{(t+1)-n}\}$, the forecast values of any exogenous inputs $\{\hat{y}_t, \hat{y}_{t+1}\}$ for the first two steps, and possibly previous $m - 1$ recorded values of the same exogenous inputs $\{y_{t-1}, y_{(t+1)-m}\}$.

In each iteration (time step), the total number of inputs remains intact. However, as the algorithm moves forward in time, the recorded (real) values used as input in the first steps are gradually replaced by the outputs of the ANN of the previous steps (distorted forecast values) and the forecast values of the exogenous inputs accordingly. The input shifting is illustrated in Figure 4.1.

The above process is repeated until the desired scenario generation horizon is reached. The entire process is also repeated as many times as the desired number of scenarios. Due to the random generation of error values, a different path is created in each iteration. However, all paths follow the statistical properties of the initial error time series.

The algorithm is also presented step by step in Figure 4.2, where N_T denotes the desired number of forecast periods (steps) and N_Ω denotes the desired number of scenarios.

4.2.3.3 Scenario Generation for Cross-Correlated Stochastic Processes

The management of power systems with high renewable penetration involves many stochastic processes that are statistically dependent. For instance, the energy injection from adjacent wind farms or PV stations frequently follows similar patterns. Modeling this statistical correlation is crucial for the system operator (SO) who is responsible for the scheduling and real-time operation of the power system. In this chapter, the procedure presented in [24] is embedded in the proposed ANN-based scenario generation methodology for the creation of spatially and temporally cross-correlated scenarios regarding the energy injection from neighboring RES plants.

In this context, in case two stochastic processes Y_a and Y_b are statistically dependent, the dependency is transferred to the series of residuals ε_a and ε_b and, therefore, ε_a and ε_b should be cross correlated.

In order to determine the degree of dependency of the series of the residuals, a cross-correlogram that represents the cross-correlation coefficients for

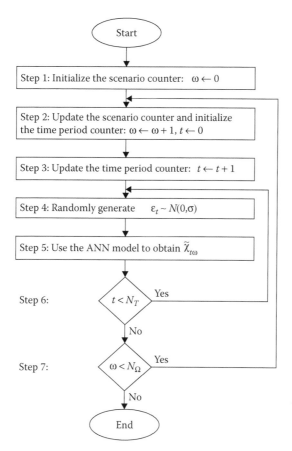

FIGURE 4.2
Scenario generation algorithm.

different time lags can be derived [3]. According to the shape of the residual cross-correlogram, three different types of dependent stochastic processes can be distinguished, namely, contemporaneous, quasi-contemporaneous, and noncontemporaneous stochastic processes. For further details on the distinction and the features of cross-correlated stochastic processes, the interested reader can refer [24].

In general, a variance–covariance matrix **G** can be used to identify the cross-correlations between ε_a and ε_b. This matrix is symmetric by definition and, therefore, can always be diagonalized. In other words, an orthogonal transformation can be used to model such a series of errors, as follows:

$$\varepsilon = \begin{pmatrix} \varepsilon^a \\ \varepsilon^b \end{pmatrix} = \mathbf{B} \cdot \xi = \mathbf{B} \cdot \begin{pmatrix} \xi^a \\ \xi^b \end{pmatrix} \tag{4.1}$$

In this context, white noises (independent standard normal errors ξ) can be generated and then cross-correlated according to the variance–covariance matrix $\mathbf{G}$ using the orthogonal transformation. In most engineering applications, matrix $\mathbf{G}$ besides being symmetric is also positive definite and as such can be decomposed through the computationally advantageous *Cholesky decomposition*, which avoids the calculation of eigenvalues and eigenvectors. The Cholesky decomposition can be stated as follows:

$$\mathbf{G} = \mathbf{B} \cdot \mathbf{B}^{\mathrm{T}} \tag{4.2}$$

where:

> $\mathbf{B}$ is an inferior triangular matrix that turns out to be the orthogonal matrix required for transformation (4.1)

The analytical description of the scenario generation methodology for cross-correlated stochastic processes is outside the scope of this chapter, and the interested reader can refer [24] for further details. Indicative results from the incorporation of the aforementioned methodology into the proposed ANN-based scenario generation algorithm presented in Section 4.2.3.2 are given in Section 4.2.5.

4.2.3.4 Scenario Reduction Techniques

More than often, the computational performance of stochastic programming optimization models is highly dependent on the size of the scenario set. For this reason, a compromise between the necessary number of scenarios and the computational burden of the associated stochastic programming model needs to be made, so that the problem can be solved using acceptable computational resources. For this purpose, scenario reduction techniques are usually applied. Various scenario reduction techniques have been reported in the literature so far.

The scenario reduction methodology adopted in this chapter is based on the concept of the probability distance [24]. In general, the probability distance quantifies how *close* two different sets of scenarios representing the same stochastic process are. In this context, if a large scenario set is close enough to a reduced one in terms of the probability distance, the optimal solution of the simpler problem (which is formulated and solved using the reduced set of scenarios) is expected to be close to the optimal value of the original problem (which is formulated and solved with the extended set of scenarios). An overview of the theoretical background underlying the concept of probability distance is thoroughly presented in [25], whereas its application to scenario reduction is discussed in detail in [26]. In this chapter, the scenario reduction methodology based on the probability distance criterion is used to effectively reduce the initially extended number of scenarios created by the proposed ANN-based scenario generation methodology.

4.2.4 Scenario Generation for Unit Availability

One of the main sources of uncertainty to be taken into account for the modeling of the short-term operation of the power system is the availability of generating units, that is, the ability of the unit to produce at its nominal capacity or its inability to fully or partially produce electricity due to a forced outage.

In order to model the availability or unavailability of generating units, availability scenarios are usually created. In this work, the generation of availability scenarios for units is based on a widely used technique, which involves creating *availability* and *nonavailability* states using a two-state Markov model.

The availability history of a generating unit is usually represented by a two-state time series, where $Up = available$ and $Down = unavailable$, as shown in Figure 4.3.

This dynamic sequence of states is constructed using a two-state Markov model, as illustrated in Figure 4.4.

According to Figure 4.4, each generating unit is characterized by a failure rate λ (times per year), according to which it leaves the state Up and enters the state $Down$. Similarly, it leaves the state $Down$ with a repair rate μ (times per year), according to which it enters the state Up again. Considering the unit mean time to failure $t_{mean}^{Up} = 8760/\lambda$ (in hour) and the unit mean time to repair $t_{mean}^{Down} = 8760/\mu$ (in hour), the time t^{Up}, for which the unit remains available until the next failure, and the time t^{Down}, for which the unit is offline until it becomes available again, can be considered as random variables following exponential distribution, with time constants t_{mean}^{Up} and t_{mean}^{Down}, respectively, as follows:

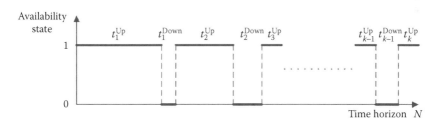

FIGURE 4.3
Unit availability state sequence.

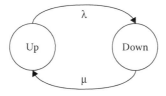

FIGURE 4.4
A two-state Markov model.

$$F\left(t^{\text{Up}}\right) = 1 - e^{\lambda \cdot t^{\text{Up}}/8760} \tag{4.3}$$

$$F\left(t^{\text{Down}}\right) = 1 - e^{\mu \cdot t^{\text{Down}}/8760} \tag{4.4}$$

The time constants $t_{\text{mean}}^{\text{Up}}$ and $t_{\text{mean}}^{\text{Down}}$ result through the assessment of the unit availability during a long study period in the past. It is common for the constant $t_{\text{mean}}^{\text{Up}}$ not to be given explicitly but to be calculated indirectly using the EFOR$_{\text{D}}$ (equivalent forced outage rate—demand) factor. In this case, the time constant $t_{\text{mean}}^{\text{Up}}$ is calculated as follows:

$$t_{\text{mean}}^{\text{Up}} = \frac{\left(1 - \text{EFOR}_{\text{D}}\right)}{\text{EFOR}_{\text{D}}} \cdot t_{\text{mean}}^{\text{Down}} \tag{4.5}$$

Using the inverse transform method [27], the random time to the next failure (status *Up*) and the random time to the next repair (status *Down*) are given, respectively, by the following expressions:

$$t^{\text{Up}} = F^{-1}(y) = -\frac{8760}{\lambda} \cdot \ln(1-y) = -t_{\text{mean}}^{\text{Up}} \cdot \ln(1-y) \tag{4.6}$$

$$t^{\text{Down}} = F^{-1}(y) = -\frac{8760}{\mu} \cdot \ln(1-y) = -t_{\text{mean}}^{\text{Down}} \cdot \ln(1-y) \tag{4.7}$$

where:
y is a random variable uniformly distributed in [0,1]

Therefore, expressions 4.6 and 4.7 allow for the generation of k random samples t_k^{Up}, t_k^{Down} through the generation of random samples y_k from the uniform distribution in [0,1].

The detailed algorithm for the generation of unit availability scenarios is as follows:

- *Step 1*: Consider a time horizon of N hours, a number of N_I generating units, and a desired number of availability scenarios N_Ω. For long-term reliability assessment studies, the time horizon can be extended to one year (8760 hours), whereas in short-term studies, it covers a shorter period (e.g., 24 hours).
- *Step 2*: Initialize the scenario counter: $\omega \leftarrow 0$.
- *Step 3*: Update the scenario counter and initialize the unit counter: $\omega \leftarrow \omega + 1, i \leftarrow 0$.
- *Step 4*: Update the unit counter: $i \leftarrow i + 1$.
- *Step 5*: Consider unit i to be available at the beginning of the time horizon.

- *Step 6*: Initialize the sample counter: $k \leftarrow 0$.
- *Step 7*: Update the sample counter: $k \leftarrow k + 1$.
- *Step 8*: Create a random value y_k from the uniform distribution [0,1]. The time to next failure k of unit i is computed as follows:

$$t_k^{\text{Up}} = F^{-1}(y) = -\frac{8760}{\lambda} \cdot \ln(1 - y_k) \tag{4.8}$$

- *Step 9*: Create a random value y_k from the uniform distribution [0,1]. The time to next repair k of unit i is computed as follows:

$$t_k^{\text{Down}} = F^{-1}(y) = -\frac{8760}{\mu} \cdot \ln(1 - y_k) \tag{4.9}$$

- *Step 10*: For very short-term studies (e.g., 24-hour time horizon), the time to next repair t_k^{Down} can be neglected, given that when the unit becomes unavailable during a single hour of the 24-hour horizon, it is considered to remain unavailable up to the end of the 24-hour period. For longer time horizons, steps 7–9 are repeated until the time series of the successively produced t_k^{Up}, t_k^{Down} covers the time period of N hours.
- *Step 11*: If $i < N_I$ go to step 4, else go to step 12.
- *Step 12*: If $\omega < N_\Omega$ go to step 3, else the scenario generation process is completed.

Finally, the availability state $av_{i,t,\omega}$ of unit i for hour t and scenario ω is defined as follows (see Figure 4.3).

$$av_{i,t,\omega} = 1 \quad \forall i, t_{k-1}^{\text{Down}} < t < t_k^{\text{Up}}, \omega \tag{4.10}$$

$$av_{i,t,\omega} = 0 \quad \forall i, t_k^{\text{Up}} < t < t_k^{\text{Down}}, \omega \tag{4.11}$$

In order to efficiently reduce the large number of scenarios generated by the procedure already described, a similar scenario reduction technique, with that already described in Section 4.2.3.4, was implemented.

4.2.5 Numerical Application

In this section, the application of the ANN-based scenario generation methodology for three distinct stochastic variables, namely, electric load, PV, and wind production, is analytically presented. The implementation took place for the system load of the real-world insular power system of Crete for spatially close real-life wind farms located in Crete and for adjacent PV plants located in Greece. In each one of the three cases, different number of scenarios and scenario reduction techniques are applied on purpose, in order to highlight the flexibility of the proposed methodology.

4.2.5.1 Scenario Generation for System Load

In order to create scenarios for the system load of Crete, the ANN was trained with (1) the hourly load time series and (2) the maximum and minimum daily temperature values of the insular power system of Crete for 2011 and 2012. The recall stage took place for 2013. The hourly load time series (real measurements) for the three-year period (2011–2013) is presented in Figure 4.5.

There are many combinations of time lags that can lead to a successful training. After extensive testing, 29 time lags were chosen, extending back to one week prior to the forecast hour. All ANN inputs are presented in Table 4.2. The maximum and minimum daily temperature values in Heraklion, Crete, were also used as ANN exogenous inputs. The time series of the residuals derived from the 1-hour-ahead training for 2011–2012 is presented in Figure 4.6, whereas the autocorrelation diagram of the same time series is presented in Figure 4.7.

It is obvious that the autocorrelation coefficients beyond the zero lag are negligible. In Figure 4.8, the histogram of the residual time series is illustrated, which is approximated by a normal distribution with zero mean and standard deviation $\sigma = 4.83$ MW.

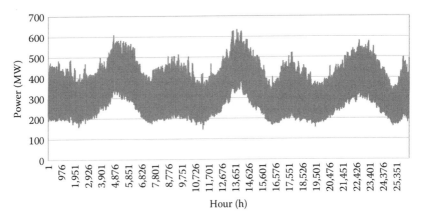

FIGURE 4.5
Hourly load of the insular power system of Crete for 2011–2013.

TABLE 4.2

ANN Inputs—One Hour-Ahead Electric Load Forecasting

Past Values	$t - k$; $k = 1,...,7$, $23,...,26$, $48,...,50$, $72,...,74$, $96,...$, 98, $120,...,122$, $144,...,146$, $168,...,170$

Exogenous Inputs	• Maximum and minimum daily temperature values
	  • Previous day $d - 1$ (historical data)
	  • Next day (forecast data)

Time Indices	HOD, DOW, and DOY

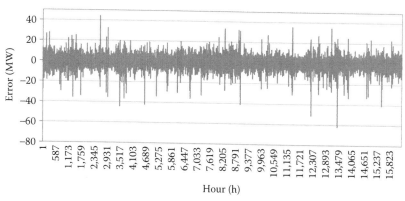

FIGURE 4.6
One-hour-ahead residuals (errors) time series of the Crete system load yielded by the training stage for 2011–2012.

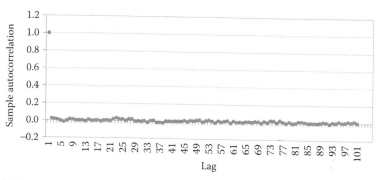

FIGURE 4.7
Autocorrelation function of the residual time series of the Crete system load yielded by the training stage for 2011–2012.

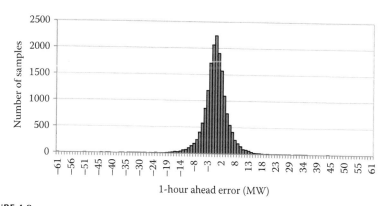

FIGURE 4.8
Histogram of the residual time series of the Crete system load yielded by the training stage for 2011–2012 and normal distribution fitting.

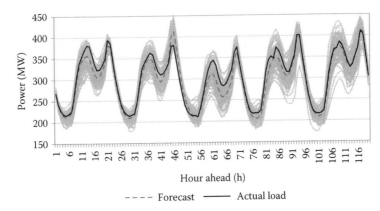

FIGURE 4.9
Generation of 100 load scenarios for 5 days ahead.

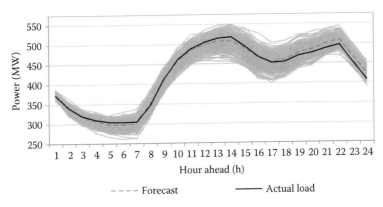

FIGURE 4.10
Generation of 1000 load scenarios for 24 hours ahead.

As described above, the scenario generation methodology can be extended up to many steps ahead, provided that forecasts of the exogenous inputs are available up to the desired horizon. Figure 4.9 shows the proposed methodology used to create 100 scenarios for 5 days ahead with hourly discretization, beginning Friday the 10th of May 2013. The real and forecast system load values are also presented. In addition, 1000 scenarios for 24 hours ahead are created for the 18th of July 2013 and shown in Figure 4.10. The 1000 initially created equiprobable scenarios are then reduced to five scenarios (reduced set) using the scenario reduction methodology outlined in Section 4.2.3.4. The final set of scenarios along with their correspondent probabilities is presented in Figure 4.11.

4.2.5.2 Scenario Generation for PV Production

The proposed ANN-based scenario generation methodology is also used to create scenarios for PV production. In general, PV production is usually

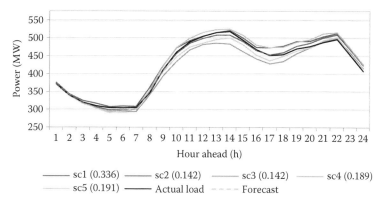

FIGURE 4.11
Reduced set of five final scenarios with their correspondent probabilities.

highly correlated between adjacent PV stations. Therefore, the proposed methodology was applied for the creation of cross-correlated scenarios for the production of two PV stations located in the adjacent prefectures of Attica and Viotia in central mainland Greece. The installed capacities of the PV stations are 0.15 and 1 MW, respectively. After numerous tests, 11 time lags were chosen. In addition, exogenous factors were also used to improve both the ANN training and the scenario generation procedure. Irradiation measurements during the training phase and irradiation forecasts during scenario generation were used as exogenous inputs. All inputs of the related ANN are given in Table 4.3.

The training phase includes the period from January 2011 to July 2012. Results are presented for a single day of August 2012. The scenario generation procedure comprises the generation of an initial set of 50 cross-correlated scenarios (original set) that was finally reduced to a set of 20 scenarios per PV station (reduced set). For the reader's convenience, in all figures, the PV production is normalized with the installed capacity of each PV station and, therefore, per unit (pu) values are used.

Figure 4.12 illustrates the original sets of 50 cross-correlated scenarios for each PV station. Both sets were created according to the methodology

TABLE 4.3

ANN Inputs—One Hour-Aheads PV Production Forecasting

Past Values	$t - k; k = 1,...,5, 24,...,26, 48,...,50$
Exogenous Inputs	• Total daily solar irradiation
	• Previous day $d - 1$ (historical data)
	• Next day (forecast data)
	• Hourly solar irradiation
	• Previous hour $t - 1$ (historical data)
	• Next hour (forecast data)
Time Indices	HOD and DOY

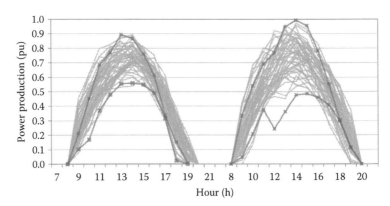

FIGURE 4.12
Initial set of cross-correlated scenarios (50 scenarios per PV stations—Viotia: left curves, Attica: right curves).

described in Section 4.2.3.3. It is noted that the initial 50 scenarios are equiprobable, and, therefore, each pair of cross-correlated scenarios of the two PV stations is assigned a probability of $1/50 = 0.02$.

Once the initial sets of scenarios are generated, the scenario reduction technique already described in Section 4.2.3 is applied, keeping the stochastic information contained in the extreme scenarios of each PV station intact. Extreme scenarios are defined as those scenarios where the maximum daily production has the maximum and the minimum value, respectively, among all scenarios. These scenarios, despite their low probability of occurrence, can have a significant impact on the short-term scheduling of the power system. In order to maintain this information, the extreme scenarios of each PV station are excluded from the scenario reduction procedure that follows. In Figure 4.12 the extreme scenarios of each PV station are illustrated in bold line along with an associated marker (Viotia: left curves, Attica: right curves).

Given that the scenarios for the two PV stations have been generated with the aforementioned cross-correlation procedure, for each scenario that is selected for each PV station, the corresponding cross-correlated scenario of the other PV station is also selected, in order to be excluded from the scenario reduction procedure. In Figure 4.13, the four scenarios that have been selected as extreme ones and are excluded from the scenario reduction process are presented. In this way, the modified initial set of scenarios, in which the scenario reduction process applies, is determined and presented in Figure 4.14. These sets include $50 - 4 = 46$ scenarios each.

The implementation of the scenario reduction algorithm on the modified initial sets of scenarios results in the reduced scenario sets that comprise 16 scenarios for each PV station. The reduced scenario sets are presented in Figure 4.15, where a grayscale is used to denote scenarios with higher (darker shade) to lower (lighter shade) probability, accordingly. The final sets of scenarios comprising 20 scenarios per PV station are illustrated in Figure 4.16.

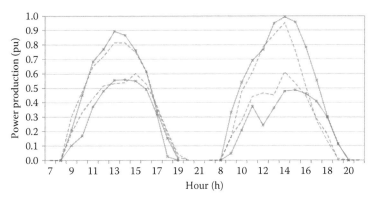

FIGURE 4.13
Sets of the extreme scenarios (four scenarios per PV station—Viotia: left curves, Attica: right curves—solid line: extreme scenario of each PV station, dotted line: corresponding cross-correlated scenario of the other PV station).

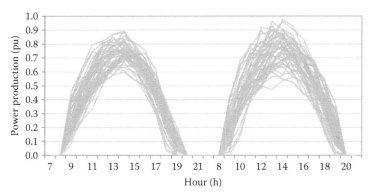

FIGURE 4.14
Modified initial sets of scenarios to be reduced (46 scenarios per PV station—Viotia: left curves, Attica: right curves).

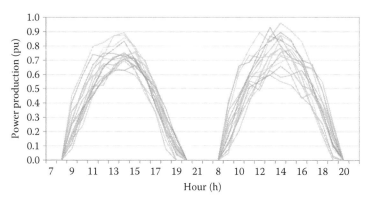

FIGURE 4.15
Reduced sets of scenarios (16 scenarios per PV station—Viotia: left curves, Attica: right curves).

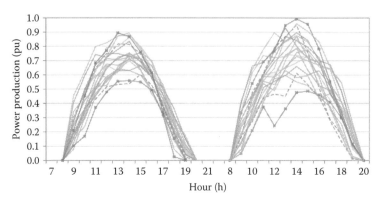

FIGURE 4.16
Final sets of scenarios including the reduced and the extreme scenarios set (20 scenarios per PV station—Viotia: left curves, Attica: right curves).

It is evident that the final sets result from the composition of the reduced sets (16 scenarios per PV station) and the four extreme scenarios initially selected.

4.2.5.3 Scenario Generation for Wind Power Production

Following the same logic with the PV scenario generation, indicative results from the process of creating cross-correlated scenarios for the electricity production of four wind farms, namely, Aiolos, Rokas, Honos Iweco, and Rokas-Modi, located in the prefecture of Lasithi, Crete, are presented next. Their installed capacities are 10, 12.90, 4.50, and 4.80 MW, respectively. The related ANN inputs are presented in Table 4.4.

The scenario generation process follows the same steps with the previous case. Once the initial sets of 50 scenarios are generated, the extreme scenarios are extracted, which are now defined as those scenarios where the maximum daily wind power production (in MWh) has the maximum and the minimum value, respectively, among all scenarios. The modified initial sets of scenarios, in which the scenario reduction process applies, is determined and presented in Figure 4.17. These sets include 50 (initial scenarios) – 8 (extreme scenarios) = 42 scenarios each. The implementation of the scenario reduction algorithm on the modified initial sets of scenarios results in the reduced scenario sets that comprise 12 scenarios for each wind farm. The final sets of

TABLE 4.4

ANN Inputs—One Hour-Ahead Wind Production Forecasting

Past Values	$t - k; k = 1,\ldots,5, 23,\ldots,25$
Exogenous Inputs	• Hourly wind Speed
	• Previous hours t (−1, −2, −3) (historical data)
	• Next hour (forecast data)
Time Indices	HOD and DOY

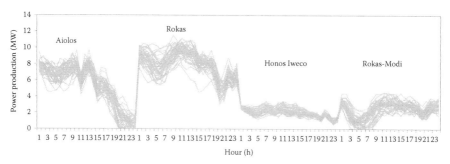

FIGURE 4.17
Modified initial sets of scenarios to be reduced (42 scenarios per wind farm).

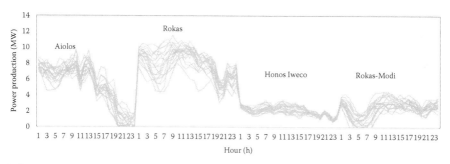

FIGURE 4.18
Final sets of scenarios including the reduced and the extreme scenarios set (20 scenarios per wind farm).

20 scenarios composed of the reduced sets and the extreme scenarios initially selected are presented in Figure 4.18.

4.3 Scheduling Models for the Short-Term Operation of the Insular Power System

4.3.1 Introduction

4.3.1.1 Short-Term Power System Operations

The short-term operations scheduling of power systems has been tradition-ally based on a two-level—UC/ED—hierarchy paradigm [28]: day-ahead UC scheduling is first performed at around 12:00 noon of the day preceding the dispatch day in order to determine the commitment status of all dispatchable units during all dispatch periods of the dispatch day, with an hourly time resolution. Real-time economic dispatch (RTED) is performed every 5 min-utes and determines the active power output of all committed dispatchable units (unit base points) for the next 5-minute interval. Current UC practice assumes deterministic knowledge (perfect forecast) of system conditions for the next day, typically load demand and component availability. Reserve requirements and $N - 1$ contingency analysis ensure a certain degree of robustness against forecast errors. Until recently, the perfect forecast was a reasonable assumption since both the load demand and the availability of the conventional, dispatchable units could be fairly accurately forecast, with component failures within the next day considered a rather rare event. When system conditions deviate substantially from forecasts, forward (day-ahead) or intraday revised UC schedules are computed.

The critical problem of UC in a power system has been addressed by many different novel approaches and algorithms so far. These approaches range from simple exhaustive enumeration [29–30] and priority listing methodologies [31–33] to dynamic programming [34–40], artificial intelligence [41–47], and Lagrangian relaxation techniques [48–53]. Lately, more advanced MILP models have been adopted [53–60], following the rapid growth of state-of-the-art hardware and optimization software.

The growth of variable (intermittent) renewable energy in the generation mix of many power systems has challenged the traditional paradigm of the short-term power system operation [9,10,61–67]. Variable generation (VG) technologies deliver energy on an as-available basis, and increase the level of variability and uncertainty in power system operations. As the VG penetration further increases, current short-term operating practices will be inadequate and will need to be revised.

In the literature, there are several approaches to cope with the increased uncertainty in the short-term operation of the power system, including advanced forecasting tools [63], maintaining increased amounts of reserves [64], and use of stochastic [9,10,65–67] or robust [68–69] optimization models. Although various advanced stochastic and robust optimization models have been proposed in the literature, SOs are still reluctant to use them in operations practice. Apart from the complexity and the high computational requirements of stochastic optimization, the main reason is institutional: scenario generation and weighting may raise market transparency issues.

Therefore, up to now, SOs rely on implementing more accurate deterministic models and facing uncertainty and variability by maintaining increased levels of reserves and introducing frequently updated forecasts and additional intraday system operations. Maintaining high reserve levels can be uneconomical and could also render the scheduling infeasible. In this context, advanced markets in the United States have already begun to restructure their short-term operation and market models by adding and modifying operation functions based on frequently updated forecasts. Some of the most advanced techniques to face increased uncertainty include the following:

- Frequently revised reliability unit commitment (RUC), with hourly granularity, to adapt the commitment decisions based on most recent information on changing system conditions [61–62].

- Intraday rolling UC with subhourly time resolution and scheduling horizons up to several hours in order to recommit fast-start units [70].
- Real-time dispatch with 5-minute resolution and look-ahead capabilities (e.g., next hour) in order to capture the forthcoming wind energy variations [71–72]. The benefits of this approach have been explored in [73].
- Real-time dispatch and fast-start UC [74].
- Flexible ramp constraints and new ramp products [71].

In the literature, several deterministic models have been presented to cope with increased uncertainty. In [75] and [76], the authors developed deterministic UC models that are executed on a rolling basis. In [77], the authors examine the effect of RES variability and uncertainty in a one-day simulation across all multiple timescales down to automatic generation control (AGC). The advantage of [77] lies in that they examine the effect of RES generation in multiple time frames of the short-term power system operations simultaneously, whereas the majority of other works focus on a single timescale (usually day ahead or real time). The authors in [78] have presented a novel deterministic model that unifies the UCED functions in a real-time tool that uses variable time resolution and a scheduling horizon of up to 36 hours to better accommodate large amounts of RES generation.

4.3.1.2 Short-Term Insular System Operations

Owing to their small size and autonomous nature, insular power systems are simpler to monitor and control than large, multiarea interconnected power systems. There are no seams issues and no low frequency inter-area oscillations threaten the integrity of insular power systems. However, owing to the same reasons, insular power systems are more vulnerable to high RES penetration. They cannot import flexible generation from neighboring power systems when wind is not blowing and sun is not shining, and they cannot export the excess renewable generation of windy and sunny days to their neighbors through the interconnections. In addition, they cannot take advantage of the renewable generation *portfolio effect*, that is, the fact that renewable generation becomes more predictable and less variable when aggregated over a wide geographic area. Finally, the effect of increased renewable penetration on the system inertia and the system primary frequency response characteristic are harder to manage in the absence of the help from neighboring systems.

Whichever the nature of the power system is (i.e., insular or interconnected), an important criterion for the optimal short-term operation of a power system is to meet the variable load demand with minimum operating cost using an optimal mix of generating units, according to their operating characteristics. SOs regularly use short-term scheduling models in order to fulfill this objective, among which UC is considered to be one of the best available options for longer time horizons (e.g., one day or several hours ahead). For near to real-time horizons (e.g., 5 or 15 minutes ahead), ED is usually used. In this case, the commitment status of the conventional generating units for each dispatch interval has been already determined by the most recent UC solution, and ED aims only at dispatching the online units solely respecting their ramp rates and technical limits to meet the system load at least cost.

In general, UC is a complex optimization problem, where the SO aims at minimizing the total production cost over the scheduling horizon. In general, the total production cost comprises fuel costs, which are primarily associated with the operating status and the production level of the generating units, start-up costs, and shutdown costs. As a result, the UC problem has been traditionally solved in centralized power systems to determine the best possible commitment status, the start-up/shutdown sequences, and the respective power outputs for all available generating units, so as to satisfy the forecast demand and the system-wide reserve requirements in all time intervals of the scheduling horizon. Moreover, the SO has to respect various generating unit constraints (such as the minimum up/down times, the start-up/shutdown trajectories), which further reduce his flexibility to select which generating units to start-up and/or shut down.

In this sense, the UC optimization problem has the following form:

$$\text{Minimize (Total production costs)} = \begin{pmatrix} \text{Fuel costs + Start-up costs} \\ \text{+ Shutdown costs} \end{pmatrix}$$

This condition depends on the following constraints:

1. System power balance
2. System reserve requirements
3. Unit capacity limits
4. Unit minimum up/down times
5. Unit initial conditions and status restrictions (must-run, fixed-MW, and unit availability)
6. Unit ramp-rate limits

Constraints 1 and 2 are the system-wide coupling constraints, whereas constraints 3–6 are local unit-wise constraints. The specific nature of the UC problem has been exploited by SOs through various solution algorithms, in order to achieve a feasible and as close to optimal solution as possible. It should be noted that for interconnected power systems, the UC system constraints must be accordingly modified to take into account the interchange schedules and the tie-line constraints.

The ED problem formulation is simpler than the UC formulation, since the objective function comprises only the fuel costs and the problem is subject only to the system-wide coupling constraints 1 and 2 and the unit-wise constraints 3, 6.

The mathematical formulation for both problems as designed and implemented in this project is analytically described in the following sections.

4.3.2 Mathematical Formulation of the Proposed Scheduling Models

As already mentioned in Section 4.3.1.1, the MILP approach has been proposed in the last decade as a viable and efficient alternative methodology for solving various optimization problems associated with the short-term operation of power systems, such as the traditional or the competitive UC problem. MILP models have been widely applied, since most independent SOs (ISOs), along with the research community, recognized that critical decisions associated with the operation of the power system can be effectively represented by integer (more specifically binary) variables and, therefore, classical linear programming (LP) approaches could not be used to explicitly model and solve such complex problems. In MILP formulations, the commitment decisions denoting practically the on/off status of the generating units in various operating phases (e.g., offline, start-up, dispatchable, and shutdown) are modeled using binary variables, whereas the power output, reserve contribution, and flow decisions are modeled using continuous variables.

4.3.2.1 Multiple-Level Scheduling

4.3.2.1.1 Introduction

In this section, the short-term operation of the insular power system is modeled using the following three distinct MILP-based optimization models:

1. A rolling day-ahead scheduling (RDAS) model
2. An intraday dispatch scheduling (IDS) model
3. An RTED model

Figure 4.19 illustrates the general input–output data structure of the aforementioned MILP models. It is mentioned that the RDAS and IDS models have been developed either in a deterministic framework (considering all unit and system parameters deterministically known for the respective scheduling period) or in a scenario-based stochastic programming framework to account for uncertainties associated with the random unit and system parameters (e.g., system load, RES production), as already discussed in Section 4.2.

The solutions of all the above models are coordinated aiming at the minimization of the total operating cost of the conventional (thermal) generating units in the insular power system and the maximization of the zero variable cost RES injection.

Specifically, the RDAS model solves the short-term UC problem with an hourly time resolution, where a simultaneous multiple-hour co-optimization of energy and reserve resources is performed under a large set of unit and system constraints (e.g., unit start-up and shutdown procedures, minimum up/down-time constraints, minimum/maximum power output restrictions, ramp-rate

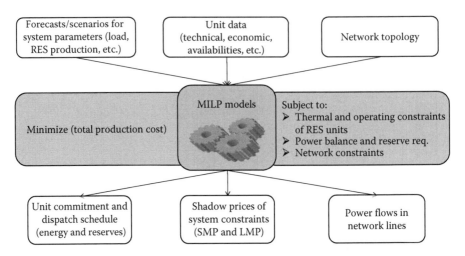

FIGURE 4.19
General input–output data structure of MILP models.

limits, system reserve requirements, and transmission limits), further described in the following paragraphs. RDAS model is typically solved twice for each dispatch day (day D): once prior to the beginning of the dispatch day (e.g., at 21:00 of day D − 1), covering the entire day-ahead scheduling period (24 hours of day D), and once few hours prior to noon of day D (e.g., at 10:00 a.m.) covering the second half of day D (i.e., hours 13–24), as updated forecasts regarding the system and unit parameters (e.g., system load, availability of units, and RES injection) are made available during the progress of day D.

The solution of the RDAS model provides the commitment status and dispatch scheduling for all generating units (conventional and RES) during the respective scheduling period (24 or 12 hours) under study. In this context, the optimal commitment status of slow base-load units is considered as binding (fixed decisions) for the formulation and solution of the intermediate IDS models that follow, whereas the optimal commitment status and dispatch scheduling for all other units (fast units and RES plants) provided by RDAS are indicative.

The IDS model follows, in general, the formulation of the RDAS model. The main difference lies in that the UC problem is now solved every 30 minutes on a multiple-hour basis (typically 4 hours) with a 30-minute time resolution covering the entire time range of the dispatch day D. Since the solution of the RDAS models provides the commitment of the slow units for day D (binding start-up and/or shutdown decisions), which is considered to remain unchanged unless a unit forced outage takes place, the committed slow units are modeled as must-run units in the IDS runs during day D. In this sense, the IDS model is solved successively during day D deciding only on the commitment status of the fast thermal units and RES plants (if applicable) as well as the dispatch scheduling of all (slow and fast) units.

Finally, the RTED model is implemented for the determination of the optimal generation dispatch levels of all generating units in real time (e.g., every 15 minutes) to satisfy the system load demand and the spinning reserves requirements. The main feature of RTED model is that no commitment decisions are taken, since the commitment status of all generating units for each dispatch interval has already been determined by the most recent RDAS and IDS models solutions. Therefore, RTED can be considered as a special case of the aforementioned MILP UC problem, since it is, in fact, an LP problem (no binary decision variables are used) and aims at dispatching the online units respecting only their ramp rates and technical limits to meet the system load at least cost. Following the current trend of advanced electricity markets in the United States and Europe that recently adopted RTED models with a look-ahead horizon in order to deal more efficiently with the unpredictability and variability of RES, the proposed RTED model has been formulated for a 15-minute dispatch period in conjunction with a look-ahead horizon of one hour.

More details on the features and the mathematical formulation of the proposed scheduling models are given in the following paragraphs. Finally, it is noted that both the time horizon and the time resolution of all three scheduling models are parametric and can be easily adapted to the needs of every single insular power system.

Figure 4.20 illustrates the main characteristics of the three distinct scheduling models, whereas the coordination sequence of the three scheduling models along with an indicative scheduling timeline is illustrated in Figure 4.21.

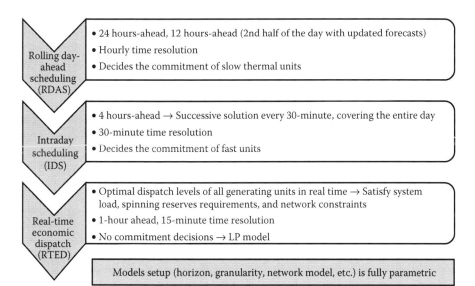

FIGURE 4.20
Characteristics of scheduling models.

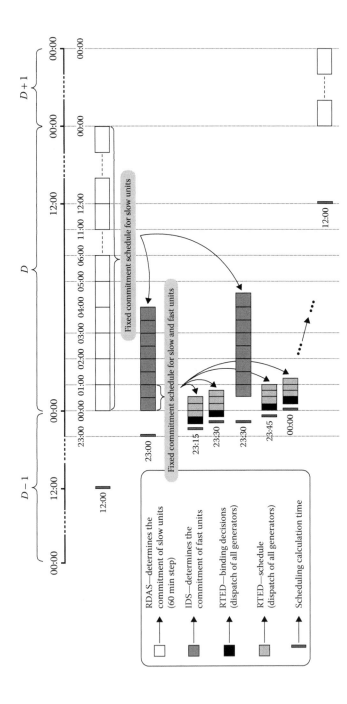

FIGURE 4.21
Coordination sequence of scheduling processes—Multiple-level scheduling.

The UC models are formulated and solved considering all unit and system parameters deterministically known. Specifically, the availability/non-availability of all generating units for the entire scheduling horizon is considered to be known a priori (no unit forced outages are modeled), whereas the system load and the RES injection values over the entire scheduling period are considered to be equal to their forecast values made available prior to the implementation of the scheduling optimization models.

The deterministic formulation of the UC models can be considered as a special case of the respective stochastic formulation further described in the following sections, where only a single scenario comprising the forecast values of the random unit and system parameters is considered.

4.3.2.1.2 Nomenclature

For the reader's convenience, an explanation of all symbols that appear in the detailed mathematical formulation of the optimization problem that follows is provided below:

Sets

$b \in B$	Transmission system branches, denoting either transmission lines or transformers
$f \in F$	Steps of the marginal cost function of thermal unit i
$g \in G$	Generating units (conventional and RES)
$i \in I$	Conventional generating units ($I \subseteq G$)
$j \in J$	RES units ($J \subseteq G$ and $J = J^w \cup J^{PV} \cup \dots J^k$), where J^w is the set of wind parks, J^{PV} is the set of PV units, and J^k is the set of RES technology k units
$\ell \in L$	Types of generating unit start-up $L = \{h, w, c\}$, where h: hot, w: warm, and c: cold start-up
$m \in M$	Reserves types $M = \{1+, 1-, 2+, 2-, 3\}$, where $m = 1+$: primary-up, $m = 1-$: primary-down, $m = 2+$: secondary-up, $m = 2-$: secondary-down, $m = 3$: tertiary (spinning—3S and nonspinning—3NS)
$n \in N$	Transmission system nodes
$t \in T$	Scheduling time intervals
T^-	Scheduling horizon extended to the past, where $T^- \supseteq T$
T^+	Scheduling horizon extended to the future, where $T^+ \supseteq T$

Parameters

$\mathbf{A}_{nb}$	Node-to-branch incidence matrix, where $\mathbf{A}_{nb} = \mathbf{A}_{nb}^+ - \mathbf{A}_{nb}^-$
$\mathbf{A}_{nb}^-$	Node-to-branch incidence matrix whose element A_{nb}^- is equal to 1 if branch b ends to node n and 0 otherwise

$\mathbf{A}_{nb}^{+}$ Node-to-branch incidence matrix whose element A_{nb}^{+} is equal to 1 if branch b begins from node n and 0 otherwise

$\mathbf{A}_{ng}$ Node-to-generator incidence matrix whose element A_{ng} is equal to 1 if generating unit g is connected to node n and 0 otherwise

b_b Branch b susceptance, in pu

$\mathbf{B}_{n,n'}$ Node susceptance matrix, in pu

C_{it}^{m} Additional cost for the procurement of reserve type m from thermal unit i in time interval t, in €/MWh

D_{nt} Load demand in node n and time interval t, in MW

D_t System load demand in time interval t, in MW

DT_i Minimum down time of unit i, in hours

E_{if} Size of step f of thermal unit i marginal cost function, in MW

F_b^{max} Branch b rating, in MW

NLC_i No-load cost of unit i (for one-hour operation), in €/hour

P_i^{max} Maximum power output of thermal unit i, in MW

$P_i^{\mathrm{max,AGC}}$ Maximum power output of thermal unit i while operating under AGC, in MW

P_i^{min} Minimum power output of thermal unit i, in MW

$P_i^{\mathrm{min,AGC}}$ Minimum power output of thermal unit i while operating under AGC, in MW

P_i^{syn} Synchronization load of unit i, in MW

P_{is}^{soak} Power output of thermal unit i during interval s of the soak phase, in MW

P_j^{forecast} Forecast power output of RES unit j, in MW

$\mathbf{PTDF}_{bn}$ Power transfer distribution factor matrix of branch b for power transfer from node n to the reference node, in p.u.

R_i^{m} Maximum contribution of thermal unit i in reserve type m, in MW

RD_i Ramp-down rate of thermal unit i, in MW/minute

$\mathrm{RD}_i^{\mathrm{AGC}}$ Ramp-down rate of thermal unit i while operating under AGC, in MW/minute

RR_t^{m} System requirement in reserve type m during time interval t, in MW

RU_i Ramp-up rate of thermal unit i, in MW/minute

$\mathrm{RU}_i^{\mathrm{AGC}}$ Ramp-up rate of thermal unit i while operating under AGC, in MW/minute

SDC_i Shutdown cost of thermal unit i, in €

SUC_i^{ℓ}	Start-up cost of thermal unit i from type-ℓ standby until load with synchronization, in €
T_i^{des}	Time from technical minimum power output to desynchronization of thermal unit i, in time intervals
T_i^{ℓ}	Time off-load before going into longer standby conditions ($\ell = w$: hot to warm and $\ell = c$: hot to cold) of thermal unit i, in time intervals
T_i^{off}	Reservation time of thermal unit i (prior to start-up)
$T_i^{\mathrm{soak},\ell}$	Soak time of thermal unit i under type-ℓ start-up, in time intervals
$T_i^{\mathrm{syn},\ell}$	Time to synchronize thermal unit i under type-ℓ start-up, in time intervals
T_{step}	Duration of time intervals (variable for different scheduling models), in minutes
UT_i	Minimum up-time of unit i, in time intervals
VLL_{nt}	Value of lost load in node n and time interval t, in €/MW
VLL_t	Value of lost load in time interval t, in €/MW

Continuous variables

d_{nt}^{LNS}	Load not served in node n and time interval t, in MW
d_t^{LNS}	Load not served in time interval t, in MW
e_{ift}	Portion of step f of the ith unit's marginal cost function loaded in time interval t, in MW
p_{bt}^{f}	Power flow in branch b during time interval t, in MW
p_{it}	Power output of thermal unit i dispatched by the ISO in time interval t, in MW
p_{it}^{des}	Power output of thermal unit i during the desynchronization phase in time interval t, in MW
p_{it}^{soak}	Power output of thermal unit i during the soak phase in time interval t, in MW
p_{jt}	Power output of RES unit j in time interval t, in MW
p_{nt}^{inj}	Total power injection in node n during time interval t, in MW
r_{it}^{m}	Contribution of thermal unit i in reserve type m during time interval t, in MW
θ_{nt}	Voltage angle of node n during time interval t, in rad

Binary variables

u_{it}	Binary variable which is equal to 1 if thermal unit i is committed during time interval t

u_{it}^{AGC}	Binary variable which is equal to 1 if thermal unit i operates under AGC and provides secondary reserve during time interval t
u_{it}^n	Binary variable which is equal to 1 if thermal unit i is in operating phase n during time interval t, where n = syn: synchronization, n = soak: soak, n = disp: dispatchable, n = des: desynchronization
u_{it}^{3NS}	Binary variable which is equal to 1 if thermal unit i provides tertiary nonspinning reserve during time interval t
u_{jt}	Binary variable which is equal to 1 if RES unit j is committed during time interval t
y_{it}	Binary variable which is equal to 1 if thermal unit i is started-up during time interval t
y_{it}^ℓ	Binary variable which is equal to 1 if a type-ℓ start-up of thermal unit i is initiated during time interval t
z_{it}	Binary variable which is equal to 1 if thermal unit i is shut-down during time interval t

4.3.2.1.3 Deterministic UC Model Without Network Representation

In this section, the deterministic UC model is mathematically formulated, comprising only the system and the generating units (conventional and RES) operating constraints. The incorporation of the transmission network representation under a DC power flow modeling is analytically described in the following paragraphs.

The objective is the minimization of the total cost for the procurement of energy and reserves from the generating units, simultaneously respecting the operating constraints of all systems and generating units.

As already mentioned, the UC problem is formulated as an MILP optimization problem, where binary variables are used to properly model the commitment, start-up, and shutdown decisions of generating units. These binary variables are hereinafter denoted by u_{it}, y_{it}, and z_{it}, respectively. Moreover, a detailed representation of the operating phases of generating units is provided. For this reason, additional binary variables u_{it}^{syn}, u_{it}^{soak}, u_{it}^{disp}, and u_{it}^{des}, are introduced for the modeling of the synchronization, soak, dispatch, and desynchronization phases, respectively. Finally, for each unit operating under the automatic generation control (AGC), an additional binary variable, u_{it}^{AGC}, is used.

4.3.2.1.3.1 Operating Phases of a Thermal Unit

Figure 4.22 shows the different operating phases of a thermal generating unit. After being reserved ($u_{it} = 0$) for T_i^{off} time intervals ($T_i^{off} \geq DT_i$), the unit starts-up at time interval t_1 ($y_{it} = 1$) and remains committed ($u_{it} = 1$) until it is shut down at time interval $t5$ ($z_{it} = 1$).

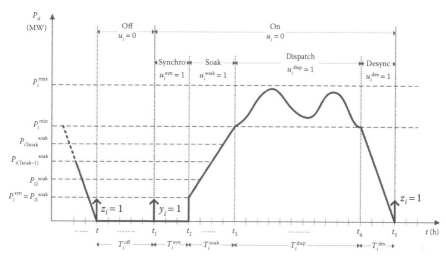

FIGURE 4.22
Operating phases of a thermal unit.

Once committed, the unit follows, in general, four consecutive operation phases, namely, (1) synchronization, (2) soak, (3) dispatchable, and (4) desynchronization, denoted by the aforementioned binary variables. The first two phases—namely, the synchronization and soak procedures—comprise the unit start-up phase. During the dispatchable phase, the unit can receive dispatch instructions to vary its power output between its technical minimum P_{it}^{min} and its nominal power output P_{it}^{max} according to its ramp-rate limits, and contributes to the system reserves. During the desynchronization phase the unit power output follows a predefined shutdown sequence.

The accurate modeling of the unit start-up sequence requires special attention. Three start-up types are modeled, $\ell \in L = \{h, w, c\}$ (h: hot, w: warm, and c: cold), each with distinct synchronization time, soak time, and start-up cost, as shown in Table 4.5. Both $T_i^{syn,\ell}$ and $T_i^{soak,\ell}$ depend on the thermal unit's prior reservation time, T_i^{off}. In Table 4.5, parameter T^- represents a large number of time intervals of the past (larger than the maximum reservation time to cold start of all units).

TABLE 4.5

Thermal Unit Start-Up Modeling

Start-Up Type	Prior Reservation Time (Time Intervals) $\underline{T}_i^\ell \leq T_i^{off} < \overline{T}_i^\ell$	Synchronization Time (Time Intervals)	Soak Time (Time Intervals)	Start-Up Cost (€)
Hot	$0 \leq T_i^{off} < T_i^w$	$T_i^{syn,h}$	$T_i^{soak,h}$	SUC_i^h
Warm	$T_i^w \leq T_i^{off} < T_i^c$	$T_i^{syn,w}$	$T_i^{soak,w}$	SUC_i^w
Cold	$T_i^c \leq T_i^{off} < T^-$	$T_i^{syn,c}$	$T_i^{soak,c}$	SUC_i^c

The unit start-up sequence consists of two phases: (1) synchronization phase and (2) soak phase. Once a type-ℓ start-up decision is taken, the thermal unit enters the synchronization phase, which lasts for $T_i^{\text{syn},\ell}$ time intervals and during which the power output of the unit is 0 MW. Subsequently, the unit enters the soak phase, which lasts for $T_i^{\text{soak},\ell}$ time intervals and during which the unit operates between the synchronization load, P_i^{syn}, and the technical minimum power output, P_i^{min}. A detailed description of the unit's soak phase ramp-up sequence is given in the following paragraphs.

Once a thermal unit enters a hot, warm, or cold start-up phase, it should complete the start-up sequence, and enter the dispatchable phase as long as needed to satisfy the minimum up-time requirement before shutting down.

It is clarified that the model input parameters $T_i^{\text{off}}, T_i^{\ell}, T_i^{\text{des}}, T_i^{\text{syn},\ell}, T_i^{\text{soak},\ell}, UT_i,$ and DT_i presented in the aforementioned analysis represent hourly values. Depending on the granularity of the examined problem regarding the time resolution, these parameters should be converted to the appropriate time frame prior to the optimization procedure. For instance, if generating unit g requires 3 hours for its desynchronization phase ($\tilde{T}_g^{\text{des}} = 3h$, where the tilde denotes the raw input data) and the time granularity of the optimization problem is 15 minutes, then the variable time step ΔT is equal to $\Delta T = \left(T^{\text{step}}/60\right) = (15/60) = 0.25h$ and the required desynchronization time of the generating unit g for this time resolution is equal to $T_g^{\text{des}} = \left(\tilde{T}_g^{\text{des}}/0.25\right) = (3/0.25) = 12$ time intervals. Apparently, if the problem time resolution is equal to 1 hour, then $\Delta T = \left(T^{\text{step}}/60\right) = (60/60) = 1$ and, therefore, $T_g^{\text{des}} = \tilde{T}_g^{\text{des}} = 3$ time intervals.

In the following sections, all above mentioned time constants are expressed as number of time intervals according to the selected granularity of the optimization model.

Objective function

The objective function of the short-term scheduling problem is formulated as follows:

$$
\text{Min} \left\{ \begin{aligned}
& \sum_{t \in T} \sum_{i \in I} \left[\left(\sum_{f \in F} C_{if} \cdot e_{ift} + \text{NLC}_i \cdot \left(u_{it} - u_{it}^{\text{syn}} \right) + \sum_{m \in M} C_{it}^m \cdot r_{it}^m \right) \cdot \left(\frac{T_{\text{step}}}{60} \right) \right. \\
& \left. + \sum_{\ell \in L} \text{SUC}_i^{\ell} \cdot y_{it}^{\ell} + \text{SDC}_i \cdot z_{it} \right] + \left(\frac{T_{\text{step}}}{60} \right) \cdot \sum_{t \in T} \text{VLL}_t \cdot d_t^{\text{LNS}}
\end{aligned} \right\}
\tag{4.12}
$$

The objective function (4.12) to be minimized includes the following: (1) the energy production cost, (2) the no-load cost, (3) the cost for the procurement of the required reserves (ancillary services), (4) the start-up and shutdown costs, and (5) the cost of the unserved load.

It is noted that the start-up cost of generating unit i depends on the start-up type (hot, warm, and cold). In addition, since the energy production costs, the no-load costs and the reserves provision costs are generally expressed per hour (€/h), the respective components in the objective function are multiplied by the ratio $(T_{step}/60)$ to allow for the accurate representation of the variable duration of the time intervals. On the contrary, the start-up and shutdown costs are independent of the duration of the time intervals chosen.

In the following subsections, the generating units and system constraints are analytically presented.

4.3.2.1.3.2 Thermal Unit Operating Constraints   The operating constraints of a thermal unit may be classified as follows:

Start-up type constraints

$$y_{it}^\ell \leq \sum_{\tau=t-\overline{T}_i^\ell+1}^{t-\underline{T}_i^\ell} z_{i\tau} \qquad \forall i \in I,\ t \in T,\ \ell \in L, \quad \text{where } \tau \in T^- \tag{4.13}$$

$$y_{it} = \sum_{\ell \in L} y_{it}^\ell \quad \forall i \in I,\ t \in T \tag{4.14}$$

Constraints 4.13 and 4.14 select the appropriate start-up type of the thermal unit i, depending on the unit's prior reservation time, as described by the first two columns of Table 4.5. This is achieved in Equation 4.13 by constraining the type-ℓ start-up variable of unit i during time interval t, y_{it}^ℓ, to be zero, unless there was a prior shutdown of the unit in the time interval $\left(t-\overline{T}_i^\ell, t-\underline{T}_j^\ell\right]$. Constraint 4.14 ensure that only one of the start-up type binary variables y_{it}^ℓ may be equal to 1.

Synchronization phase constraints

$$u_{it}^{syn,\ell} = \sum_{\tau=t-T_i^{syn,\ell}+1}^{t} y_{i\tau}^\ell \quad \forall i \in I,\ t \in T,\ \ell \in L, \quad \text{where } \tau \in T^- \tag{4.15}$$

$$u_{it}^{syn} = \sum_{\ell \in L} u_{it}^{syn,\ell} \quad \forall i \in I,\ t \in T \tag{4.16}$$

Constraint 4.15 ensure that unit i enters the synchronization phase immediately following its start-up. The synchronization binary variable becomes equal to 1 whenever there is a type-ℓ start-up of the unit in the prior $T_i^{syn,\ell}$ hours, where the duration $T_i^{syn,\ell}$ of the synchronization phase depends on the unit start-up type. Moreover, constraint 4.16 ensure that only one synchronization type per start-up is selected.

Soak phase constraints

$$u_{it}^{\text{soak},\ell} = \sum_{\tau=t-T_i^{\text{syn},\ell}-T_i^{\text{soak},\ell}+1}^{t-T_i^{\text{syn},\ell}} y_{i\tau}^{\ell} \quad \forall i \in I, t \in T, \ell \in L, \quad \text{where } \tau \in T^{-} \qquad (4.17)$$

$$u_{it}^{\text{soak}} = \sum_{\ell \in L} u_{it}^{\text{soak},\ell} \quad \forall i \in I, t \in T \qquad (4.18)$$

Constraint 4.17 ensure that thermal unit i should enter a soak phase following its synchronization (see Figure 4.22). The duration of the soak phase depends on the unit i start-up type (hot, warm, cold). Constraint 4.18 ensure that only one soak phase type per start-up is selected.

In general, during the soak phase, the power output of a unit follows a predefined sequence of MW values, depending on the start-up type, ℓ, as shown in Figure 4.23:

$$\left\{ P_{is}^{\text{soak},\ell}, s = 1,\dots, T_i^{\text{soak},\ell} \right\} \qquad (4.19)$$

Thus, during the soak phase, the unit power output is constrained by the following:

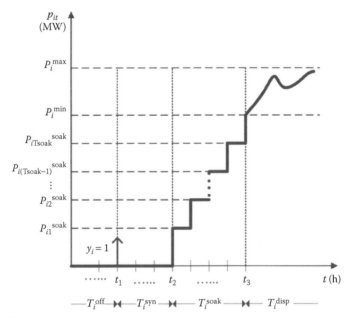

FIGURE 4.23
Soak phase of a thermal unit with predefined sequence of MW values.

$$p_{it}^{\text{soak}} = \sum_{\ell \in L} \sum_{\tau = t - T_i^{\text{syn},\ell} - T_i^{\text{soak},\ell} + 1}^{t - T_i^{\text{syn},\ell}} y_{i\tau}^{\ell} \cdot P_{i, t - T_i^{\text{syn}} - \tau + 1}^{\text{soak},\ell} \qquad \forall i \in I, t \in T, \ell \in L, \qquad (4.20)$$

where $\tau \in T^-$

Desynchronization phase constraints

$$u_{it}^{\text{des}} = \sum_{\tau = t+1}^{t + T_i^{\text{des}}} z_{i\tau} \qquad \forall i \in I, t \in T, \quad \text{where } \tau \in T^+ \qquad (4.21)$$

$$p_{it}^{\text{des}} = \sum_{\tau = t+1}^{t + T_i^{\text{des}}} z_{i\tau} \cdot (\tau - t) \cdot \frac{P_i^{\text{min}}}{T_i^{\text{des}}} \qquad \forall i \in I, t \in T, \quad \text{where } \tau \in T^+ \qquad (4.22)$$

Constraint 4.21 ensure that unit i should operate in a desynchronization phase lasting T_i^{des} time intervals before its shutdown (see Figure 4.22). The unit power output during the desynchronization process decreases linearly from its technical minimum power output to 0 MW, as enforced by Equation 4.22. T^+ denotes the planning horizon extended to the future.

Minimum up/down-time constraints

$$\sum_{\tau = t - \text{UT}_i + 1}^{t} y_{i\tau} \leq u_{it} \qquad \forall i \in I, t \in T, \quad \text{where } \tau \in T^- \qquad (4.23)$$

$$\sum_{\tau = t - \text{DT}_i + 1}^{t} z_{i\tau} \leq 1 - u_{it} \qquad \forall i \in I, t \in T, \quad \text{where } \tau \in T^- \qquad (4.24)$$

Equations (4.23) and (4.24) enforce the minimum up- and down-time constraints, respectively; that is, unit i must remain committed (de-committed) at time interval t if its start-up (shutdown) occurred during the previous $\text{UT}_i - 1$ ($\text{DT}_i - 1$) time intervals.

Logical status of commitment

$$u_{it} = u_{it}^{\text{syn}} + u_{it}^{\text{soak}} + u_{it}^{\text{disp}} + u_{it}^{\text{des}} \qquad \forall i \in I, t \in T \qquad (4.25)$$

$$y_{it} - z_{it} = u_{it} - u_{i(t-1)} \qquad \forall i \in I, t \in T \qquad (4.26)$$

$$y_{it} + z_{it} \leq 1 \qquad \forall i \in I, t \in T \qquad (4.27)$$

Equations 4.25 through 4.27 represent the logical status of commitment variables. Constraint 4.25 ensure that only one, at most, of the binary variables corresponding to the different commitment states of unit i can equal be equal to 1 in every time interval. Constraint 4.26 model the logic of the start-up and shutdown status change, whereas constraint 4.27 enforce that unit i may not be started-up and shutdown simultaneously in a given period.

Ramp-up/down constraints

$$p_{it} - p_{i(t-1)} \le RU_i \cdot T_{\text{step}} + P_i^{\min} \cdot \left(u_{it}^{\text{syn}} + u_{it}^{\text{soak}}\right) \qquad \forall i \in I, t \in T \tag{4.28}$$

$$p_{i(t-1)} - p_{it} \le RD_i \cdot T_{\text{step}} + P_i^{\min} \cdot u_{i(t-1)}^{\text{des}} \qquad \forall i \in I, t \in T \tag{4.29}$$

Constraint 4.28 and 4.29 introduce the effect of ramp-rate limits on the power output of units. Since RU and RD are generally expressed in MW/minute, the first part of the right-hand side of the abovementioned inequalities is multiplied by the variable time interval duration T_{step}, so that these two constraints model the ramp-up/-down capability of thermal units for any time interval duration selected. Moreover, the second part of the right-hand side of inequalities and Equations 4.28 and 4.29 are added, so that these constraints are relaxed when unit i is in the synchronization, soak, or desynchronization phase.

Power output constraints

$$u_{it}^{\text{AGC}} \le u_{it}^{\text{disp}} \qquad \forall i \in I, t \in T \tag{4.30}$$

$$0 \le r_{it}^{1+} \le R_i^1 \cdot u_{it}^{\text{disp}} \qquad \forall i \in I, t \in T \tag{4.31}$$

$$0 \le r_{it}^{1-} \le R_i^1 \cdot u_{it}^{\text{disp}} \qquad \forall i \in I, t \in T \tag{4.32}$$

$$0 \le r_{it}^{2+} \le 15 \cdot RU_i^{\text{AGC}} \cdot u_{it}^{\text{AGC}} \qquad \forall i \in I, t \in T \tag{4.33}$$

$$0 \le r_{it}^{2-} \le 15 \cdot RD_i^{\text{AGC}} \cdot u_{it}^{\text{AGC}} \qquad \forall i \in I, t \in T \tag{4.34}$$

$$0 \le r_{it}^{3S} \le R_i^{3S} \cdot u_{it}^{\text{disp}} \qquad \forall i \in I, t \in T \tag{4.35}$$

$$r_{it}^{3NS} \le R_i^{3NS} \cdot u_{it}^{3NS} \qquad \forall i \in I, t \in T \tag{4.36}$$

$$r_{it}^{3NS} \ge P_i^{\min} \cdot u_{it}^{3NS} \qquad \forall i \in I, t \in T \tag{4.37}$$

$$u_{it}^{3NS} \le 1 - u_{it} \qquad \forall i \in I, t \in T \tag{4.38}$$

$$p_{it} - r_{it}^{2-} \geq 0 \cdot u_{it}^{\text{syn}} + p_{it}^{\text{soak}} + p_{it}^{\text{des}} + P_i^{\min} \cdot \left(u_{it}^{\text{disp}} - u_{it}^{\text{AGC}} \right)$$
$$+ P_i^{\min,\text{AGC}} \cdot u_{it}^{\text{AGC}} \qquad \forall i \in I, t \in T \tag{4.39}$$

$$p_{it} + r_{it}^{2+} \leq 0 \cdot u_{it}^{\text{syn}} + p_{it}^{\text{soak}} + p_{it}^{\text{des}} + P_i^{\max} \cdot \left(u_{it}^{\text{disp}} - u_{it}^{\text{AGC}} \right) + P_i^{\max,\text{AGC}} \cdot u_{it}^{\text{AGC}} \tag{4.40}$$
$$\forall i \in I, t \in T$$

$$p_{it} - r_{it}^{1-} - r_{it}^{2-} \geq 0 \cdot u_{it}^{\text{syn}} + p_{it}^{\text{soak}} + p_{it}^{\text{des}} + P_i^{\min} \cdot u_{it}^{\text{disp}} \qquad \forall i \in I, t \in T \tag{4.41}$$

$$p_{it} + r_{it}^{1} + r_{it}^{2+} + r_{it}^{3S} \leq 0 \cdot u_{it}^{\text{syn}} + p_{it}^{\text{soak}} + p_{it}^{\text{des}} + P_i^{\max} \cdot u_{it}^{\text{disp}} \qquad \forall i \in I, t \in T \tag{4.42}$$

Constraint 4.30 state that unit i may provide AGC if and only if it is on dispatch. Constraints 4.31 through 4.36 set the upper limits of primary up/down, secondary-up/down, and tertiary spinning/nonspinning reserves, respectively. As shown by constraints 4.30 through 4.35, unit i may contribute in synchronized reserves if and only if it is dispatchable, while during the synchronization, soak and desynchronization phases the synchronized reserves contribution is equal to zero. Constraint 4.37 enforce the tertiary nonspinning reserve contribution to be greater than the minimum power output of unit i. Constraint 4.38 state that unit i may provide tertiary nonspinning reserve if and only if it is offline.

Constraints 4.39 through 4.42 define the limits of the power output of thermal unit i in every commitment state. The first three terms of the right-hand side of Equations 4.39 through 4.42 constrain the output of the unit during synchronization, soak, and desynchronization. If unit i is on synchronization in time interval t (i.e., $u_{it}^{\text{syn}} = 1$), the power output will be equal to 0, whereas the terms p_{it}^{soak} and p_{it}^{des} are defined in Equations 4.20 and 4.22, respectively.

Total production cost of a thermal unit

The total production cost of a thermal unit includes the unit's start-up type, dependent start-up cost, SUC_i^f (see Figure 4.24), shutdown cost, SDC_i, and the hourly operating cost defined by the unit's no-load cost, NLC_i, and the step-wise marginal cost function, $\left(Q_{if}, C_{if} \right)$, $f \in F^i$, shown in Figure 4.25.

The general case of a step-wise marginal cost function of the units is modeled with the addition of new variables mc_{ift} and e_{ift}, defined as follows:

F^i set of steps of the step-wise marginal cost function of unit i, $F^i = \left\{ 1, 2, \ldots, F^i \right\}$

e_{ift} portion of step f of the ith unit's marginal cost function loaded in time interval t, in MWh.

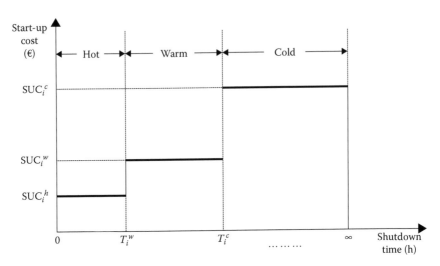

FIGURE 4.24
Start-up cost from hot, warm, and cold standby until load with synchronization.

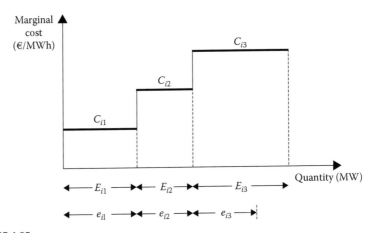

FIGURE 4.25
Thermal generating unit step-wise marginal cost function (convex cost function).

mc$_{ift}$ binary variable indicating unit i output higher than, or equal to, step f during time interval t (needed for nonconvex cost function only)

The total production cost of a thermal unit is modeled as follows:

$$c_{it}(p_{it}) = \sum_{\ell \in L} \text{SUC}_i^\ell \cdot y_{it}^\ell + \text{SDC}_i \cdot z_{it} + \text{NLC}_i \cdot \left(u_{it} - u_{it}^{\text{syn}}\right)$$
$$+ \sum_{f \in F^i} C_{if} \cdot e_{ift} \qquad \forall i \in I, t \in T \tag{4.43}$$

In the general case of nonconvex cost functions, the following constraints 4.44 through 4.46 must be added to the model:

$$\sum_{f \in F^i} e_{ift} = p_{it} \qquad \forall i \in I, t \in T \tag{4.44}$$

$$mc_{i(f+1)t} \cdot E_{if} \le e_{ift} \le mc_{ift} \cdot E_{if} \qquad \forall i \in I, f \in F^i, t \in T \tag{4.45}$$

$$mc_{i(f+1)t} \le mc_{ift} \qquad \forall i \in I, f \in F^i, t \in T \tag{4.46}$$

If the unit's marginal cost function is nondecreasing (convex cost function; see Figure 4.25), the use of the additional binary variables mc_{ift} is avoided and constraints 4.45 and 4.46 should be replaced by constraint 4.47 in the model.

$$0 \le e_{ift} \le E_{if} \qquad \forall i \in I, f \in F^i, t \in T \tag{4.47}$$

The start-up cost used in (4.43) is discretized in three levels, one for each start-up type (see Figure 4.24).

It should be noted that the start-up cost, SUC_i^ℓ, includes expenses up to the end of the synchronization phase; the operating cost of a unit during the soak phase is computed separately, as the sum of the unit's no-load cost and the integral of the unit's marginal cost function, as shown in the last two terms of Equation 4.43. Similarly, the shut-down cost, SDC_i, involves all expenses needed for the unit's shutdown besides the unit operating cost during the desynchronization phase.

Operating constraints of RES plants

Different types of RES plants are considered in the framework of the short-term scheduling models, namely, wind, PV, small hydro, geothermal, and biomass. Among all, wind and PV units cover the largest share of the installed capacity, as well as the associated energy, produced in the insular power systems involved in the project.

It is common that RES plants are considered to operate with zero variable cost, thus being assigned a dispatch priority by the SO. In fact, the short-term operation of these plants is much simpler than that of the thermal units, since they are not subject to the respective complex inter-temporal and cost constraints already described above. Therefore, in general, their power output is only restricted by the available power output [see constraints (Equation 4.48)], which is calculated with the use of appropriate forecasting tools and inserted as input (known parameter) in the deterministic optimization model.

$$p_{jt} \leq P_j^{\text{forecast}} \cdot u_{jt} \qquad \forall j \in J^{\text{w}}, t \in T \tag{4.48}$$

However, the increased variability and uncertainty of the wind production allows for an exception to the above rule: the wind energy production is allowed to be curtailed as much as needed to ensure that a maximum renewable energy penetration level is maintained in the system without simultaneously violating the technical minima of the online conventional generating units.

In the present formulation, the curtailment of the wind production is organized based on specific rules currently applied by the SO in the power system of Crete Island. According to these rules, the mathematical formulation of the wind production curtailment procedure consists of the following group of equations, which are also included in the formulation of the optimization problem:

$$p_{jt} \leq P_{jt}^{\text{SP}} \qquad \forall j \in J^{\text{w}}, t \in T \tag{4.49}$$

$$P_{jt}^{\text{SP}} = \frac{P_j^{\max}}{\sum_{j \in J^{\text{w}}} P_j^{\max}} \cdot P_t^{\text{SP,tot}} \qquad \forall j \in J^{\text{w}}, t \in T \tag{4.50}$$

$$P_t^{\text{SP,tot}} \leq \lambda \cdot D_t \qquad \forall t \in T \tag{4.51}$$

$$P_t^{\text{SP,tot}} \leq D_t - \sum_{i \in I} \left[P_i^{\min} \cdot u_{it}^{\text{disp}} + p_{it}^{\text{soak}} + p_{it}^{\text{des}} \right] \qquad \forall t \in T \tag{4.52}$$

$$P_t^{\text{SP,tot}} = \sum_{j \in J^{\text{w}}} P_{jt}^{\text{SP}} \qquad \forall t \in T \tag{4.53}$$

where:

P_{jt}^{SP} is the maximum allowed output (set point) of wind farm j in time interval t (in MW)

$P_t^{\text{SP,tot}}$ is the total maximum allowed output (total set point) of all wind farms in time interval t (in MW)

$P_j^{\max}$ is the nominal (installed) capacity of wind farm j (in MW)

λ is a coefficient (in pu) determining the desired wind power penetration level of the power system and it currently lies in the range of 30%–40%

p_{it}^{soak} is the unit power output during the soak phase that is calculated according to Equation 4.20

p_{it}^{des} is the unit power output during the desynchronization phase, determined by Equation 4.22

Constraint 4.49 define the power output limits of each wind farm j in time interval t with respect to its respective set point. Constraint 4.50

define the set point of each wind farm j in time interval t according to the ratio of the installed capacity of the wind farm to the total wind installed capacity of the power system (prorata calculation). Constraints 4.51 and 4.52 define the total set point, calculated on the basis of the desired wind power penetration level λ, the system load demand, and the constrained power output of the committed conventional units. Finally, constraints 4.53 ensure that the summation of the set points of all wind farms is equal to the total set point of the insular power system.

System constraints

$$\sum_{i \in I} r_{it}^{1+} \geq RR_t^{1+} \qquad \forall t \in T \tag{4.54}$$

$$\sum_{i \in I} r_{it}^{1-} \geq RR_t^{1-} \qquad \forall t \in T \tag{4.55}$$

$$\sum_{i \in I} r_{it}^{2+} \geq RR_t^{2+} \qquad \forall t \in T \tag{4.56}$$

$$\sum_{i \in I} r_{it}^{2-} \geq RR_t^{2-} \qquad \forall t \in T \tag{4.57}$$

$$\sum_{i \in I} r_{it}^{3S} + \sum_{i \in I} r_{it}^{3NS} \geq RR_t^{3} \qquad \forall t \in T \tag{4.58}$$

Constraints 4.54 through 4.58 enforce the deterministic system requirements for the provision of the primary, secondary, and tertiary reserves. In all cases, the sum of all contributions of thermal units to a single reserve type must be at least equal to the respective reserve requirement.

$$\left(\frac{T_{\text{step}}}{60}\right) \cdot \left(\sum_{i \in I} p_{it} + \sum_{j \in J} p_{jt}\right) = \left(\frac{T_{\text{step}}}{60}\right) \cdot \left(D_t - d_t^{\text{LNS}}\right) \qquad \forall t \in T \tag{4.59}$$

Finally, Equation 4.59 describes the power balance, where the total power output of all generating units (both thermal and RES) must be equal to the forecast load in each time interval t. It should be noted that the additional term $T_{\text{step}}/60$ is added, so that the Lagrange multiplier of Equation 4.59 has an economical interpretation, in terms of the marginal clearing price (€/MWh), in all time intervals.

4.3.2.1.4 Deterministic UC Model with DC Power Flow

In this section, the deterministic UC model taking into account the transmission network modeling under a DC power flow formulation is described. It should be noted that the DC power flow formulation derives from the actual AC power flow formulation, assuming that all line resistances are equal to zero, voltage magnitudes are equal to 1 pu, and voltage angle differences between interconnected nodes (i,j) are sufficiently small, so that $\sin\theta_{ij} \approx \theta_{ij}$ and $\cos\theta_{ij} \approx 1$ [79]. Many works in the literature have shown that the DC power flow model provides a good approximation of the actual flows and its linearity can been exploited in the scheduling tools.

$$p_{bt}^f = -b_b \cdot \sum_{n\in N} A_{nb} \cdot \theta_{nt} \qquad \forall b \in B, t \in T \tag{4.60}$$

$$p_{nt}^{inj} = \sum_{n'\in N} B_{n,n'} \cdot \theta_{n't} \qquad \forall n \in N, t \in T \tag{4.61}$$

where:

A_{nb} is the (n,b) element of the node-to-branch incidence matrix $\mathbf{A}_{nb}$ matrix

$B_{n,n'}$ is the (n,n') element of the $\mathbf{B}_{n,n'}$ node susceptance matrix, which is defined according to the following equation

$$\mathbf{B}_{n,n'} = \mathbf{A}_{nb} \cdot diag(-b_b) \cdot \mathbf{A}_{nb}^{\mathsf{T}} \tag{4.62}$$

Equation 4.60 describes the DC power flow equation with regard to the angle formulation. According to this equation, the power flow in branch b and during the time interval t is equal to the susceptance of the examined branch multiplied by the angle difference between the beginning and the ending nodes of the examined branch b. Equation 4.61 describes the overall power injection in node n. Combining both Equations 4.60 and 4.61 yields the following:

$$p_{bt}^f = \sum_{n\in N} PTDF_{bn} \cdot p_{nt}^{inj} \qquad \forall b \in B, t \in T \tag{4.63}$$

The term $PTDF_{bn}$ is the (b,n) element of the power transfer distribution factor matrix $\mathbf{PTDF}_{bn}$, which is computed by

$$\mathbf{PTDF}_{bn} = diag(-b_b) \cdot \mathbf{A}_{nb}^{\mathsf{T}} \cdot [\mathbf{B}_{n,n'}]^{-1} \tag{4.64}$$

and describes the fraction of the amount of a power transfer from node n to the reference node that flows over the branch b. Equation 4.63 actually links the power injection in node n with the power flow in branch b. The power injection in node n and in time interval t is equal to the following:

$$p_{nt}^{inj} = \sum_{i\in I} A_{ni} \cdot p_{it} + \sum_{j\in J} A_{nj} \cdot p_{jt} - \left(D_{nt} - d_{nt}^{LNS}\right) \qquad \forall n \in N, t \in T \tag{4.65}$$

where:

A_g^n is the element of the respective $\mathbf{A}_g^n$ incidence matrix that is equal to 1 if generating unit g (thermal unit i or RES unit j) is connected to node n and 0 otherwise

D_{nt} is the load in node n and in time interval t

Taking into consideration the aforementioned analysis and the equations, modeling the thermal units and operating constraints of RES units, along with the system reserves constraints, presented in Sections 4.3.2.1.3 and 4.3.2.1.4, the overall short-term UC problem incorporating the DC network representation is mathematically formulated as follows:

$$
\text{Min}\left\{
\begin{array}{l}
\sum_{t\in T}\sum_{i\in I}\left[\left(\sum_{f\in F}C_{if}\cdot e_{ift}+\text{NLC}_i\cdot\left(u_{it}-u_{it}^{\text{syn}}\right)+\sum_{m\in M}C_{it}^m\cdot r_{it}^m\right)\cdot\left(\frac{T_{\text{step}}}{60}\right)\right. \\
\left.+\sum_{\ell\in L}\text{SUC}_i^\ell\cdot y_{it}^\ell+\text{SDC}_i\cdot z_{it}\right]+\left(\frac{T_{\text{step}}}{60}\right)\cdot\sum_{t\in T}\sum_{n\in N}\text{VLL}_{nt}\cdot d_{nt}^{\text{LNS}}
\end{array}
\right\}
\tag{4.66}
$$

subject to:

Constraints 4.13 through 4.42, 4.44, 4.47, and 4.48 through 4.58

$$
\left(\frac{T_{\text{step}}}{60}\right)\cdot\left(\sum_{n\in N}\sum_{i\in I}A_{ni}\cdot p_{it}+\sum_{n\in N}\sum_{j\in J}A_{nj}\cdot p_{jt}\right)
$$
$$
=\left(\frac{T_{\text{step}}}{60}\right)\cdot\sum_{n\in N}\left(D_{nt}-d_{nt}^{\text{LNS}}\right)\qquad \forall t\in T
\tag{4.67}
$$

$$
p_{bt}^{\text{f}}=\sum_{n\in N}\text{PTDF}_{bn}\cdot p_{nt}^{\text{inj}}\qquad \forall b\in B,\,t\in T
\tag{4.68}
$$

$$
p_{nt}^{\text{inj}}=\sum_{i\in I}A_{ni}\cdot p_{it}+\sum_{j\in J}A_{nj}\cdot p_{jt}-\left(D_{nt}-d_{nt}^{\text{LNS}}\right)\qquad \forall n\in N,\,t\in T
\tag{4.69}
$$

$$
-F_b^{\max}\le p_{bt}^{\text{f}}\le F_b^{\max}\qquad \forall b\in B,\,t\in T
\tag{4.70}
$$

where Equation 4.67 is the system-wide power balance equation, constraints 4.68 and 4.69 are the power flow equations with respect to the PTDF-based formulation and finally Equation 4.70 enforces the power flow limit in branch b with regard to its thermal limit.

4.3.2.1.5 RTED Model with AC Power Flow

As already mentioned, the RTED is implemented for the determination of the optimal generation dispatch levels of all online generating units in real time (e.g., every 15 minutes) to satisfy the system load demand and the spinning reserves requirements.

In the present RTED model, the time interval used is 15 minutes and a look-ahead horizon of one hour is adopted. In addition, in this stage, no commitment decisions are taken, since the commitment status of all generating units for each dispatch interval has already been determined by the most recent RDAS and IDS models solutions, according to the UC formulations already presented in the Sections 4.3.2.1.3 and 4.3.2.1.4. Therefore, the RTED model comprises all constraints of the UC formulation with all binary variables corresponding to the commitment decisions being fixed to 0 (offline) or 1 (online), accordingly. The present formulation also includes full AC network representation with equations constraining the branch apparent power and voltage magnitude, turning the optimization problem into a non-linear problem owning to the physics of the electric network.

The proposed RTED with AC load flow formulation is suitable for meshed transmission networks and radial distribution networks with fixed network configuration (it does not optimize distribution network configuration).

For the reader's convenience, only the symbols not defined yet in the previous UC models are defined next and the additional network constraints follow.

4.3.2.1.3.1 Additional Nomenclature   Additional nomenclature includes the following parameters and continuous variables:

Parameters

b_b	Series susceptance of branch b, in pu
b_b^{sh}	Total shunt susceptance of branch b, in pu
b_n	Shunt susceptance of node n, in pu
g_b	Series conductance of branch b, in pu
g_b^{sh}	Total shunt conductance of branch b, in pu
g_n	Shunt conductance of node n, in pu
i_{nt}^p	Active component of constant current load of node n at time interval t, in pu
i_{nt}^q	Reactive component of constant current load of node n at time interval t, in pu
Q_i^{max}, Q_i^{min}	Reactive power limits of generator i, in MVAr
Q_{nt}	Reactive component of constant power load in node n and time interval t, in MVAr
S_b^{max}	Apparent power rating of branch b, in MVA
S_{base}	Base value for power, in MVA
V_n^{max}, V_n^{min}	Voltage magnitude limits of node n, in pu
$\Theta_n^{max}, \Theta_n^{min}$	Voltage angle limits of node n, in pu

Continuous variables

p_{bt}^+ Active power flow of branch b at *from bus* side, in time interval t, in pu

p_{bt}^- Active power flow of branch b at to bus side, in time interval t, in MW

q_{bt}^+ Reactive power flow of branch b at from bus side, in time interval t, in pu

q_{bt}^- Reactive power flow of branch b at *to bus* side, in time interval t, in MVAr

q_{it} Reactive power output of generator i and time interval t, in MVAr

v_{nt} Voltage magnitude of node n, at time interval t, in pu

θ_{nt} Voltage angle of node n, at time interval t, in rad

4.3.2.1.3.2 Mathematical Formulation—Additional Network Constraints The overall RTED problem with the AC network representation is a deterministic nonlinear optimization problem comprising the single objective function (Equation 4.71) to be minimized subject to specific groups of constraints, further described as follows:

$$
\text{Min}\left\{
\begin{array}{l}
\displaystyle\sum_{t\in T}\sum_{i\in I}\left[\left(\sum_{f\in F}C_{if}\cdot e_{ift}+\text{NLC}_i\cdot\left(\bar{u}_{it}-\bar{u}_{it}^{-\text{syn}}\right)+\sum_{m\in M}C_{it}^m\cdot r_{it}^m\right)\cdot\left(\frac{T_{\text{step}}}{60}\right)\right. \\[4mm]
\displaystyle\left.+\sum_{\ell\in L}\text{SUC}_i^\ell\cdot\bar{y}_{it}^{-\ell}+\text{SDC}_i\cdot\bar{z}_{it}\right]+\left(\frac{T_{\text{step}}}{60}\right)\cdot\sum_{t\in T}\text{VLL}_t\cdot d_t^{\text{LNS}}
\end{array}
\right\}
\tag{4.71}
$$

The above equation is subject to constraints 4.13 through 4.42, 4.44, 4.47, and 4.48 through 4.58 (with fixed binary variables) and constraints 4.72 through 4.83 representing the full network constraints to be included in the optimization model.

In Equation 4.71, the binary variables $\bar{u}_{it},\bar{u}_{it}^{\text{syn}},\bar{y}_{it},\bar{z}_{it}$ are fixed to 0 or 1, according to the respective commitment decisions already determined by the solution of the most recent RDAS and IDS models.

Active and reactive branch power flow at *from bus* side

$$
p_{bt}^+=\left(g_b+g_b^{\text{sh}}\right)\cdot\sum_{n\in N}A_{bn}^+\cdot v_{nt}^2
$$

$$
-\sum_{n\in N}A_{bn}^+\cdot v_{nt}\sum_{n\in N}A_{bn}^-\cdot v_{nt}\left[
\begin{array}{l}
g_b\cdot\cos\left(\displaystyle\sum_{n\in N}A_{bn}^+\cdot\theta_{nt}-\sum_{n\in N}A_{bn}^-\cdot\theta_{nt}\right) \\[4mm]
+b_b\cdot\sin\left(\displaystyle\sum_{n\in N}A_{bn}^+\cdot\theta_{nt}-\sum_{n\in N}A_{bn}^-\cdot\theta_{nt}\right)
\end{array}
\right]
\tag{4.72}
$$

$$
\forall b\in B,\, t\in T
$$

$$q_{bt}^+ = -\left(b_b + b_b^{sh}\right) \cdot \sum_{n \in N} A_{bn}^+ \cdot v_{nt}^2$$

$$-\sum_{n \in N} A_{bn}^+ \cdot v_{nt} \sum_{n \in N} A_{bn}^- \cdot v_{nt} \left[\begin{matrix} g_b \cdot \sin\left(\sum_{n \in N} A_{bn}^+ \cdot \theta_{nt} - \sum_{n \in N} A_{bn}^- \cdot \theta_{nt} \right) \\ - b_b \cdot \cos\left(\sum_{n \in N} A_{bn}^+ \cdot \theta_{nt} - \sum_{n \in N} A_{bn}^- \cdot \theta_{nt} \right) \end{matrix} \right] \quad (4.73)$$

$$\forall b \in B, t \in T$$

Equations 4.72 and 4.73 represent the active and reactive power flow at the *from bus* side of branch b at time interval t as a function of the voltages and angles of the *from bus* and *to bus* sides.

Active and reactive branch power flow at *to bus* side

$$p_{bt}^- = \left(g_b + g_b^{sh}\right) \cdot \sum_{n \in N} A_{bn}^- \cdot v_{nt}^2$$

$$-\sum_{n \in N} A_{bn}^- \cdot v_{nt} \sum_{n \in N} A_{bn}^+ \cdot v_{nt} \left[\begin{matrix} g_b \cdot \cos\left(\sum_{n \in N} A_{bn}^- \cdot \theta_{nt} - \sum_{n \in N} A_{bn}^+ \cdot \theta_{nt} \right) \\ + b_b \cdot \sin\left(\sum_{n \in N} A_{bn}^- \cdot \theta_{nt} - \sum_{n \in N} A_{bn}^+ \cdot \theta_{nt} \right) \end{matrix} \right] \quad (4.74)$$

$$\forall b \in B, t \in T$$

$$q_{bt}^- = -\left(b_b + b_b^{sh}\right) \cdot \sum_{n \in N} A_{bn}^- \cdot v_{nt}^2$$

$$-\sum_{n \in N} A_{bn}^- \cdot v_{nt} \sum_{n \in N} A_{bn}^+ \cdot v_{nt} \left[\begin{matrix} g_b \cdot \sin\left(\sum_{n \in N} A_{bn}^- \cdot \theta_{nt} - \sum_{n \in N} A_{bn}^+ \cdot \theta_{nt} \right) \\ - b_b \cdot \cos\left(\sum_{n \in N} A_{bn}^- \cdot \theta_{nt} - \sum_{n \in N} A_{bn}^+ \cdot \theta_{nt} \right) \end{matrix} \right] \quad (4.75)$$

$$\forall b \in B, t \in T$$

Equations 4.74 and 4.75 represent the active and reactive power flow at the *to bus* side of branch b at time interval t, as a function of the voltages and angles of the *from bus* and *to bus* sides.

Active and reactive node power injection

$$\sum_{i \in I} A_i^n \cdot p_{it} + \sum_{j \in J} A_j^n \cdot p_{jt} - D_{nt}$$

$$= S_{base} \cdot \left(v_{nt}^2 \cdot g_n + v_{nt} \cdot i_{nt}^p + \sum_{b \in B} A_{bn}^+ \cdot p_{bt}^+ + \sum_{b \in B} A_{bn}^- \cdot p_{bt}^- \right) \quad \forall n \in N, t \in T \quad (4.76)$$

$$\sum_{i \in I} A_i^n \cdot q_{it} + \sum_{j \in J} A_j^n \cdot q_{jt} - Q_{nt}$$

$$= S_{\text{base}} \cdot \left(v_{nt}^2 \cdot b_n + v_{nt} \cdot i_{nt}^q + \sum_{b \in B} A_{bn}^+ \cdot q_{bt}^+ + \sum_{b \in B} A_{bn}^- \cdot q_{bt}^- \right) \qquad \forall n \in N, t \in T$$

(4.77)

Equations 4.76 and 4.77 represent the active and reactive power injection in node *n*, at time interval *t*, as the sum of the power flows of the branches connected to the node, as well as the power of the shunt admittance, and constant current load connected to the node.

Branch apparent power constraints

$$\left[S_{\text{base}} \cdot \left(p_{bt}^+ + q_{bt}^+ \right) \right]^2 \le \left(S_b^{\text{max}} \right)^2 \qquad \forall b \in B, t \in T \tag{4.78}$$

$$\left[S_{\text{base}} \cdot \left(p_{bt}^- + q_{bt}^- \right) \right]^2 \le \left(S_b^{\text{max}} \right)^2 \qquad \forall b \in B, t \in T \tag{4.79}$$

Equations 4.78 and 4.79 constrain the apparent power flow of branch *b*, at time interval *t*, at the *from bus* and *to bus* sides, respectively.

Voltage-angle constraints

$$V_n^{\text{min}} \le v_{nt} \le V_n^{\text{max}} \qquad \forall n \in N, t \in T \tag{4.80}$$

$$\Theta_n^{\text{min}} \le \theta_{nt} \le \Theta_n^{\text{max}} \qquad \forall n \in N, t \in T \tag{4.81}$$

$$\theta_{nt} = 0 \quad n = \text{Slack}, \quad t \in T \tag{4.82}$$

Equations 4.80 and 4.81 limits the voltage magnitude and angle within permissible bounds. Equation 4.82 fixes the slack node angle to zero.

Generator reactive power constraints

$$Q_i^{\text{min}} \cdot \bar{u}_{it} \le q_{it} \le Q_i^{\text{max}} \cdot \bar{u}_{it} \qquad \forall i \in I, t \in T \tag{4.83}$$

Equation 4.83 limits the reactive power of generating unit *i*, at time interval *t*, to its technical limits, only when generating unit *i* is online.

4.3.2.2 Unified UCED Model

4.3.2.2.1 Introduction

An alternative approach for the short-term operation of insular power systems is the unified UCED (UUCED) model, presented in [78]. The variable time resolution UUCED model is an integrated tool that can smoothly bridge the short-term scheduling with real-time decisions and better accommodate the uncertain nature of renewable generation. The model unifies the DAS, IDS, and RTD functions into a single tool that uses a scheduling horizon of up to 36 hours. The model uses variable time resolution: the first hours are modeled with finer time resolution (i.e., 15- and 30-minute intervals), whereas coarser time resolution (i.e., 1-hour intervals) is adopted for the following hours of the scheduling horizon. This approach is adopted since wind forecasts with shorter lead times tend to be more accurate. Consequently, it is crucial to use frequently updated wind forecasts in a powerful real-time tool. The commitment and dispatch decisions of the first scheduling interval are binding, while the remaining schedule is advisory. This unified approach increases the flexibility of the generation fleet by allowing unit recommitment and re-dispatch for the entire scheduling horizon. The binding dispatch and commitment decisions of the first intervals are very robust, since they are taken in anticipation of the system conditions for an extended scheduling period.

4.3.2.2.2 Length of the Scheduling Horizon

The scheduling horizon length varies from 12 to 36 hours ahead, depending on the starting point during the day, as shown in Figure 4.26. The horizon length is set so that it respects the conventional DAS timeline. The length could increase/decrease depending on the specific generation system needs. The length adopted for the test case in Section 4.3.3 is sufficient for the generation fleet characteristics of Crete island, where the start-up time of the slowest conventional unit does not exceed 10 hours.

4.3.2.2.3 Variable Time Resolution

Adopting finer time resolution for real-time functions is crucial, since load and wind forecasts are more accurate for short lead times and, therefore, a lower amount of reserves can be scheduled for the intrastep variations. However, using fine resolution over the entire scheduling horizon (i.e., 36 hours) would cumber the optimization with unnecessary computational burden, since wind forecasts for longer lead times tend to be rather inaccurate. These opposing needs lead to a variable time resolution compromise. The basic idea of modeling the variable time resolution is to use a 15-minute time step for the first scheduling hour, a 30-minute time step for the second hour, and an hourly time step for the remaining scheduling horizon. In order to align the variable time intervals with clock hours, in case the scheduling does not begin at an exact clock hour, the following rules apply: (1) use a 15-minute time step for at least 1 hour, followed by a 30-minute time step for at least one hour and an hourly time step thereafter; (2) align intervals of a specific duration with intervals of the immediately longer duration; and (3) span the scheduling horizon with the minimum number of intervals. Table 4.6 illustrates all possible combinations of the variable time resolution horizon depending on the specific 15-minute interval of hour *h* that the

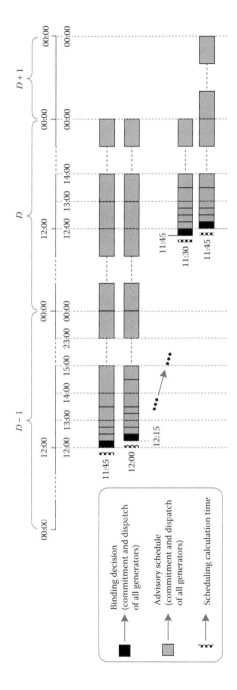

FIGURE 4.26
Timeline of the UUCED operation model.

TABLE 4.6

Definition of the Variable Time Intervals

	h					h+1		h+2	h+3	h+4	...
	h:00	h:15	h:30	h:45							
1	15	15	15	15	30		30	60	60	60	60
2		15	15	15	15	15	30	30	30	60	60	60
3			15	15	15	15	30	30	30	60	60	60
4				15	15	15	15	15	30	30	60	60	60

scheduling begins. It is noted that the selected time granularity is system specific.

In order to modify the scheduling horizon from a constant time framework to a variable time framework, the time constants that are usually used in a typical UC model (e.g., unit synchronization and soak time and minimum up/down time) need to be converted from minutes (or hours) to variable time intervals. The detailed formulation has already been analyzed in [78], but, for the reader's convenience, an illustrative example is presented next.

The minimum up time (MUT) of a single generating unit can be simulated in a constant time resolution model with the constraint in Equation 4.84.

$$u_t \geq \sum_{\tau=t-\tilde{T}^{up}+1}^{t} y_\tau \tag{4.84}$$

where:

$\tilde{T}^{up}$ is the value of the unit MUT expressed in hours

y_t is a binary variable which is equal to 1 if the unit starts-up during hour t

u_t is a binary variable equal to 1 if the unit is online during hour t and 0 otherwise.

An MUT of 4 hours is equal to four time intervals in a model that uses hourly granularity and equal to 20 time intervals in a model that uses 15-minute granularity. However, in a model with variable time resolution, the value of the MUT expressed in time intervals depends on the time interval t at which it terminates, since a backward-looking modeling of the intertemporal constraints is adopted [78]. Figure 4.27 illustrates that an MUT of 4 hours is equal to four time intervals for $t = 13$, five time intervals for $t = 12$, and seven time intervals for $t = 11$.

Consequently, the MUT of each generating unit has to be calculated separately at each time interval of the scheduling horizon and constraint in Equation 4.84 must be transformed to Equation 4.85:

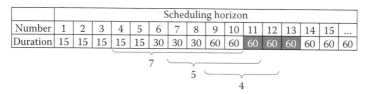

FIGURE 4.27
Example that shows how a minimum up time of 4 hours is converted to be used in a variable time resolution modeling.

$$u_t \geq \sum_{\tau=t-T_t^{up}+1}^{t} y_\tau \qquad (4.85)$$

It is evident that the MUT is now indexed on time intervals and, in general, it depends on its terminal time interval. It is important to follow the above transformation procedure for all time constants and the respective constraints of the optimization model. The readers may refer [78] for further details on the relevant techniques employed.

4.3.2.2.4 Model Complexity

The UUCED model incorporates all the equations presented in Section 4.3.2.1 after the necessary parameter conversion, as described in Section 4.3.2.2.3 and in more detail in [78].

4.3.2.2.5 Pros and Cons

The advantage of using the variable time resolution UUCED model is that we avoid the complex interaction between different scheduling models, as presented in Section 4.3.2.1.

The main drawback of the UUCED model is its computational tractability. The implementation of this model in practice requires the solution of an MILP optimization problem with look-ahead horizon of up to 36 hours (divided into subintervals of variable length) within the time window of the real-time interval (e.g., within 5–15 minutes). However, it is perfectly applicable for small insular systems with a few tens of generators and buses.

4.3.2.3 Stochastic Scheduling Models

4.3.2.3.1 Introduction

As already mentioned, in the last decade, the integration of variable RES injection led the industry and the academic community to the implementation of sophisticated mathematical models in order to deal with the uncertainty of the intermittent energy sources.

Stochastic programming has been proposed as a viable approach for solving the uncertain UC problem. Stochastic programming is the mathematical field that aims at taking optimal decisions in problems that involve uncertain data [80]. Contributions from many disciplines, including operations research, mathematics, and probability theory, have led to its implementation in various problems, ranging from financial planning to agriculture and computer network applications and, of course, to energy-related problems.

Many researchers have utilized two-stage stochastic programming methodologies in order to solve the uncertain UC problem [9,10,65–67]. In the existing literature, two variations for the mathematical formulation of the stochastic UC problem are currently used.

In the variable-node formulation, the problem variables are associated with decision points of the stochastic optimization problem [24]. Under this formulation, the first-stage variables (scenario-independent) and the second-stage variables (scenario-dependent) are linked together through coupling constraints. As a result, in the stochastic UC problem, the commitment decisions taken in the first-stage take explicitly into consideration the realization of uncertainty in all scenarios of the second-stage. The variable-node formulation is a compact formulation that is suitable for a direct solution approach [67].

In the scenario-based formulation, the binary and continuous decisions are all scenario-dependent variables, meaning that the commitment status and the power output/reserve contribution of each generating unit are calculated for each scenario. The various scenarios are linked together through the well-known *nonanticipativity constraints*, which enforce specific decision variables to be common in all scenarios, for example, the commitment status of the slow generating units. This formulation requires more decision variables and equations than the node-variable formulation, but its structure allows for its efficient exploitation by various decomposition algorithms [10].

The application of the scenario-based stochastic programming approach to the classical deterministic UC problem, as already discussed in Section 4.3.2.1, is presented in the following paragraphs.

4.3.2.3.2 Nomenclature

It should be noted that most symbols used in the equations of the stochastic programming model have been already defined in Section 4.3.2.1.2 and, for the reader's convenience, are not repeated here. Only symbols defined for the first time and scenario-dependent symbols are presented.

Sets

$i \in I^s$ Slow-start conventional generating units ($I^s \subseteq I$)

$s \in S$ Scenarios

Parameters

D_{nts} Load demand in node n, in time interval t, and scenario s, in MW

D_{ts} System load demand in time interval t and scenario s, in MW

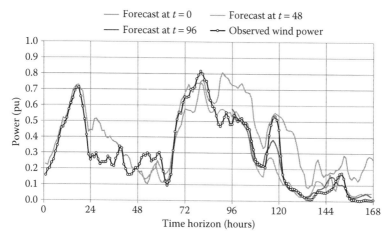

FIGURE 2.1
Influence of the forecast refreshments along time. The forecasts are provided at 00 UTC and refreshed after 48 and 96 hours.

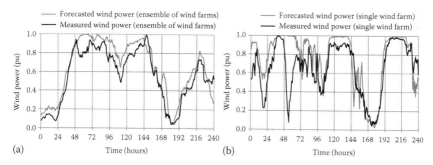

(a)

(b)

FIGURE 2.2
Example of forecast (day ahead) and observed wind power of a group of wind farms (a) and a single wind farm (b) for the same period.

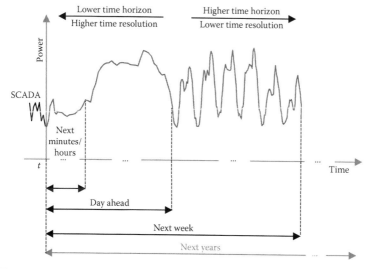

FIGURE 2.4
Scheduling diagram over the period t up to a week.

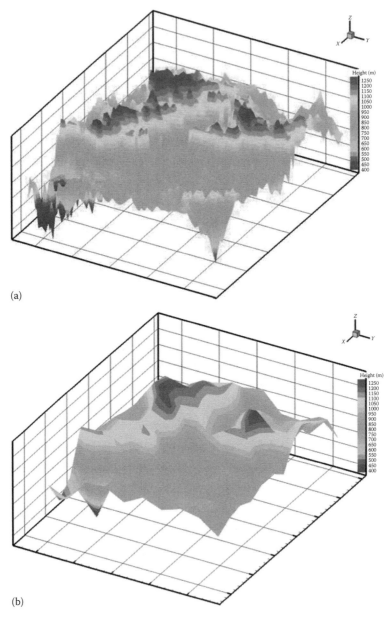

(a)

(b)

FIGURE 2.5
Digital elevation model of the terrain surface with spatial resolutions of 5 km (a) and 25 km (b).

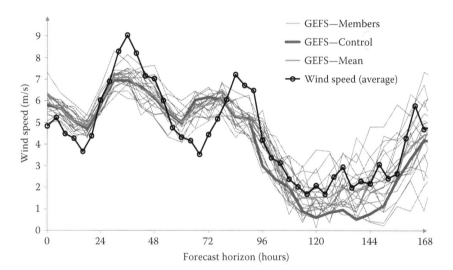

FIGURE 2.6
Example of a forecast for the next 168 hours of wind speed at a height of 10 meters and mean measured value.

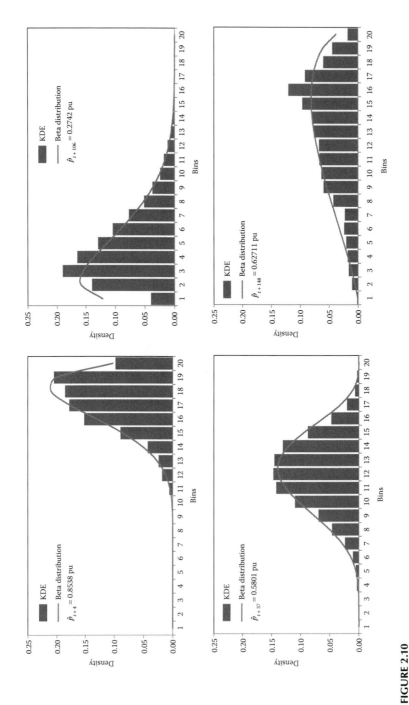

FIGURE 2.10

A comparison between KDE and beta distribution obtained.

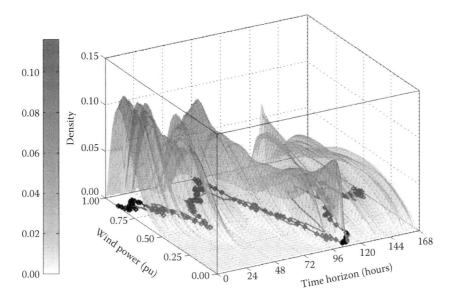

FIGURE 2.11
Representation of the obtained probability density function of each step ahead of the forecast (for 168 hours ahead). The black dotted line represents the observed wind power and the red line the deterministic forecast.

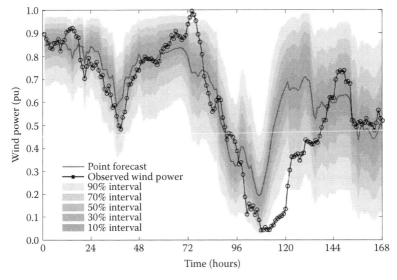

FIGURE 2.12
Example of a point and a probabilistic wind power forecast for the next 168 hours for prediction intervals of 10%, 30%, 50%, 70%, and 90%.

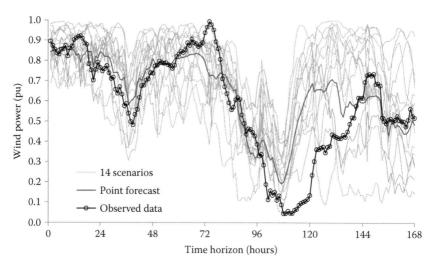

FIGURE 2.13

Wind power forecast scenarios calculated based on the methodology presented in the literature. (Adapted from P. Pinson et al., *Wind Energy*, 2008.)

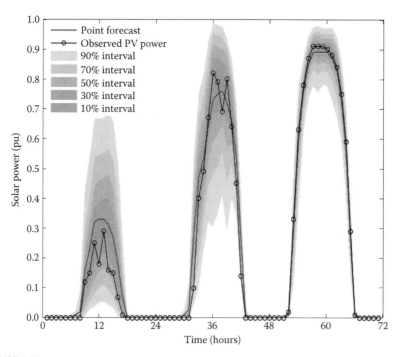

FIGURE 2.15

Example of a probabilistic PV power forecast. Each forecast is provided at 00 UTC for the next 24 hours.

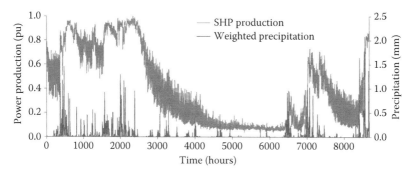

FIGURE 2.16
Temporal evolution of the power production of the aggregated SHPs and the average precipitation.

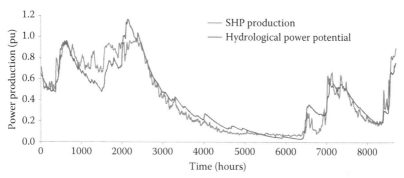

FIGURE 2.17
Temporal evolution of the daily moving averaged power production of the aggregated SHPs and the corresponding calculated HPP.

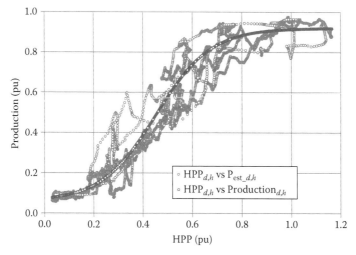

FIGURE 2.18
Relation between HPP with the parameterized sigmoid and the measured values (average).

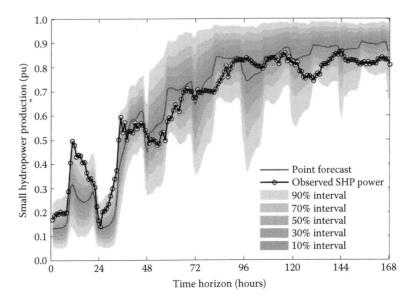

FIGURE 2.19
Prediction intervals of a probabilistic forecast of hydro production obtained from an ensemble of small hydropower plants provided at 00 UTC for the next 168 hours.

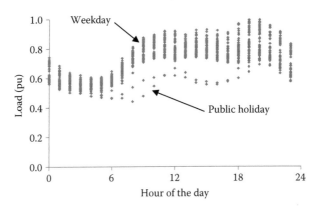

FIGURE 2.20
Relationship between electric load consumption (pu) and time of the day in hours for a weekday (Wednesday).

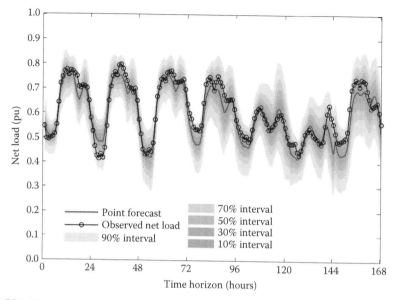

FIGURE 2.25
Representation of the net load uncertainty by a set of intervals forecast, based on methodology explained in Figure 2.24.

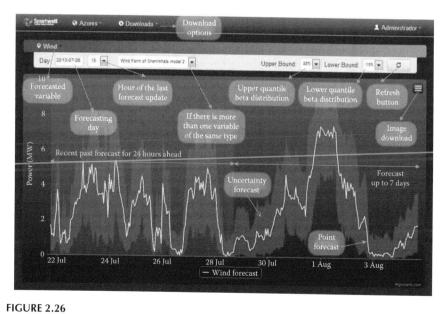

FIGURE 2.26
Options and information provided by the forecasting platform developed for the wind farm of Graminhais in São Miguel Islands in the Azores.

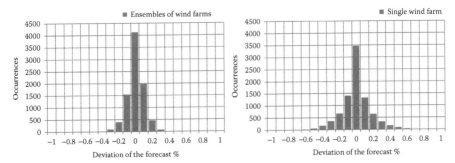

FIGURE 2.29
Scheduling solutions provided by: (a) the methodology developed and (b) system operator (SO).

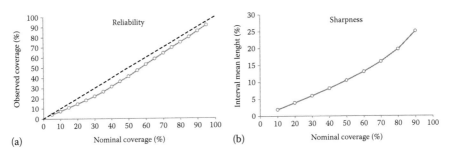

(a)

(b)

FIGURE 2.30
Reliability and sharpness of probabilistic net load forecast.

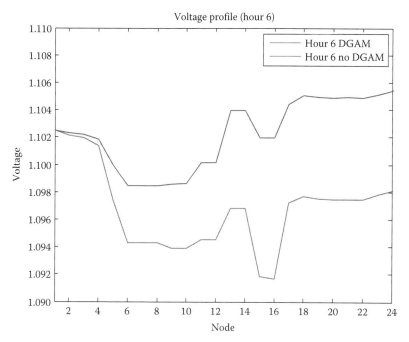

FIGURE 6.18
Impact of DGAM on voltage profile during valley hours.

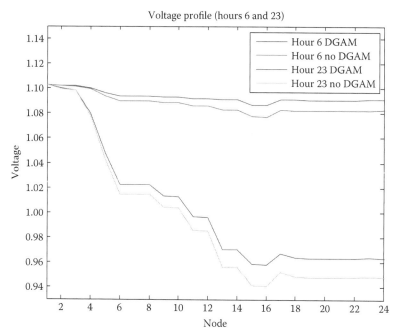

FIGURE 6.31
Impact of DGAM on voltage profile during peak and valley hours.

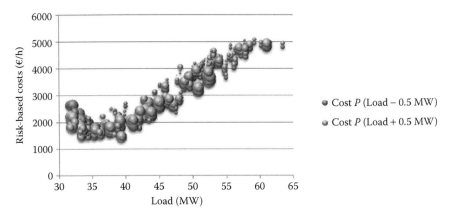

FIGURE 6.32
Cost P (Load − 0.5 MW) and Cost P (Load + 0.5 MW) charts (the higher sized points represent higher wind generation).

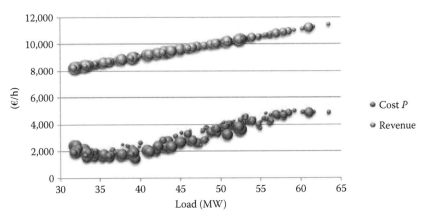

FIGURE 6.33
Cost and revenue per load (the large-sized points represent higher wind generation).

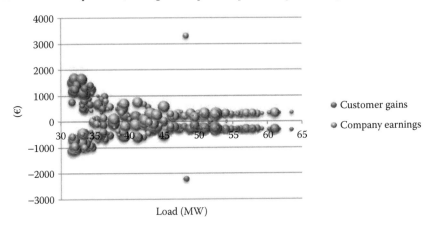

FIGURE 6.38
Company earnings and customer gains per load chart (the higher sized points represent higher wind generation).

P_{js}^{forecast}	Forecast power output of RES unit j in scenario s, in MW
VLL_{nt}	Value of lost load in node n and time interval t in €/MW
VLL_t	Value of lost load in time interval t in €/MW
π_s	Probability of scenario s

Continuous variables

d_{nts}^{LNS}	Load not served in node n, time interval t, and scenario s, in MW
d_{ts}^{LNS}	Load not served in time interval t and scenario s, in MW
e_{ifts}	Portion of step f of the ith unit's marginal cost function loaded in time interval t and in scenario s, in MW.
p_{its}	Power output of thermal unit i dispatched by the ISO in time interval t and scenario s, in MW
p_{bts}^{f}	Power flow in branch b in time interval t and scenario s, in MW
p_{its}^{des}	Power output of thermal unit i during the desynchronization phase in time interval t and scenario s, in MW
p_{its}^{soak}	Power output of thermal unit i during the soak phase in time interval t and scenario s, in MW
p_{jts}	Power output of RES unit in time interval t and scenario s, in MW
p_{nts}^{inj}	Total power injection in node n, in time interval t, and scenario s, in MW
r_{its}^{m}	Contribution of thermal unit i in reserve type m in time interval t and scenario s, in MW
θ_{nts}	Voltage angle of node n, in time interval t, and scenario s, in rad

Binary variables

u_{it}	Binary variable which is equal to 1 if slow-start thermal unit i is committed during time interval t
u_{its}	Binary variable which is equal to 1 if thermal unit i is committed in time interval t and scenario s
u_{its}^{3NS}	Binary variable which is equal to 1 if thermal unit i provides tertiary nonspinning reserve in time interval t and scenario s
u_{its}^{AGC}	Binary variable which is equal to 1 if thermal unit i operates under AGC and provides secondary reserve in time interval t and scenario s
u_{its}^{n}	Binary variable which is equal to 1 if thermal unit i is in operating phase n in time interval t and scenario s, where n = syn: synchronization, n = soak: soak, n = disp: dispatchable, n = des: desynchronization

u_{jts} Binary variable which is equal to 1 if RES unit j is committed in time interval t and scenario s

y_{it} Binary variable which is equal to 1 if slow-start thermal unit i is started-up during time interval t

y_{its} Binary variable which is equal to 1 if thermal unit i is started-up in time interval t and scenario s

y_{its}^{ℓ} Binary variable which is equal to 1 if a type-ℓ start-up of thermal unit i is initiated in time interval t and scenario s

z_{it} Binary variable which is equal to 1 if slow-start thermal unit i is shutdown during time interval t

z_{its} Binary variable which is equal to 1 if thermal unit i is shutdown in time interval t and scenario s

4.3.2.3.3 Stochastic UC Model without Network Representation

In this section, the stochastic scenario-based UC model is mathematically formulated, comprising only the system and the unit operating constraints. The incorporation of the transmission network representation under a DC power flow model is analytically described in Section 4.3.2.3.4.

In general, the problem functions and constraints follow those of Section 4.3.2.1.3, with the main difference that they are now defined for each scenario. For this reason, an additional explanation is given only when it is required, since they all have already been explained previously.

All continuous power output (i.e., e_{ifts}, p_{its}, p_{its}^{des}, p_{its}^{soak}, and p_{jts}) and reserve contribution (i.e., r_{its}^m) decision variables, along with all binary commitment decision variables (i.e., u_{its}, u_{its}^{AGC}, u_{its}^{syn}, u_{its}^{soak}, u_{its}^{disp}, u_{its}^{des}, u_{jts}, y_{its}^{ℓ}, y_{its}, and z_{its}), for each unit i are now scenario-dependent. Additional parameters are also defined, namely, the scenario-based load demand D_{nts} and the RES forecast power output $P_{js}^{forecast}$, along with the probability π_s of scenario s.

Objective function

The objective function of the day-ahead scheduling problem is now formulated as follows:

$$\text{Min} \sum_{s \in S} \pi_s \cdot \sum_{t \in T} \left[\sum_{i \in I} \left\{ \left[\sum_{f \in F} C_{if} \cdot e_{ifts} + \text{NLC}_i \cdot \left(u_{its} - u_{its}^{syn} \right) + \sum_{m \in M} C_{it}^m \cdot r_{its}^m \right] \cdot \left(\frac{T_{step}}{60} \right) + \sum_{\ell \in L} \text{SUC}_i^{\ell} \cdot y_{its}^{\ell} + \text{SDC}_i \cdot z_{its} \right\} + \left(\frac{T_{step}}{60} \right) \cdot \text{VLL}_t \cdot d_{ts}^{LNS} \right] \quad (4.86)$$

The objective function in Equation 4.86 is almost identical to that of the deterministic UC problem and comprises the sum of the total operating cost of each scenario, multiplied by the probability for the realization of the respective scenario.

Constraints 4.87 through 4.130 enforce the unit operating and the system constraints of the stochastic optimization problem.

4.3.2.3.3.1 Operating Constraints of a Thermal Unit   Operating constraints of a thermal unit may be classified as follows:

Start-up type constraints

$$y_{its}^{\ell} \leq \sum_{\tau=t-\bar{T}_i^{\ell}+1}^{t-\bar{T}_i^{\ell}} z_{i\tau s} \qquad \forall i \in I, \, t \in T, \, s \in S, \, \ell \in L, \quad \text{where } \tau \in T^- \qquad (4.87)$$

$$y_{its} = \sum_{\ell \in L} y_{its}^{\ell} \qquad \forall i \in I, \, t \in T, \, s \in S \qquad (4.88)$$

Synchronization phase constraints

$$u_{its}^{\text{syn},\ell} = \sum_{\tau=t-T_i^{\text{syn},\ell}+1}^{t} y_{i\tau s}^{\ell} \qquad \forall i \in I, \, t \in T, \, s \in S, \, \ell \in L, \quad \text{where } \tau \in T^- \qquad (4.89)$$

$$u_{its}^{\text{syn}} = \sum_{\ell \in L} u_{its}^{\text{syn},\ell} \qquad \forall i \in I, \, t \in T, \, s \in S \qquad (4.90)$$

Soak phase constraints

$$u_{its}^{\text{soak},\ell} = \sum_{\tau=t-T_i^{\text{syn},\ell}-T_i^{\text{soak},\ell}+1}^{t-T_i^{\text{syn},\ell}} y_{i\tau s}^{\ell} \qquad \forall i \in I, \, t \in T, \, s \in S, \, \ell \in L, \quad \text{where } \tau \in T^- \qquad (4.91)$$

$$u_{its}^{\text{soak}} = \sum_{\ell \in L} u_{its}^{\text{soak},\ell} \qquad \forall i \in I, \, t \in T, \, s \in S \qquad (4.92)$$

Desynchronization phase constraints

$$u_{its}^{\text{des}} = \sum_{\tau=t+1}^{t+T_i^{\text{des}}} z_{i\tau s} \qquad \forall \, i \in I, \, t \in T, \, s \in S, \quad \text{where } \tau \in T^+ \qquad (4.93)$$

$$p_{its}^{\text{des}} = \sum_{\tau=t+1}^{t+T_i^{\text{des}}} z_{i\tau s} \cdot (\tau - t) \cdot \frac{P_i^{\min}}{T_i^{\text{des}}} \qquad \forall \, i \in I, \, t \in T, \, s \in S, \quad \text{where } \tau \in T^+ \qquad (4.94)$$

Minimum up-/down-time constraints

$$\sum_{\tau=t-UT_i+1}^{t} y_{i\tau s} \leq u_{its} \qquad \forall i \in I, t \in T, s \in S, \quad \text{where } \tau \in T^- \qquad (4.95)$$

$$\sum_{\tau=t-DT_i+1}^{t} z_{i\tau s} \leq 1 - u_{its} \qquad \forall i \in I, t \in T, s \in S, \quad \text{where } \tau \in T^- \qquad (4.96)$$

Logical status of commitment

$$u_{its} = u_{its}^{\text{syn}} + u_{its}^{\text{soak}} + u_{its}^{\text{disp}} + u_{its}^{\text{des}} \qquad \forall i \in I, t \in T, s \in S \qquad (4.97)$$

$$y_{its} - z_{its} = u_{its} - u_{i(t-1)s} \qquad \forall i \in I, t \in T, s \in S \qquad (4.98)$$

$$y_{its} + z_{its} \leq 1 \qquad \forall i \in I, t \in T, s \in S \qquad (4.99)$$

Ramp-up/-down constraints

$$p_{its} - p_{i(t-1)s} \leq RU_i \cdot T_{\text{step}} + P_i^{\min} \cdot \left(u_{its}^{\text{syn}} + u_{its}^{\text{soak}}\right) \qquad \forall i \in I, t \in T, s \in S \quad (4.100)$$

$$p_{i(t-1)s} - p_{its} \leq RD_i \cdot T_{\text{step}} + P_i^{\min} \cdot u_{i(t-1)s}^{\text{des}} \qquad \forall i \in I, t \in T, s \in S \qquad (4.101)$$

Power output constraints

$$u_{its}^{\text{AGC}} \leq u_{its}^{\text{disp}} \qquad \forall i \in I, t \in T, s \in S \qquad (4.102)$$

$$0 \leq r_{its}^{1+} \leq R_i^1 \cdot u_{its}^{\text{disp}} \qquad \forall i \in I, t \in T, s \in S \qquad (4.103)$$

$$0 \leq r_{its}^{1-} \leq R_i^1 \cdot u_{its}^{\text{disp}} \qquad \forall i \in I, t \in T, s \in S \qquad (4.104)$$

$$0 \leq r_{its}^{2+} \leq 15 \cdot RU_i^{\text{AGC}} \cdot u_{its}^{\text{AGC}} \qquad \forall i \in I, t \in T, s \in S \qquad (4.105)$$

$$0 \leq r_{its}^{2-} \leq 15 \cdot RD_i^{\text{AGC}} \cdot u_{its}^{\text{AGC}} \qquad \forall i \in I, t \in T, s \in S \qquad (4.106)$$

$$0 \leq r_{its}^{3S} \leq R_i^{3S} \cdot u_{its}^{\text{disp}} \qquad \forall i \in I, t \in T, s \in S \qquad (4.107)$$

$$r_{its}^{3NS} \leq R_i^{3NS} \cdot u_{its}^{3NS} \qquad \forall i \in I, t \in T, s \in S \qquad (4.108)$$

$$r_{its}^{3NS} \geq P_i^{\min} \cdot u_{its}^{3NS} \qquad \forall i \in I, t \in T, s \in S \qquad (4.109)$$

$$u_{its}^{3NS} \leq 1 - u_{its} \qquad \forall i \in I, t \in T, s \in S \qquad (4.110)$$

$$p_{its} - r_{its}^{2-} \geq 0 \cdot u_{its}^{\mathrm{syn}} + p_{its}^{\mathrm{soak}} + p_{its}^{\mathrm{des}} + P_i^{\mathrm{min}} \cdot \left(u_{its}^{\mathrm{disp}} - u_{its}^{\mathrm{AGC}}\right)$$
$$+ P_i^{\mathrm{min,AGC}} \cdot u_{its}^{\mathrm{AGC}} \qquad \forall i \in I, t \in T, s \in S \tag{4.111}$$

$$p_{its} + r_{its}^{2+} \leq 0 \cdot u_{its}^{\mathrm{syn}} + p_{its}^{\mathrm{soak}} + p_{its}^{\mathrm{des}} + P_i^{\mathrm{max}} \cdot \left(u_{its}^{\mathrm{disp}} - u_{its}^{\mathrm{AGC}}\right)$$
$$+ P_i^{\mathrm{max,AGC}} \cdot u_{its}^{\mathrm{AGC}} \qquad \forall i \in I, t \in T, s \in S \tag{4.112}$$

$$p_{its} - r_{its}^{1-} - r_{its}^{2-} \geq 0 \cdot u_{its}^{\mathrm{syn}} + p_{its}^{\mathrm{soak}} + p_{its}^{\mathrm{des}} + P_i^{\mathrm{min}} \cdot u_{its}^{\mathrm{disp}}$$
$$\forall i \in I, t \in T, s \in S \tag{4.113}$$

$$p_{its} + r_{its}^{1} + r_{its}^{2+} + r_{its}^{3S} \leq 0 \cdot u_{its}^{\mathrm{syn}} + p_{its}^{\mathrm{soak}} + p_{its}^{\mathrm{des}} + P_i^{\mathrm{max}} \cdot u_{its}^{\mathrm{disp}}$$
$$\forall i \in I, t \in T, s \in S \tag{4.114}$$

Total production constraints of a thermal unit

$$\sum_{f \in F^i} e_{ifts} = p_{its} \qquad \forall i \in I, t \in T, s \in S \tag{4.115}$$

$$\mathrm{mc}_{i(f+1)ts} \cdot E_{if} \leq e_{ifts} \leq \mathrm{mc}_{ifts} \cdot E_{if} \qquad \forall i \in I, f \in F^i, t \in T, s \in S,$$
$$\text{where } \mathrm{mc}_{ifts} \in \{0,1\}^{ifts} \tag{4.116}$$

$$\mathrm{mc}_{i(f+1)ts} \leq \mathrm{mc}_{ifts} \qquad \forall i \in I, f \in F^i, t \in T, s \in S \tag{4.117}$$

$$0 \leq e_{ifts} \leq E_{if} \qquad \forall i \in I, f \in F^i, t \in T, s \in S \tag{4.118}$$

Operating constraints of RES plants

$$p_{jts} \leq P_{js}^{\mathrm{forecast}} \cdot u_{jts} \qquad \forall j \in J^{\mathrm{w}}, t \in T, s \in S \tag{4.119}$$

$$p_{jts} \leq P_{jt}^{\mathrm{SP}} \qquad \forall j \in J^{\mathrm{w}}, t \in T, s \in S \tag{4.120}$$

$$P_{jts}^{\mathrm{SP}} = \frac{P_j^{\mathrm{max}}}{\sum\limits_{j \in J^{\mathrm{w}}} P_j^{\mathrm{max}}} \cdot P_{ts}^{\mathrm{SP,tot}} \qquad \forall j \in J^{\mathrm{w}}, t \in T, s \in S \tag{4.121}$$

$$P_{ts}^{\mathrm{SP,tot}} \leq \lambda \cdot D_{ts} \qquad \forall t \in T, s \in S \tag{4.122}$$

$$P_{ts}^{\mathrm{SP,tot}} \leq D_{ts} - \sum_{i \in I}\left[P_i^{\mathrm{min}} \cdot u_{its}^{\mathrm{disp}} + p_{its}^{\mathrm{soak}} + p_{its}^{\mathrm{des}} \right] \qquad \forall t \in T, s \in S \tag{4.123}$$

$$P_{ts}^{\mathrm{SP,tot}} = \sum_{j \in J^w} P_{jts}^{\mathrm{SP}} \qquad \forall t \in T, s \in S \tag{4.124}$$

System constraints

$$\sum_{i \in I} r_{its}^{1+} \geq RR_t^{1+} \qquad \forall t \in T, s \in S \tag{4.125}$$

$$\sum_{i \in I} r_{its}^{1-} \geq RR_t^{1-} \qquad \forall t \in T, s \in S \tag{4.126}$$

$$\sum_{i \in I} r_{its}^{2+} \geq RR_t^{2+} \qquad \forall t \in T, s \in S \tag{4.127}$$

$$\sum_{i \in I} r_{its}^{2-} \geq RR_t^{2-} \qquad \forall t \in T, s \in S \tag{4.128}$$

$$\sum_{i \in I} r_{its}^{3S} + \sum_{i \in I} r_{its}^{3NS} \geq RR_t^3 \qquad \forall t \in T, s \in S \tag{4.129}$$

$$\left(\frac{T_{\text{step}}}{60}\right) \cdot \left(\sum_{i \in I} p_{its} + \sum_{j \in J} p_{jts}\right) = \left(\frac{T_{\text{step}}}{60}\right) \cdot \left(D_{ts} - d_{ts}^{\text{LNS}}\right) \qquad \forall t \in T, s \in S \tag{4.130}$$

Nonanticipativity constraints

$$u_{its} = u_{it} \qquad \forall i \in I^s, t \in T, s \in S \tag{4.131}$$

$$y_{its} = y_{it} \qquad \forall i \in I^s, t \in T, s \in S \tag{4.132}$$

$$z_{its} = z_{it} \qquad \forall i \in I^s, t \in T, s \in S \tag{4.133}$$

Constraints 4.131 through 4.133 denote the nonanticipativity constraints for all slow-start units. These constraints practically denote that scenarios with a common history must have the same decisions. In this case, constraints 4.131 through 4.133 enforce that the commitment status of all slow-start thermal generating units must be identical in all scenarios s. It should be noted that variables u_{it}, y_{it}, and z_{it} are the first-stage variables, whereas all other variables are the second-stage variables of the two-stage stochastic programming problem with recourse.

4.3.2.3.4 *Stochastic UC Model with DC Power Flow*

In this section, the stochastic UC model with DC power flow constraints is briefly described. Similar to the network-less stochastic UC problem, the formulation presented follows the respective deterministic model with the addition of the nonanticipativity constraints 4.131 through 4.133. In this case, the power flow related variables p_{bts}^{f} and p_{nts}^{inj} are also defined for each scenario s.

As a result, the overall day-ahead stochastic scheduling problem with the DC network representation is formulated as follows:

$$\text{Min} \sum_{s \in S} \pi_s \cdot \sum_{t \in T} \left[\begin{matrix} \sum_{i \in I} \left[\begin{matrix} \left[\sum_{f \in F} C_{if} \cdot e_{ifts} + NLC_i \cdot \left(u_{its} - u_{its}^{syn} \right) + \sum_{m \in M} C_{it}^{m} \cdot r_{its}^{m} \cdot \right] \\ \left(\dfrac{T_{step}}{60} \right) + \sum_{\ell \in L} SUC_i^{\ell} \cdot y_{its}^{\ell} + SDC_i \cdot z_{its} \end{matrix} \right] \\ + \left(\dfrac{T_{step}}{60} \right) \cdot \sum_{n \in N} VLL_{nt} \cdot d_{nts}^{LNS} \end{matrix} \right] \tag{4.134}$$

subject to:

Constraints 4.87 through 4.114, 4.115, 4.118, 4.125 through 4.129, and 4.131 through 4.133

$$\left(\frac{T_{step}}{60} \right) \cdot \left(\sum_{n \in N} \sum_{i \in I} A_{ni} \cdot p_{its} + \sum_{n \in N} \sum_{j \in J} A_{nj} \cdot p_{jts} \right) = \left(\frac{T_{step}}{60} \right) \cdot \sum_{n \in N} \left(D_{nts} - d_{nts}^{LNS} \right) \tag{4.135}$$

$$\forall t \in T, s \in S$$

$$p_{bts}^{f} = \sum_{n \in N} PTDF_{bn} \cdot p_{nts}^{inj} \qquad \forall b \in B, t \in T, s \in S \tag{4.136}$$

$$p_{nts}^{inj} = \sum_{i \in I} A_{ni} \cdot p_{its} + \sum_{j \in J} A_{nj} \cdot p_{jts} - \left(D_{nts} - d_{nts}^{LNS} \right) \qquad \forall n \in N, t \in T, s \in S \tag{4.137}$$

$$-F_b^{max} \le p_{nts}^{f} \le F_b^{max} \qquad \forall b \in B, t \in T, s \in S \tag{4.138}$$

where Equation 4.135 is the system-wide power balance equation, constraints 4.136 and 4.137 are the power flow equations with respect to the PTDF-based formulation, and, finally, Equation 4.138 enforces the power flow limit in branch b with regard to its thermal limit. Constraints are all defined for each scenario s and for each time period t of the scheduling horizon.

As it can be observed from the above equations, the stochastic UC problem with DC network constraints is identical to the respective deterministic problem described in Section 4.3.2.1.4, with the sole difference that all variables are now scenario-dependent and the nonanticipativity constraints 4.131 and 4.133 are also enforced to the overall optimization problem.

4.3.3 Case Study

The developed MILP models have been tested on the power system of Crete, which is a large insular system, currently comprising 25 conventional thermal units of different technologies with a total installed thermal capacity of 765 MW. In addition, there are 31 wind parks and around 2800 PV installations (990 ground-mounted and 1806 rooftop systems) with total installed capacities of 186 and 94 MW, respectively. Table 4.7 presents an overview of the generation mix of the insular power system of Crete at the end of 2013.

Figures 4.28 through 4.33 illustrate indicative results from the application of the scheduling models in the power system of Crete. Two different cases have been examined, which are differentiated only in terms of the wind energy production distribution during the day. It is noted that the total available wind generation (in MWh) is identical in both cases. In this sense, the effect that the wind energy generation profiles may have on the short-term scheduling of the entire generation mix is analyzed. Specifically, in case 1 (morning wind), wind blows strongly during the first hours of the day and gradually decreases during the evening and night hours. On the contrary, in case 2 (afternoon wind), wind generation is weak during the early morning hours and gradually increases and reaches its peak during the final hours of the 24-hour scheduling horizon.

Figures 4.28 and 4.29 illustrate the day-ahead scheduling per unit type for both cases. In both cases, combined cycle gas turbine (CCGT) is considered a must-run unit, since its continuous operation is strictly required by the SO in order not only to alleviate voltage stability issues on the west side of the island, where the CCGT unit is located, but also to contribute significantly to different types of spinning reserves (i.e., primary, secondary, and tertiary). In the case of high morning wind (case 1), the low system load during the early morning hours in conjunction with the fact that base-load thermal units cannot operate below their minimum power output, as well as they cannot be shutdown and subsequently start-up immediately due to critical technical

TABLE 4.7

Crete Power System Overview (December 2013)

Unit Technology	Unit Fuel	Number of Units	Installed Capacity (MW)
Steam	Fuel oil	7	198
Combined cycle gas turbine	Diesel	1	132
Open cycle gas turbine	Diesel	11	290
Internal combustion engine	Fuel oil	6	145
Wind	–	31	186
PV	–	2800	94
Total			1046

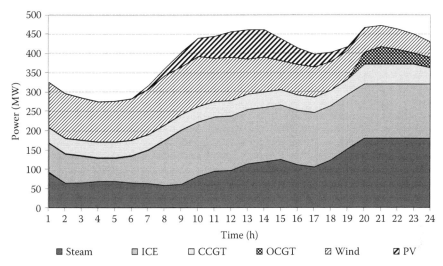

FIGURE 4.28
Day-ahead scheduling—morning wind (case 1).

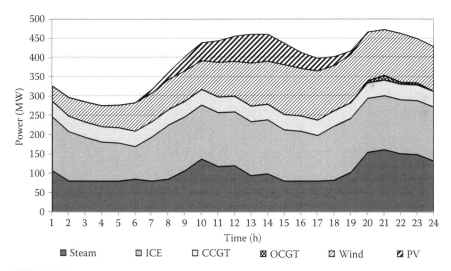

FIGURE 4.29
Day-ahead scheduling—afternoon wind (case 2).

and economic constraints (e.g., long minimum up/down times, high start-up costs, and provision of reserves) lead to a notable wind curtailment of 76.3 MWh/day, which correspond to 3.3% of the total daily available wind production for this case (see Figure 4.28).

On the contrary, in the case of high afternoon wind (case 2), these issues are eliminated (wind curtailment falls to 4.3 MWh/day or 0.002%), since the wind generation contributes mainly to the shaving of the noon and evening peak load, also restricting significantly the operation of high-cost thermal units, such as the open cycle gas turbine (OCGT) units (see Figure 4.29), which in case 1 are deemed necessary to serve the evening peak-load (see Figure 4.28).

Figures 4.30 and 4.31 illustrate the total energy and reserves contribution provided by all conventional thermal units for the two cases examined. The system reserve requirements are determined on a regular basis by the SO according to the specific needs of the power system taking also into account

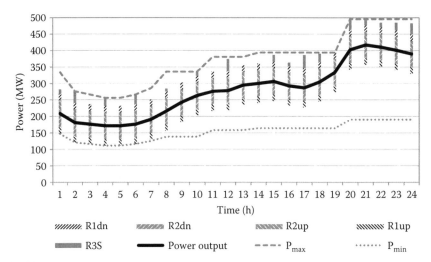

FIGURE 4.30
Energy and reserves contribution from conventional units—morning wind (case 1).

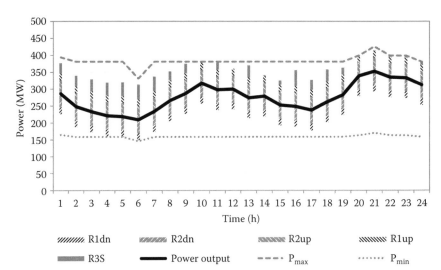

FIGURE 4.31
Energy and reserves contribution from conventional units—afternoon wind (case 2).

the RES share. The dashed and dotted red lines denote the total maximum and minimum power output of all conventional units that are online in each hour of the scheduling period, respectively. It is clear that the power output of all thermal units plus/minus the corresponding total contribution in spinning reserves (i.e., R1up: primary-up, R1dn: primary-down, R2up: secondary-up, R2dn: secondary-down, R3S: tertiary spinning) do not exceed the corresponding technical limits (i.e., maximum/minimum power output) of the thermal generation system.

Figures 4.32 and 4.33 illustrate the distribution of the total daily production cost in the different conventional thermal unit types for the two cases studied. As expected, in case 1, where wind generation substitutes mainly low-cost thermal units [i.e., steam and internal combustion engine (ICE) units] during the early morning hours (see Figure 4.28), the total system

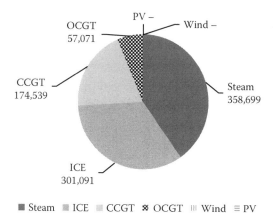

FIGURE 4.32
Total daily production cost—morning wind (case 1).

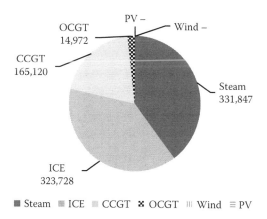

FIGURE 4.33
Total daily production cost—afternoon wind (case 2).

production cost is equal to €907,838 with the share of the production cost of OCGT units being equal to 6.3% (= €57,031/€907,838). On the contrary, in case 2, where wind generation substitutes mainly the energy production of medium-/high-cost thermal units (i.e., CCGT and OCGT units) during the evening hours (see Figure 4.29), the total system production cost falls to €835,668 (–7.95% as compared to case 1) and the share of the production cost of OCGT units falls to 1.8% (= €14,972/€835,668).

References

1. European Commission (EU), Energy roadmap 2050, 2012. Available at: http://ec.europa.eu/energy/publications/doc/2012_energy_roadmap_2050_en.pdf.
2. United States Environmental Protection Agency (EPA), Roadmap for incorporating energy efficiency/renewable energy policies and programs into state and tribal implementation plans, July 2012. Available at: http://epa.gov/airquality/eere/manual.html.
3. G. E. Box and G. Jenkins, *Time Series Analysis, Forecasting and Control*, Holden Day, San Francisco, CA, 1976.
4. Y. Li, Y. Su, and L. Shu, An ARMAX model for forecasting the power output of a grid connected photovoltaic system, *Renewable Energy*, vol. 66, pp. 78–89, 2014.
5. J. Juban, L. Fugon, and G. Kariniotakis, Probabilistic short-term wind power forecasting based on kernel density estimators, in *Proceedings of the European Wind Energy Conference*, Milan, Italy, May 2007.
6. A. G. Bakirtzis, V. Petridis, S. J. Kiartzis, M. C. Alexiadis, and A. H. Maissis, A neural network short term load forecasting model for the Greek power system, *IEEE Transactions on Power Systems*, vol. 11, no. 2, pp. 858–863, 1996.
7. M. C. Alexiadis, P. S. Dokopoulos, H. S. Sahsamanoglou, and I. M. Manousaridis, Short-term forecasting of wind speed and related electric power, *Solar Energy*, vol. 63, no. 1, pp. 61–68, 1998.
8. F. Wang, Z. Mi, and S. Su, Short-term solar irradiance forecasting model based on artificial neural network using statistical feature parameters, *Energies*, vol. 5, no. 5, pp. 1355–1370, 2012.
9. A. Tuohy, P. Meidom, E. Deny, and M. O'Malley, Unit commitment for systems with significant wind penetration, *IEEE Transactions on Power Systems*, vol. 24, no. 2, pp. 592–601, 2009.
10. A. Papavasiliou, S. S. Oren, and R. P. O'Neill, Reserve requirements for wind power integration: A scenario-based stochastic programming framework, *IEEE Transactions on Power Systems*, vol. 26, no. 4, pp. 2197–2206, 2011.
11. P. A. Ruiz, C. R. Philbrick, E. Zak, K. W. Cheung, and P. W. Sauer, Uncertainty management in the unit commitment problem, *IEEE Transactions on Power Systems*, vol. 24, no. 2, pp. 643–651, 2009.
12. K. Hoyland and S. W. Wallace, Generating scenario trees for multistage decision problems, *Management Science*, vol. 47, no. 2, pp. 295–307, 2001.
13. K. Hoyland, M. Kaut, and S. W. Wallace, A heuristic for moment-matching scenario generation, *Computational Optimization and Applications*, vol. 24, no. 2–3, pp. 169–185, 2003.

14. P. Beraldi, F. De Simone, and A. Violi, Generating scenario trees: A parallel integrated simulation-optimization approach, *Journal of Computational and Applied Mathematics*, vol. 233, no. 9, pp. 2322–2331, 2010.

15. E. Deniz and J. T. Luxhoj, A scenario generation method with heteroskedasticity and moment matching, *Engineering Economist*, vol. 56, no. 3, pp. 231–253, 2011.

16. J. Dupacova, G. Consigli, and S. W. Wallace, Scenarios for multistage stochastic programs, in *Proceedings of the 8th International Conference on Stochastic Programming*, Vancouver, Canada, August 1998.

17. S. Mehrota and D. Papp, Generating moment matching scenarios using optimization techniques, *Siam Journal on Optimization*, vol. 23, no. 2, pp. 963–999, 2013.

18. N. Gulpinar, B. Rustem, and R. Settergren, Simulation and optimization approaches to scenario tree generation, *Journal of Economic Dynamics & Control*, vol. 28, no. 7, pp. 1291–1315, 2004.

19. G. C. Pflug, Scenario tree generation for multiperiod financial optimization by optimal discretization, *Mathematical Programming*, vol. 89, no. 2, pp. 251–271, 2001.

20. P. Pinson, H. Madsen, H. A. Nielsen, G. Papaefthymiou, and B. Klockl, From probabilistic forecasts to statistical scenarios of short-term wind power production, *Wind Energy*, vol. 12, no. 1, pp. 51–62, 2009.

21. R. Barth, L. Soder, C. Weber, H. Brand, and D. J. Swider, Methodology of the scenario tree tool tech. rep. D6.2 (d), January 2006. Available at: http://www.wilmar.risoe.dk/Results.htm.

22. V. S. Pappala, I. Erlich, K. Rohrig, and J. Dobschinski, A stochastic model for the optimal operation of a wind-thermal power system, *IEEE Transactions on Power Systems*, vol. 24, no. 2, pp. 940–950, 2009.

23. X.-Y. Ma, Y.-Z. Sun, and H.-L. Fang, Scenario generation of wind power based on statistical uncertainty and variability, *IEEE Transactions on Sustainable Energy*, vol. 4, no. 4, pp. 894–904, 2013.

24. A. J. Conejo, *Decision Making Under Uncertainty in Electricity Markets*, Springer, New York, 2010.

25. S. T. Rachev. *Probability Metrics and the Stability of Stochastic Models*, John Wiley and Sons, Chichester, 1991.

26. J. Dupačová, N. Gröwe-Kuska, and W. Römisch, Scenario reduction in stochastic programming: An approach using probability metrics, *Mathematical Programming Series A*, vol. 95, pp. 493–511, 2003.

27. L. Devroye, *Non-Uniform Random Variate Generation*, Springer-Verlag, New York, 1986. Available at: http://luc.devroye.org/rnbookindex.html.

28. S. Stoft, *Power Systems Economics: Designing Markets for Electricity*, John Wiley and Sons, New York, 2002.

29. K. Hara, M. Kimura, and N. Honda, A method for planning economic unit commitment and maintenance of thermal power systems, *IEEE Transactions on Power Apparatus and Systems*, vol. PAS-85, no. 5, pp. 427–436, 1966.

30. R. H. Kerr, J. L. Scheidt, A. J. Fontana, and J. K. Wiley, Unit commitment, *IEEE Transactions on Power Apparatus and Systems*, vol. PAS-85, no. 5, pp. 417–421, 1966.

31. R. R. Shoults, S. K. Chang, S. Helmick, and W. M. Grady, A practical approach to unit commitment, economic dispatch and savings allocation for multiple-area pool operation with import/export constraints, *IEEE Transactions on Power Apparatus and Systems*, vol. PAS-99, no. 2, pp. 625–635, 1980.

32. F. N. Lee, Short-term thermal unit commitment—A new method, *IEEE Transactions on Power Systems*, vol. 3, no. 2, pp. 421–428, 1988.
33. T. Senjyu, K. Shimabukuro, K. Uezato, and T. Funabashi, A fast technique for unit commitment problem by extended priority list, *IEEE Transactions on Power Systems*, vol. 18, no. 2, pp. 882–888, 2003.
34. P. G. Lowery, Generating unit commitment by dynamic programming, *IEEE Transactions on Power Apparatus and Systems*, vol. PAS-85, no. 5, pp. 422–426, 1966.
35. C. K. Pang, G. B. Sheble, and F. Albuyeh, Evaluation of dynamic programming based methods and multiple area representation for thermal unit commitments, *IEEE Transactions on Power Apparatus and Systems*, vol. PAS-100, no. 3, pp. 1212–1218, 1981.
36. W. L. Snyder, H. D. Powell, Jr., and J. C. Rayburn, Dynamic programming approach to unit commitment, *IEEE Transactions on Power Systems*, vol. PWRS-2, no. 2, pp. 339–350, 1987.
37. W. J. Hobbs, G. Hermon, S. Warner, and G. B. Sheble, An enhanced dynamic programming approach for unit commitment, *IEEE Transactions on Power Systems*, vol. 3, no. 3, pp. 1201–1205, 1988.
38. Z. Ouyang and S. M. Shahidehpour, An intelligent dynamic programming for unit commitment application, *IEEE Transactions on Power Systems*, vol. 6, no. 3, pp. 1203–1209, 1991.
39. C. L. Chen and S. L. Chen, Short-term unit commitment with simplified economic dispatch, *Electric Power Systems Research*, vol. 21, pp. 115–120, 1991.
40. C. Li, R. B. Johnson, A. J. Svoboda, C. Tseng, and E. Hsu, A robust unit commitment algorithm for hydro-thermal optimization, *IEEE Transactions on Power Systems*, vol. 13, no. 3, pp. 1051–1056, 1998.
41. I. Boussaïd, J. Lepagnot, and P. Siarry, A survey on optimization metaheuristics, *Information Sciences*, vol. 237, pp. 82–117, 2013.
42. H. Mantawy, Y. L. Abdel-Magid, and S. Z. Selim, A new genetic-based tabu search algorithm for unit commitment problem, *Electric Power Systems Research*, vol. 49, pp. 71–78, 1999.
43. B. Zhao, C. X. Guo, B. R. Bai, and Y. J. Cao, An improved particle swarm optimization algorithm for unit commitment, *Electrical Power and Energy Systems*, vol. 28, pp. 482–490, 2006.
44. S. A. Kazarlis, A. G. Bakirtzis, and V. Petridis, A genetic algorithm solution to the unit commitment problem, *IEEE Transactions on Power Systems*, vol. 11, no. 1, pp. 83–92, 1996.
45. T. T. Maifeld and G. B. Sheble, Genetic-based unit commitment algorithm, *IEEE Transactions on Power Systems*, vol. 11, no. 3, pp. 1359–1370, 1996.
46. A. Rudolf and R. Bayrleithner, A genetic algorithm for solving the unit commitment problem of a hydro-thermal power system, *IEEE Transactions on Power Systems*, vol. 14, no. 4, pp. 1460–1468, 1999.
47. G. Damousis, A. G. Bakirtzis, and P. S. Dokopoulos, A solution to the unit-commitment problem using integer-coded genetic algorithm, *IEEE Transactions on Power Systems*, vol. 19, no. 2, pp. 1165–1172, 2004.
48. I. Cohen and V. R. Sherkat, Optimization based methods for operations scheduling, *Proceedings of the IEEE*, vol. 75, no. 12, pp. 1574–1591, 1987.
49. A. Merlin and P. Sandrin, A new method for unit commitment at electricité de France, *IEEE Transactions on Power Apparatus and Systems*, vol. PAS-102, no. 5, pp. 1218–1225, 1983.

50. R. Baldick, The generalized unit commitment problem, *IEEE Transactions on Power Systems*, vol. 10, no. 1, pp. 465–475, 1995.

51. J. Svoboda, C. Tseng, C. Li, and R. B. Johnson, Short-term resource scheduling with ramp constraints, *IEEE Transactions on Power Systems*, vol. 12, no. 1, pp. 77–83, 1997.

52. C. P. Cheng, C. W. Liu, and C. C. Liu, Unit commitment by Lagrangian relaxation and genetic algorithms, *IEEE Transactions on Power Systems*, vol. 15, no. 2, pp. 707–714, 2000.

53. Y. Fu, M. Shahidehpour, and Z. Li, Security-constrained unit commitment with AC constraints*, *IEEE Transactions on Power Systems*, vol. 20, no. 3, pp. 1538–1550, 2005.

54. T. S. Dillon, K. W. Edwin, H. D. Kochs, and R. J. Taud, Integer programming approach to the problem of optimal unit commitment with probabilistic reserve determination, *IEEE Transactions on Power Apparatus and Systems*, vol. PAS-97, no. 6, pp. 2154–2166, 1978.

55. D. Streiffert, R. Philbrick, and A. Ott, A mixed integer programming solution for market clearing and reliability analysis, in *Proceedings of the IEEE Power Engineering Society General Meeting*, San Francisco, CA, 2005, June 12–16.

56. T. Li and M. Shahidehpour, Price-based unit commitment: A case of Lagrangian relaxation versus mixed integer programming, *IEEE Transactions on Power Systems*, vol. 20, no. 4, pp. 2015–2025, 2005.

57. M. Carrion and J. M. Arroyo, A computationally efficient mixed-integer linear formulation for the thermal unit commitment problem, *IEEE Transactions on Power Systems*, vol. 21, no. 3, pp. 1371–1378, 2006.

58. A. Frangioni, C. Gentile, and F. Lacalandra, Tighter approximated MILP formulations for unit commitment problems, *IEEE Transactions on Power Systems*, vol. 24, no. 1, pp. 105–113, 2009.

59. J. Ostrowski, M. F. Anjos, and A. Vannelli, Tight mixed integer linear programming formulations for the unit commitment problem, *IEEE Transactions on Power Systems*, vol. 27, no. 1, pp. 39–46, 2012.

60. G. Morales-España, J. M. Latorre, and A. Ramos, Tight and compact MILP formulation of start-up and shut-down ramping in unit commitment, *IEEE Transactions on Power Systems*, vol. 28, no. 2, pp. 1288–1296, 2013.

61. H. Hui, C. Yu, and S. Moorty, Reliability unit commitment in the new ERCOT nodal electricity market, in *Proceedings of the IEEE Power and Energy Society General Meeting*, Calgary, Canada, July 26–30, 2009.

62. H. Hui, C. N. Yu, R. Surendran, F. Gao, and S. Moorty, Wind generation scheduling and coordination in ERCOT nodal market, in *Proceedings of the IEEE Power and Energy Society General Meeting*, San Diego, CA, July 22–26, 2012.

63. C. Monteiro, R. Bessa, V. Miranda, A. Botterud, J. Wang, and G. Conzelmann, *Wind Power Forecasting: State-of-the-art 2009*, ANL/DIS-10-1, Argonne National Laboratory, November 2009. Available at: http://www.dis.anl.gov/pubs/65613.pdf.

64. H. Holttinen et al., Methodologies to determine operating reserves due to increased wind power, *IEEE Transactions on Sustainable Energy*, vol. 3, no. 4, pp. 713–723, 2012.

65. F. Bouffard and F. D. Galiana, Stochastic security for operations planning with significant wind power generation, *IEEE Transactions on Power Systems*, vol. 23, no. 2, pp. 306–316, 2008.

66. L. Wu, M. Shahidehpour, and T. Li, Stochastic security-constrained unit commitment, *IEEE Transactions on Power Systems*, vol. 22, no. 2, pp. 800–811, 2007.

67. J. M. Morales, A. J. Conejo, and J. Pérez-Ruiz, Economic valuation of reserves in power systems with high penetration of wind power, *IEEE Transactions on Power Systems*, vol. 24, no. 2, pp. 900–910, 2009.

68. R. Jiang, J. Wang, M. Zhang, and Y. Guan, Two-stage minimax regret robust unit commitment, *IEEE Transactions on Power Systems*, vol. 28, no. 3, pp. 2271–2282, 2013.

69. D. Bertsimas et al., Adaptive robust optimization for the security constrained unit commitment problem, *IEEE Transactions on Power Systems*, vol. 28, no. 1, pp. 52–63, 2013.

70. J. E. Price and M. Rothleder, Recognition of extended dispatch horizons in California's energy markets, in *Proceedings of the IEEE Power and Energy Society General Meeting*, Detroit, MI, July 24–28, 2011.

71. K. H. Abdul-Rahman, H. Alarian, and M. Rothleder, Enhanced system reliability using flexible ramp constraint in CAISO market, in *Proceedings of the IEEE Power and Energy Society General Meeting*, San Diego, CA, July 22–26, 2012.

72. H. Hui, C. Yu, R. Surendran, F. Gao, S. Moorty, and X. Xu, Look ahead to the unforeseen: ERCOT's nonbinding look-ahead SCED study, in *Proceedings of the IEEE Power and Energy Society General Meeting*, Vancouver, Canada, July 21–25, 2013.

73. V. M. Zavala, A. Botterud, E. Constantinescu, and J. Wang, Computational and economic limitations of dispatch operations in the next-generation power grid, in *Proceedings of the Innovative Technologies for an Efficient and Reliable Electricity Supply Conference*, Waltham, MA. September 2010.

74. A. Ott, Development of smart dispatch tools in the PJM market, in *Proceedings of the IEEE Power and Energy Society General Meeting*, Detroit, MI, July 24–28, 2011.

75. B. C. Ummels, M. Gibescu, E. Pelgrum et al., Impacts of wind power on thermal generation unit commitment and dispatch, *IEEE Transactions on Energy Conversion*, vol. 22, no. 1, pp. 44–51, 2007.

76. A. Tuohy, E. Denny, and M. O'Malley, Rolling unit commitment for systems with significant installed wind capacity, in *Proceedings of the IEEE Lausanne PowerTech 2007*, Lausanne, Switzerland, July 1–5, 2007.

77. E. Ela and M. O'Malley, Studying the variability and uncertainty impacts of variable generation at multiple timescales, *IEEE Transactions on Power Systems*, vol. 27, no. 3, pp. 1324–1333, 2012.

78. E. A. Bakirtzis, P. N. Biskas, D. P. Labridis, and A. G. Bakirtzis, Multiple time resolution unit commitment for short-term operations scheduling under high renewable penetration, *IEEE Transactions on Power Systems*, vol. 29, no. 1, pp. 149–159, 2014.

79. R. D. Christie, B. F. Wollenberg, and I. Wangensteen, Transmission management in the deregulated environment, *Proceedings of the IEEE*, vol. 88, no. 2, pp. 170–195, 2000.

80. J. R. Birge, *Introduction to Stochastic Programming*, Springer Series in Operations Research and Financial Engineering, Springer, New York, 1997.

5

Reserves and Demand Response Coping with Renewable Energy Resources Uncertainty

Nikolaos G. Paterakis, Ozan Erdinç, Miadreza Shafie-khah, and Ehsan Heydarian-Forushani

CONTENTS

ABSTRACT This chapter introduces several two-stage stochastic programming models in order to procure energy and reserve services from the demand side to cope with uncertainty introduced by increasing renewable energy sources (RESs) penetration into the power system. The variable and uncertain nature of the leading RESs, such as wind power generation, calls for the development of a sophisticated balance mechanism between supply and demand to maintain the consistency of a power system. Especially, the reliable and economic operation of insular power systems may be jeopardized due to the limited flexibility of conventional power sources. The two-stage stochastic optimization models are developed with the aim of procuring the required load-following reserves from both generation and demand-side resources under high wind-power penetration. A novel load model is introduced to procure flexible reserves from industrial clients. The economic model of responsive demand and a model describing an aggregation of plug-in electric vehicles (PEVs) are also discussed in this chapter. The proposed formulations are evaluated through various simulations and different case studies performed on the insular power system of Crete, Greece.

5.1   Introduction

5.1.1   Need for Procuring Reserves by the Demand Side

It is widely accepted that RESs are likely to represent a significant portion of the production mix in many power systems around the world, a trend expected to be increasingly followed in the coming years. Programs such as the European Union's (EU) 2020 Climate and Energy Package [1], widely known as *20–20–20* targets, which stand for the raise of the share of EU energy consumption produced from RES to 20% until 2020, confirm this trend. However, the economic and environmental benefits that arise from the integration of these resources into the power system lead to additional problems due to the fact that their production is highly variable and unpredictable. Thus, their integration into the grid requires the balancing authorities of the respective power systems to procure more ancillary services (ASs) in order to ensure that power quality, reliability, and security of the system operation are maintained [2]. In the restructured environment, ASs are treated as a commodity that has significant economic value. Traditionally, these services are procured only from the generation side.

Nevertheless, the increased needs introduced by the large-scale penetration of RES and the growth of demand for electricity, in combination with the scarcity and the increased cost of conventional fuels and the limited new investments in new units, bring upon a peculiar situation.

Various studies, including [3], argue that resources located in the demand side may also provide some types of ASs and even with the significant advantages of fast, statistically reliable response and distributed nature.

In practice, although the importance of having an active demand side contributing in system reliability efforts is recognized, only a few leading independent system operators (ISOs) and transmission system operators (TSOs), especially in North America, have established recently demand response (DR) programs that allow the participation of consumers in the provision of critical ASs. A notable example is the demand-side AS program (DSASP), initiated in 2009 by the New York ISO, which exploits the ability of DR to provide regulation, synchronous reserve, and nonsynchronous reserve [4]. In Europe, DR activities are generally in an initial status [5] despite the fact that the provision of AS by DR has been recognized as mandatory [6] to support greater future integration of RES, mainly due to regulatory barriers.

5.1.2 Reserve Procurement Under Uncertainty: An Overview

There are several works dealing with reserve allocation from both production and demand side sources to cope with uncertainties in system operation. A comprehensive description of AS needed for the reliable operation of a power system can be found in [7]. In [8,9], existing ancillary mechanisms for 11 power systems have been summarized and their economic features underlined. A detailed evaluation of DR activities for AS can be found in [10]. Apart from the commonly met AS types, a new type was recently proposed by Midwest ISO [11] and California ISO [12] as flexible ramping products to increase the robustness of the load following reserves under uncertainty, such as high wind-power fluctuation. Jafari et al. [13] proposed a stochastic programming-based multiagent market model incorporating day-ahead and several intraday markets, as well as a spot real-time energy-operating reserve market in order to adjust wind fluctuation-based system uncertainties. In [13], no other demand-side resource apart from load shedding was considered. In [14], a contingency analysis-based stochastic security-constrained-system operation under significant wind-power condition was analyzed, whereas demand-side resources were not considered. In [15], a switching operation between two separate energy markets named *conventional energy market* and *green energy market* was proposed, where profit maximization of green energy systems was formulated in a stochastic programming framework without considering demand-side facilities. Similar studies neglecting demand-side resources for reserve procurement to overcome system uncertainties were also presented in [16–18]. It is also worthy to note that studies in [16–18] considered the combination of different approaches in order to mainly provide a computationally efficient way to solve unit commitment under uncertainties. The computational efficiency of unit commitment under uncertainty was also addressed in [19]. On the other hand, there are some studies providing implicit evaluation of reserves under wind penetration. Among them, required levels or spinning and nonspinning reserves under high levels of wind penetration was the scope of the study realized by Morales et al. [20]. In [20], demand-side resources are considered by modeling some portion of load demand as elastic, capable of contributing to reserve procurement. Involuntary load shedding has been also taken into account in [20].

Wind and load uncertainties were covered by scheduling optimal hourly reserves using security-constrained unit commitment (SCUC) approach in [21]. A two-stage stochastic programming framework with a set of appropriate scenarios solved using dual decomposition algorithm was provided in [22]. The dual decomposition algorithm was employed in [22] in order to address the computational issues emerging from the use of stochastic programming incorporated in unit-commitment policies. Some further studies focusing mainly on demand and stochastic programming also take place in the literature. Parvania and Fotuhi-Firuzabad [23] proposed a short-term stochastic SCUC that jointly schedules energy and spinning reserve from units and DR resources, where the main contribution to the literature was the reserve modeling of DR programs, which actually are aggregated retail customer responses submitted as bids to the market. Also, in [24] a day-ahead market structure was presented where demand side can participate in order to provide contingency reserves by bidding an offer curve representing the cost of making their loads available for curtailment. Shan et al. [25] also considered a DR-based load-side contribution to reserves under high levels of wind penetration, where demand was a linear price-responsive function. As a different study regarding reliability consideration, load uncertainty and generation unavailability were covered in [26] without additional wind power-based uncertainty in a two-stage stochastic programming framework. Apart from the stochastic programming-based literature referred above, many studies considering different modeling frameworks, such as probabilistic, rolling stochastic, and Monte Carlo criteria, can also be found in [27–30]. There are also more studies not referred here contributing to the literature from reserve procurement perspective considering different aspects of uncertainties in the system operation.

5.1.3 Objectives

It is obvious from the overview presented in Section 5.1.2 that there is an increasing research interest in AS mechanisms under the presence of significant renewable energy penetration. However, there is a lack of studies dealing with more detailed evaluation of responsive loads in the concept of DR-based reserve allocation under both demand- and production-side uncertainties in a stochastic programming framework.

The main objective of this chapter is to provide stochastic programming optimization models that explicitly consider demand-side resources, as well as test cases on an insular power system, such as the one of Crete, Greece, illustrating the potential benefits of demand-side resources integration to accommodate significant RES integration. The models that are analyzed cover several aspects of demand-side participation in energy and reserve markets. Specifically, the load models that are presented describe the following:

- Industrial customers
- PEV aggregators
- An economic model of responsive demand

Finally, special attention is given to the rigorous treatment of various sources of uncertainty and the introduction of appropriate indices to evaluate the performance and impacts of responsive demand to the system operation.

5.2 Uncertainty Modeling

5.2.1 Introduction

The scenario-based stochastic programming method is an efficient tool to make optimal decisions in problems involving uncertainty. When it comes to making decisions under uncertainty using stochastic programming, the building of scenario sets that properly represent the uncertain input parameters constitutes a task of utmost importance. In reality, the optimal decisions derived from stochastic programming models may be indeed remarkably sensitive to the scenario characteristics of uncertain data. For this reason, a remarkable effort has been made by researchers to design efficient scenario generation methods. A brief description of the most relevant ones is presented in [31].

However, it should be considered that obtaining a good set of scenarios often requires the generation of a huge number of scenarios, which may render the underlying optimization problem intractable. Therefore, it is necessary to consider a limited subset of scenarios without losing the generality of the original set. The scenario reduction techniques can reduce the number of scenarios effectively and maximally retain the fitting accuracy of samples. Scenario reduction methods have been used in [32,33].

In this section, the uncertainties are modeled as multiple different scenarios. Then, the scenario-based stochastic programming approach is employed to handle the uncertainties. In fact, this section is mainly intended to build appropriate scenario sets required in a typical stochastic process.

5.2.2 Scenario Generation and Reduction Techniques

In this chapter, three major sets of uncertainty are considered: customers, market, and wind power generation. The uncertainties of customers include uncertain behavior of responsive demands and PEV owners. Market uncertainties include the uncertainties of market prices and activated amounts of reserve. The uncertainties of the wind power generation are related with the unpredictable and uncontrollable nature of this RES. In this section, uncertainty of customers' behavior and market-related uncertainties are treated. The modeling of the wind power generation uncertainty will be discussed later in the chapter.

5.2.2.1 Uncertainty of Customers' Behavior

The probabilistic behavior of responsive demand and PEV owners has caused the related aggregators to face a plenty of uncertainties in order to participate effectively in the market. Each customer behaves differently because of social and economic concerns. Therefore, each individual behavior will be different from others. The uncertain parameters of the customers include hourly power consumption of consumers, the number of PEVs connected to aggregator per hour, connection duration of PEVs, and state of charge (SOC) of batteries while plugging-in to the grid. The DR aggregators/providers and PEV aggregators should estimate the uncertain parameters of the probabilistic behavior of their customers using past statistics data. In this chapter, these market players utilize the statistical data of a set of customers and forecast the parameters mentioned above. To this end, the aggregators model the estimation uncertainty using a probabilistic approach. For this purpose, each of these market players uses the statistical data and generates scenarios based on time series of uncertain variables using roulette wheel mechanism (RWM) [34,35]. Since the time series of all related stochastic variables are generated using unique historical data, the correlation between stochastic variables and subsequent hours has been considered.

The PEV aggregator should decrease the risk of an unreliable forecast of the behavior of the PEV owners to efficiently participate in the market. In order to overcome the problem, this chapter proposes RWM for scenario generation to reduce the forecast error. Applying the generated scenarios causes the PEV aggregator to consider the probable deviations around the forecasted number of available PEVs. Since forecast errors often have a distribution very close to the normal [36], in this chapter, the scenario generation is accomplished using the normal distribution. Based on the normal distribution, the value and the probability of each scenario is related to the mean value, μ, and the standard deviation, σ, of the forecasted number of available PEVs. The closer forecasting is performed to the market closure time the more accurate the forecast of the number of available PEVs is; therefore the standard deviation in the intraday session is considered to be less than that in the day-ahead session. This means that the aggregator's forecast of its customers in the intraday stage has less deviation from the actual value than in the day-ahead stage ($\sigma^{Intra} < \sigma^{DA}$). Moreover, the mean value of the number of available PEVs in both the sessions mentioned is considered to be equal to the actual value that will be used in the balancing market.

The total aggregated SOC of the PEVs is also uncertain, and it depends on the number of PEVs and on the daily distance driven by their owners. The lognormal distribution function is utilized to generate the probabilistic daily traveled distance [37]. The lognormal random variables are generated using standard normal random variable, N, and are computed using Equation 5.1 [38].

$$M_d = \exp(\mu_{md} + \sigma_{md} \cdot N) \tag{5.1}$$

where:

M_d is the daily distance driven by the PEVs owners

μ_m and σ_m are the lognormal distribution parameters, being calculated from mean and standard deviation of M_d based on the historical data, denoted as μ_{md} and σ_{md}, respectively

Parameters μ_m and σ_m are calculated based on Equations 5.2 and 5.3, respectively.

$$\mu_m = \ln\left(\frac{\mu_{md}^2}{\sqrt{\mu_{md}^2 + \sigma_{md}^2}}\right) \tag{5.2}$$

$$\sigma_m = \sqrt{\ln\left(1 + \frac{\sigma_{md}^2}{\mu_{md}^2}\right)} \tag{5.3}$$

The general assumption for generating the scenarios adopted in this chapter is based on [39], where an average of 2.4 trips/day, an average daily distance of 75 km, and an average daily travel time of 1.55 hours are considered. On the other hand, a PEV consumes approximately 0.22 kWh to recharge for each kilometer traveled.

The aggregated SOC depends on the number of PEVs, the type of each PEV, and on the daily distance driven by their owners. The energy storage capacity of each PEV represents the total energy capacity that is dependent on the PEV class. In [40], 24 different classes have been considered for PEV batteries. The probability distribution of the battery capacities in each PEV class occurring in a market is illustrated in Figure 5.1.

It should be noted that although a higher number of scenarios causes a more accurate consideration of the uncertainties mentioned, it produces a significant optimization problem in terms of computational requirements. Thus, a scenario reduction technique is considered, using the *k*-means clustering technique [41], resulting in a scenario tree with independent scenarios that are applied to the case studies.

According to the aforementioned description, the scenarios related to uncertain amounts of available number of PEVs and the total aggregated SOC are generated as illustrated in Figures 5.2 and 5.3, respectively. In these figures,

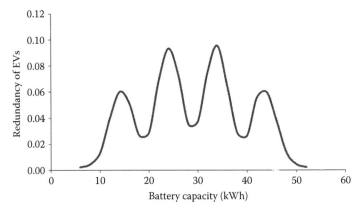

FIGURE 5.1
Distribution of battery capacity.

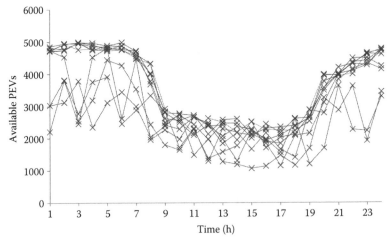

FIGURE 5.2
Number of available PEVs.

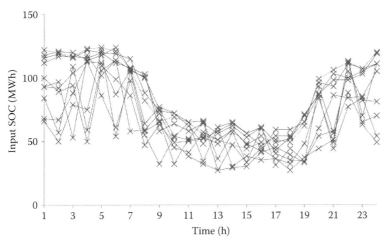

FIGURE 5.3
Available SOC of PEVs.

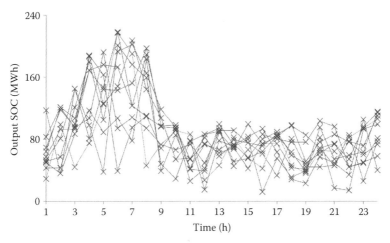

FIGURE 5.4
SOC obtained by the model proposed.

the generated scenarios and the expected value of uncertain parameters are indicated by cross-marked point and black lines, respectively. The obtained SOC of PEVs using the proposed model is shown in Figure 5.4.

5.2.2.2 Modeling the Uncertainties of Activated Amounts of Reserve

One of the uncertainties of the PEV aggregators is the probability of activated amounts of reserve by the ISO for energy provision in reserve markets. Since the uncertainty has a discrete probability distribution, can be considered as an event that occurs in a day with a known average rate, and is independent of the number of reserve activation during the previous day, it can be modeled by the Poisson distribution. On this basis, Poisson distribution is employed to model the probability of being called to provide energy in the reserve market. Thus, the probability distribution function (PDF) can be expressed by Equation 5.4:

$$f(k,\mu) = \frac{\mu^k.\exp(-\mu)}{k!}, \quad \mu > 0, \quad k = 0,1,2,\ldots \tag{5.4}$$

where:
  μ and k denote the expected value and the number of times to be called, respectively

Considering the PDF mentioned, different outcomes of ISO's behavior for calling PEV aggregator are considered by a RWM-based scenario generation procedure.

The uncertain amount of activated reserve, $\text{Act}_{t,\omega}^{\text{Res}}$, has been taken into account to be uniformly distributed between zero and PEV aggregator's quantity offered. Therefore, the PDF of quantity of activated reserve can be formulated as follows:

$$f(x) = \begin{cases} \dfrac{1}{P_{t,\omega}^{\text{Res}}}, & 0 \leq x \leq \text{Offer}_{t,\omega}^{\text{Res}} \\ 0, & \text{Otherwise} \end{cases} \tag{5.5}$$

According to Equations 5.4 and 5.5, diverse regulation requests to the aggregator by the ISO have been considered by employing RWM-based scenario generation.

5.2.2.3 Modeling the Uncertainties of Market Prices

To successfully participate in the electricity market, the aggregator has to forecast the market prices. In this chapter, three uncertain market prices are considered: day-ahead energy, reserve, and intraday energy. In order to develop an accurate and appropriate model, market prices have been characterized by lognormal distribution in each hour [42]. Thus, the PDF of market prices is represented by Equation 5.6:

$$f(\lambda,\mu,\sigma) = \frac{1}{\lambda\sigma\sqrt{2\pi}} \exp\left[-\frac{(\ln\lambda - \mu)^2}{2\sigma^2}\right] \tag{5.6}$$

where:
  μ and σ represent the mean value and the standard deviation, respectively

Considering the PDF mentioned, different realizations of markets' behavior are modeled using a scenario generation process based on RWM.

5.3 Responsive Demand Models

5.3.1 Industrial Customer Model

In this study, the industrial load is considered to comprise different groups that may work in parallel and include several individual processes, similar to real-life practice [43]. Generally, we can refer to three categories of processes, namely, totally flexible, flexible, and inflexible:

- Totally flexible processes can be considered as the ones that are not physically constrained to maintain power for a continuous interval (e.g., a set of production facilities that work as long as there is input material).

- Flexible processes are the ones that should be completed at most within a certain time interval, but with the flexibility of allocating energy consumption. Within their completion time, they can be continuous (type 1) or interruptible (type 2).

- The most rigid processes are the inflexible ones that have to be completed in a strictly specified time and energy allocation (e.g., a metallurgy process).

For the sake of simplicity, in the formulation proposed, the hourly limit of energy is considered to be uniform for each process. There are specific cases that this assumption does not cover, but this restriction is easy to overcome by defining a time varying hourly energy limit.

A process is characterized by several parameters that define the different types of flexibility in terms of energy treatment. To better illustrate the operation of the model, examples of different types of processes are presented in Figure 5.5. The totally flexible process consumes energy that can be allocated in four discrete blocks during the day. The only restriction is that no more than two blocks of energy may be allocated in a single period. The flexible process has to consume energy that can be allocated in four discrete blocks. The restrictions are that the process has to be completed in maximum three hours after it starts (no restriction in which period to start) and that no more than two energy blocks can be allocated in a single period. Also, there has to be at least one power block allocated per period (type 1). This type of process offers two degrees of freedom. First, the optimal starting period is selected, and then some parts of the consumption may be shifted in adjacent time periods. Finally, the inflexible process has to be completed in exactly two

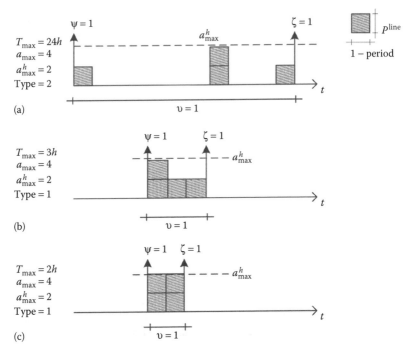

FIGURE 5.5
The types of industrial processes: (a) totally flexible, (b) flexible, and (c) inflexible.

periods after it begins (no restriction in which period to start), allocating energy blocks in a predefined manner. The only flexibility of this type of process is that the starting time can be optimally selected.

- **Operation of the industry**

    Before describing the way in which load-following reserve is procured by industrial consumers, the model of the processes described above should be mathematically expressed.

$$\sum_{t\in T} a_{p,g,d,t} = a_{p,g,d}^{max} \quad \forall p, g, d, t \tag{5.7}$$

$$P_{p,g,d,t}^{pro,S} = a_{p,g,d,t} \cdot P_{p,g,d}^{line} \quad \forall p, g, d, t \tag{5.8}$$

$$P_{d,t}^{ind,S} = D_{d,t}^{min} + \sum_{g\in G}\sum_{p\in P} P_{p,g,d,t}^{pro,S} \quad \forall d, t \tag{5.9}$$

Equation 5.7 is an energy requirement constraint. It states that all the processes should be completed within the scheduling horizon. Equations 5.8 and 5.9 define the power that a process and the industry consumer over a given period, respectively. Equation 5.9 consists of two parts. The first part is inelastic and may be characterized as minimum or mandatory and stands for must-run or uncontrollable processes of the industry. The second part consists of a controllable process load.

$$\upsilon_{p,g,d,t} \leq a_{p,g,d,t} \leq a_{p,g,d}^{max,h} \cdot \upsilon_{p,g,d,t} \quad \forall p \in P^1, g, d, t \tag{5.10}$$

$$0 \leq a_{p,g,d,t} \leq a_{p,g,d}^{max,h} \cdot \upsilon_{p,g,d,t} \quad \forall p \in P^2, g, d, t \tag{5.11}$$

Constraints in Equations 5.10 and 5.11 impose limits on the number of processes that could be scheduled for every hour by the industry. They cover both interruptible and continuous processes and they can be used in order to guarantee that limitations such as the installed power of the industry are not violated. It should be noted that the term *production line* is a general term adopted here by the authors in order to express discrete amounts of power that can be treated by processes, not necessarily referring to physical production lines.

Constraints in Equations 5.12 through 5.18 describe the logic of the commitment of a process. Especially, Equation 5.12 guarantees that a process is finished within the required completion time. Constraints in Equations 5.13 through 5.16 define the logic of operating, starting, and ending a process. Constraints in Equations 5.17 and 5.18 stipulate that a process can be run only once in the scheduling horizon.

$$\sum_{\tau=t-T_{p,g,d}^{c,max}+1}^{t} a_{p,g,d,\tau} \geq a_{p,g,d}^{max} \cdot \zeta_{p,g,d,(t+1)} \quad \forall p,g,d,t \tag{5.12}$$

$$a_{p,g,d,t} \geq \zeta_{p,g,d,(t+1)} \quad \forall p,g,d,t \tag{5.13}$$

$$\psi_{p,g,d,t} \leq a_{p,g,d,t} \quad \forall p,g,d,t \tag{5.14}$$

$$\psi_{p,g,d,t} + \zeta_{p,g,d,t} \leq 1 \quad \forall p,g,d,t \tag{5.15}$$

$$\psi_{p,g,d,t} - \zeta_{p,g,d,t} = \upsilon_{p,g,d,t} - \upsilon_{p,g,d,(t-1)} \quad \forall p,g,d,t \tag{5.16}$$

$$\sum_{t \in T} \zeta_{p,g,d,t} = 1 \quad \forall p,g,d \tag{5.17}$$

$$\sum_{t \in T} \psi_{p,g,d,t} = 1 \quad \forall p,g,d \tag{5.18}$$

Omitting constraints in Equations 5.17 and 5.18 will cause a violation of the constraint in Equation 5.12. Thus, special care should be taken when dealing with multiple cycle of a specific process within the same day.

$$\Psi_{p,g,d,t} \le \sum_{\tau=t-T^{g,max}_{(p-1),g,d}}^{t-T^{g,min}_{(p-1),g,d}} \zeta_{(p-1),g,d,\tau} \quad \forall p \in \{P \mid p > 1\}, g, d, t \tag{5.19}$$

In case of several processes that should be executed in a pre-defined order, Equation 5.19 guarantees that the next process will begin after a number of periods that can be within a minimum limit and a maximum limit. Naturally, this is a generic formulation and the appropriate values can cover any possible sequencing preferences.

- **Scheduling reserves from the industrial customer**

  These market participants can increase (down-spinning reserve) or decrease (up-spinning reserve) in a discrete amount their power consumption or even reschedule (nonspinning reserve) their production processes. It should be noted that the spinning and nonspinning reserves terminology in the case of demand-side reserves is adopted in accordance with the unit procured reserves. Spinning tends to mean *alteration of an existing consumption*, whereas nonspinning in the case of the industrial consumers stands for a time shift of a process. For instance, in [20], where a simplified responsive load was modeled, the terms *spinning* and *nonspinning* were omitted since they are not needed to distinguish the reserve mechanisms.

  Up-, down-, and nonspinning reserves that can be procured by the industrial processes are described by Equations 5.20 through 5.40.

$$R^{U,ind}_{d,t} = \sum_{p \in P} \sum_{g \in G} R^{U,pro}_{p,g,d,t} \quad \forall d, t \tag{5.20}$$

$$R^{U,pro}_{p,g,d,t} = a^{up}_{p,g,d,t} \cdot P^{line}_{p,g,d} \quad \forall p, g, d, t \tag{5.21}$$

$$0 \le a^{up}_{p,g,d,t} \le a_{p,g,d,t} \quad \forall p, g, d, t \tag{5.22}$$

The constraint in Equation 5.20 stands for the total up reserve scheduled by the industrial load during a period, whereas Equations 5.21 and 5.22 stand for the specific process reserve participation. Especially, Equation 5.22 states that no more than the number of scheduled production lines for a given interval can be scheduled for up reserve. It should be noted that Equation 5.22 is considered together with Equations 5.10 and 5.11 according to the process type.

$$R_{d,t}^{D,ind} = \sum_{p \in P} \sum_{g \in G} R_{p,g,d,t}^{D,pro} \quad \forall d,t \tag{5.23}$$

$$R_{p,g,d,t}^{D,pro} = a_{p,g,d,t}^{down} \cdot P_{p,g,d}^{line} \quad \forall p,g,d,t \tag{5.24}$$

$$0 \le a_{p,g,d,t}^{down} \le a_{p,g,d}^{max,h} \cdot \upsilon_{p,g,d,t} - a_{p,g,d,t} \quad \forall p,g,d,t \tag{5.25}$$

Similarly, Equations 5.23 through 5.25 stand for the down-reserve scheduling. Especially, Equation 5.25 states that the increase of consumption cannot overcome the hourly limit.

$$R_{d,t}^{NS,ind} = \sum_{p \in P} \sum_{g \in G} R_{p,g,d,t}^{NS,pro} \quad \forall d,t \tag{5.26}$$

$$R_{p,g,d,t}^{NS,pro} = a_{p,g,d,t}^{ns} \cdot P_{p,g,d}^{line} \quad \forall p,g,d,t \tag{5.27}$$

$$0 \le a_{p,g,d,t}^{ns} \le a_{p,g,d}^{max,h} \cdot \left(1 - \upsilon_{p,g,d,t}\right) \quad \forall p,g,d,t \tag{5.28}$$

Nonspinning reserves are defined by Equations 5.26 through 5.28. Especially, Equation 5.28 states that no more than the maximum discrete amounts of energy can be used in a given interval.

The meaning of the reserve services that can be procured from the industrial load and were described is rendered evident through constraints in Equations 5.22, 5.25, and 5.28. Up- and down-spinning reserves stand for a decrease or an increase of the consumption of a process that is scheduled to operate during an interval, whereas nonspinning reserve stands for the total shifting of the process in other periods.

5.3.2 PEV Aggregator Model

In order to consider the impact of the sources of uncertainty mentioned previously on the strategic behavior of PEV aggregator, they have been characterized as stochastic procedures, and the problem has been solved using a bi-level stochastic programming approach. It should be noted that the bi-level stochastic programming has been utilized to model the market players' behavior that consider the intraday session, in addition to the day-ahead and balancing ones [44]. In the approach mentioned, there are three stages that denote a market horizon each as illustrated in Figure 5.6.

The classification of decision variables of each stage is presented as follows:

The first stage (here-and-now) stochastic decision variables are $P_{t,\omega}^{DA}$, $P_{t,\omega}^{Res}$, $\lambda_{t,\omega}^{DA}$, and $\lambda_{t,\omega}^{Res}$. In this stage, the PEV aggregator designs its offering/bidding strategies and submits them to the day-ahead energy and reserve markets for each hour. The decisions of the here-and-now stage are made on the basis of probable realizations of the stochastic procedures including PEV owners'

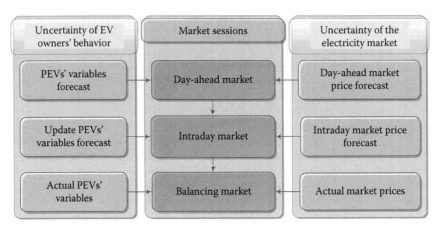

FIGURE 5.6
The proposed bi-level stochastic framework.

behavior, being called to generate, and prices of intraday and balancing markets. It should be noted that, in this stage, probable realizations of the number of available PEVs is accomplished by σ^{DA}. Consequently, scenarios of the total aggregated SOC is generated based on the achieved scenarios of available PEVs.

Stochastic variables ($P_{t,\omega}^{Intra,buy}$, $P_{t,\omega}^{Intra,sell}$, and $\lambda_{t,\omega}^{Intra}$) are the second stage (wait-and-see) variables. In this stage, the PEV aggregator submits its offering/bidding strategies to the intraday market while prices and quantities of day-ahead energy and reserve markets are known. However, prices of the intraday and balancing markets, actual behavior of PEV owners, and being called to generate are still uncertain. In the second stage, PEV aggregator can achieve new information and hence update its day-ahead scheduling between the closures of the day-ahead and intraday markets. Also, the PEV aggregator should predict intraday market clearing price in order to minimize the deviations between the latest existing forecast and the day-ahead schedule. It is notable that market prices are supposed to be stationary stochastic parameters. This means that the scenarios of market prices are generated based on the trading stage just for once. On the other hand, the PEVs' behavior is considered as a dynamic stochastic parameter that should be updated in each trading stage due to the latest obtained information in the decision-making process. On this basis, scenarios of the number of available PEVs and consequently the total aggregated SOC are generated by σ^{Intra} that is less than the σ^{DA} to reflect the higher accuracy of the forecast due to closer time to the balancing market. Other stochastic decision variables are $SOC_{i,t,\omega}$, $P_{t,\omega}^{del}$, $Act_{t,\omega}^{Res}$, $P_{i,t,\omega}^{G2V}$, $P_{i,t,\omega}^{V2G}$, $r_{i,t,\omega}^{charge}$, $r_{i,t,\omega}^{discharge}$, $\alpha_{i,t,\omega}$, $\beta_{i,t,\omega}$, $\Delta_{t,\omega}^{+}$, $\Delta_{t,\omega}^{-}$, and $\lambda_{t,\omega}^{Bal}$. These variables appertain to the balancing market. The prices and quantities of intraday market are known when these decisions are made. Also, actual prices of the balancing market, actual behavior of PEV owners, and being called to generate are known. Moreover, the deviation incurred by the PEV aggregator in each hour is known and the resulting imbalance costs are computed. When generation of PEV aggregator is higher than the forecasted value in the second stage, the system requires downward regulation services that are provided by other market participants. In such situation, the PEV aggregator has to resale the excess of its generations at a price lower than one of the day-ahead market. On the contrary, in the case of generation shortage, the system requires upward regulation services to compensate the deficit of generation. In this situation, the PEV aggregator has to cover its shortage at a price higher than that of a day-ahead market.

The PEV owners can charge their PEV batteries with or without control of PEV aggregators. However, PEV owners can take advantage, using economic benefits of connection to the grid by aggregators' control. In order to ensure the owners about desired charge of their batteries in disconnecting time, the aggregator should follow and observe the desired SOC of PEVs.

The costs of the PEV aggregator comprise fixed and variable costs. The fixed costs include the costs of installation of communication network, control, and measuring equipment. Variable costs include the purchase costs of the energy and reserve from PEVs for selling to market, cost of imbalance penalties, costs of taking part in the intraday market to compensate the imbalances, and the purchase costs of the electricity from the market for charging the batteries of PEVs.

The revenues of the PEV aggregator include incomes of participating in the energy and spinning reserve markets, incomes of participating in the intraday market, and incomes of receiving the cost of battery charging from PEV owners. According to the above discussions, the objective function of PEV aggregator can be expressed as follows:

$$
\begin{aligned}
\max_{\pi_{t,\omega}} \sum_t \Big\{ &-\text{Cost}_{\text{Infra}} + E_{\Omega 1}\Big[\text{Income}_{t,\omega}^{\text{Res}} + \text{Income}_{t,\omega}^{\text{Energy}} \\
&+ E_{\Omega 2|\Omega 1}\Big[\text{Income}_{t,\omega}^{\text{Intra}} - \text{Cost}_{t,\omega}^{\text{Intra}} \\
&+ E_{\Omega 3|\Omega 2,\Omega 1}\Big[\text{Income}_{\omega}^{\text{Charge}} + \text{Income}_{t,\omega}^{\text{Call}} \\
&+ \text{Income}_{t,\omega}^{\text{Imb}} - \text{Cost}_{t,\omega}^{\text{Imb}} - \text{Cost}_{t,\omega}^{\text{Charge}} - \text{Cost}_{t,\omega}^{\text{Obl}} - \text{Cost}_{\omega}^{\text{Res}} - \text{Cost}_{\omega}^{\text{Deg.}} \Big]\Big]\Big]\Big\}
\end{aligned}
$$

(5.29)

$$
\text{Cost}_{\text{Infra}} = \frac{1}{365} \sum_v \left[\frac{dr.\left(\text{Cost}_v^{\text{Wiring}} + \text{Cost}_v^{\text{On-board}} \right)}{1-(1+dr)^{-N_y}} \right]
$$

(5.30)

$$
\text{Income}_{t,\omega}^{\text{Res}} = P_{t,\omega}^{\text{Res}} . \lambda_{t,\omega}^{\text{Res}}
$$

(5.31)

$$
\text{Income}_{t,\omega}^{\text{Energy}} = P_{t,\omega}^{\text{DA}} . \lambda_{t,\omega}^{\text{DA}}
$$

(5.32)

$$\text{Income}_{t,\omega}^{\text{Intra}} = P_{t,\omega}^{\text{Intra,sell}}.\lambda_{t,\omega}^{\text{Intra}} \tag{5.33}$$

$$\text{Cost}_{t,\omega}^{\text{Intra}} = P_{t,\omega}^{\text{Intra,buy}}.\lambda_{t,\omega}^{\text{Intra}} \tag{5.34}$$

$$\text{Cost}_{t,\omega}^{\text{Charge}} = \sum_{v} P_{v,t,\omega}^{\text{G2V}}.\lambda_{t,\omega}^{\text{DA}} \tag{5.35}$$

$$\text{Cost}_{t,\omega}^{\text{Obl}} = \text{FOR}^{\text{Agg}}.\text{Act}_{t,\omega}^{\text{Res}}.P_{t,\omega}^{\text{del}}.\lambda_{t,\omega}^{\text{Bal}} \tag{5.36}$$

$$\text{Cost}_{t,\omega}^{\text{Imb}} = \lambda_{t,\omega}^{\text{DA}}.r_t^{-}.\Delta_{t,\omega}^{-} \tag{5.37}$$

$$\text{Income}_{t,\omega}^{\text{Call}} = \text{Act}_{t,\omega}^{\text{Res}}.\lambda_{t,\omega}^{\text{Bal}}.P_{t,\omega}^{\text{del}} \tag{5.38}$$

$$\text{Income}_{t,\omega}^{\text{Imb}} = \lambda_{t,\omega}^{\text{DA}}.r_t^{+}.\Delta_{t,\omega}^{+} \tag{5.39}$$

$$\text{Cost}_{\omega}^{\text{Res}} = \sum_{v} \left[\sum_{t=t_{\text{Connect}(v,\omega)}}^{t_{\text{Full}(v,\omega)}} P_{v,t,\omega}^{\text{G2V}}.\lambda_t^{\text{TariffRes}} \right] \tag{5.40}$$

$$\text{Income}_{\omega}^{\text{Charge}} = \sum_{v} \left[\sum_{t=t_{\text{Connect}(v,\omega)}}^{t_{\text{Full}(v,\omega)}} P_{v,t,\omega}^{\text{G2V}}.\lambda_t^{\text{TariffCharge}} \right] \tag{5.41}$$

$$\text{Cost}_{\omega}^{\text{Deg.}} = \sum_{v} \left[\sum_{t \in t_{\text{Connect}(v,\omega)}} P_v^{\text{max}}.C_d \right] \tag{5.42}$$

where:
dr is the annual discount rate
N_y is the number of years the device will last

Equation 5.29 shows the objective function of the scheduling problem and denotes the components of aggregator's profit. The objective of the aggregator is maximizing the profit in a certain period. Obviously, the profit is dependent on the behavior of the aggregator in the markets and subsequently, it is a function of uncertain variables that depend on markets sequences.

The term Cost$_{\text{Infra}}$ in Equation 5.30 denotes the infrastructure cost. The infrastructure cost includes on board incremental cost and wiring upgrade cost [45].

The aggregator income that resulted from participation in the spinning reserve market has been considered in Equation 5.31. The aggregator incomes that resulted from participation in the day-ahead energy and intra-day markets have been considered in Equations 5.32 and 5.33. Equation 5.34 represents the purchase costs from the intraday market to compensate the short on generation. Equation 5.35 denotes the purchase cost of electrical energy from the electricity market in order to charge the battery of the PEVs in scenario ω. The inability of the aggregator for energy generation at the time of being called by ISO may be caused by an error in predicting uncertain parameters. Reliability of the distribution system is modeled by FOR^{Agg}. Equation 5.36 presents the purchase cost of electrical energy in order to meet the aggregator obligations while being called for energy generation in the spinning reserve market. Equation 5.37 represents the imbalance cost due to lack of injection in comparison with day-ahead offers. Equation 5.38 considers the aggregator income resulted from being called by the ISO in order to generate electrical energy in the spinning reserve market. Equation 5.39 represents the imbalance income because of surplus of injection compared to day-ahead offers. Equation 5.40 denotes the aggregator cost of the contract with owners to persuade them to participate in the spinning reserve market. $\lambda_t^{TariffCharge}$ and $\lambda_t^{TariffRes}$ are the tariffs of participating in the energy and spinning reserve markets, respectively. Equation 5.41 shows the aggregator income resulted from receiving the batteries charge cost from PEV owners. Equation 5.42 denotes the cost of battery degradation because of participating in the electricity markets in vehicle-to-grid (V2G) mode [45]. This cost is considered as wear for V2G because of extra cycling of the battery in €/kWh. Therefore, it is related to battery capital cost and the battery lifetime as presented in Equation 5.43.

$$C_d = \frac{C_{battery}}{L_{ET}} \tag{5.43}$$

The objective function is maximized considering the constraints described below:

$$0 < SOC^{min} \le SOC_{v,t,\omega} \le SOC^{max} < 1 \tag{5.44}$$

$$SOC_{v,t,\omega} = SOC_{v,t-1,\omega} + \alpha_{v,t,\omega}.\eta_v^C.P_{v,t,\omega}^{G2V} - \beta_{v,t,\omega}.P_{v,t,\omega}^{V2G} \tag{5.45}$$

$$\alpha_{v,t,\omega} + \beta_{v,t,\omega} = 1 \tag{5.46}$$

Equation 5.44 is applied to avoid being overcharged and to take into account the depth of discharge of all connected PEVs during their connection. Equations 5.45 and 5.46 introduce changes in SOC of PEVs. Binary variables α and β ensure a PEV is not charged and discharged at the same time.

It is assumed that PEVs can be charged/discharged at any rate less than their maximum charging/discharging rates, which depend on their infrastructure [46]. The constraints are formulated as below:

$$r_{v,t,\omega}^{charge} = \frac{\left(SOC_{v,t,\omega} - SOC_{v,t-1,\omega}\right)}{\eta_v^C} \tag{5.47}$$

$$r_{v,t,\omega}^{discharge} = \left(SOC_{v,t-1,\omega} - SOC_{v,t,\omega}\right).\eta_v^D \tag{5.48}$$

$$r_{v,t,\omega}^{\text{charge}} \le r_v^{\text{charge,max}} \tag{5.49}$$

$$r_{v,t,\omega}^{\text{discharge}} \le r_v^{\text{discharge,max}} \tag{5.50}$$

The total scheduled energy of the aggregator in both day-ahead and intraday markets is given in Equation 5.51. Equation 5.52 makes the aggregator offer to the electricity markets based on the power of PEVs in V2G mode.

$$P_{t,\omega}^{\text{Sch}} + \Delta_{t,\omega}^+ - \Delta_{t,\omega}^- = P_{t,\omega}^{\text{DA}} + P_{t,\omega}^{\text{Intra,sell}} - P_{t,\omega}^{\text{Intra,buy}} \tag{5.51}$$

$$P_{t,\omega}^{\text{DA}} + P_{t,\omega}^{\text{Intra,sell}} \le \sum_v \left[\eta_v^D . P_{v,t,\omega}^{\text{V2G}} \right] \tag{5.52}$$

$$P_{p,t,\omega}^{\text{Res}} + P_{p,t,\omega}^{\text{NRes}} \le \sum_v \left[\eta_v^D . SOC_{v,t,\omega} \right] \tag{5.53}$$

Equation 5.53 ensures that the aggregator offers to the reserve market based on the power of PEVs in V2G mode.

$$\Delta_{t,\omega} = P_{t,\omega} - P_{t,\omega}^{\text{DA}} \tag{5.54}$$

$$\Delta_{t,\omega} = \Delta_{t,\omega}^+ - \Delta_{t,\omega}^- \tag{5.55}$$

Equations 5.54 and 5.55 have been employed to obtain energy deviations using the scheduled energy.

5.3.3 Economic Model of Responsive Demand

Justifying the impacts of implementing DR programs is an important issue for decision making in power system operation and planning. DR models have been developed for evaluating the impact of DR programs on power system attributes, including electricity prices, reliability, peak demands, and other power system characteristics. Obviously, more precise DR models result in more realistic estimations. DR modeling based on the concept of price elasticity of demand has been widely addressed in [47–50]. Elasticity is defined as the load's reaction to the electricity price. As the elasticity increases, the load sensitivity to price increases as well. Actually, demand can react to change in electricity tariffs in one of following ways. A set of loads are reduced without recovering it later, so-called fixed loads. Such loads have sensitivity just in a single period and it is called *self-elasticity*. Some other loads could be moved from the peak periods to off-peak periods as required, namely, transferable loads. Such behavior is called *multiperiod sensitivity* and it is evaluated by *cross elasticity*.

In order to model responsive demand, the procedure proposed in [48,49] has to be implemented. Authors showed that the following equation declares how much should be the customer's consumption to achieve maximum benefit in a 24-hour interval while participating in different DR programs.

$$d_t = d_{0,t} \cdot \left\{ \begin{array}{c} 1 + \dfrac{E(t,t)\left[\lambda_t - \lambda_{0,t} + A_t + \text{pen}_t\right]}{\lambda_{0,t}} \\[3mm] + \displaystyle\sum_{\substack{t'=1 \\ t' \neq t}}^{24} E(t,t') \dfrac{\left[\lambda_{t'} - \lambda_{0,t'} + A_{t'} + \text{pen}_{t'}\right]}{\lambda_{0,t'}} \end{array} \right\} \qquad (5.56)$$

where:

λ is related to DR program type and tariffs

Depending on incentive-based or penalty-based DR programs, A_t and pen_t can be accommodated in the above formula. The achieved formula has been employed in many recent papers to model the responsiveness of the demand. Interruptible/curtailable service (I/C) as one of the most executive DR programs is modeled using the price elasticity concept in [48]. Moreover, the impacts of other DR programs are also investigated and prioritized in [49]. An improved DR model is developed in [50], which considers the customer's behavior. The results show that the incentive-based DR programs have more effectiveness. This is due to the fact that in most communities, reward leads to a significant improvement in subjects' behavior. The study in [51] has gone a step further by proposing a flexible price elasticity concept and model. DR treated as a virtual generation resource in [52] whose marginal cost and relevant constraints are calculated due to customer information in the similar way as conventional generation units.

All of the reviewed models are applicable to short-term DR programs. While according to integrated resource portfolio planning approach, long-term DR modeling is also required and should be considered in future studies. Another aspect of DR modeling that requires more research is its capability in modeling uncertain and probabilistic parameters. Currently, uncertainties associated with DR resources are an impediment to accepting functionality of DR resources in the competitive electricity markets, where traditional resources have more deterministic features.

5.4 Supplementary Models for the Optimization Constraints

5.4.1 Model of Conventional Units

In this section, the detailed model of conventional generating units that is used in the optimization models is developed. First of all, constraints in Equations 5.57 and 5.58 formulate the generator cost function using a nondecreasing step-wise linear approximation. Constraints in Equations 5.59

and 5.60 limit the output power considering also the scheduled down and up reserves, respectively.

$$P_{i,t}^{S} = \sum_{f \in F^i} b_{i,f,t} \quad \forall i,t \tag{5.57}$$

$$0 \leq b_{i,f,t} \leq B_{i,f,t} \quad \forall i,f,t \tag{5.58}$$

$$P_{i,t}^{S} - R_{i,t}^{D} \geq P_{i}^{\min} \cdot u_{i,t} \quad \forall i,t \tag{5.59}$$

$$P_{i,t}^{S} + R_{i,t}^{U} \geq P_{i}^{\max} \cdot u_{i,t} \quad \forall i,t \tag{5.60}$$

The minimum up- and down-time constraints are enforced by Equations 5.61 and 5.62.

$$\sum_{\tau=t-\mathrm{UT}_i+1}^{t} y_{i,t} \leq u_{i,t} \quad \forall i,t \tag{5.61}$$

$$\sum_{\tau=t-\mathrm{DT}_i+1}^{t} z_{i,t} \leq 1 - u_{i,t} \quad \forall i,t \tag{5.62}$$

The start-up and shutdown status change logic is described by Equation 5.63, whereas Equation 5.64 states that it is not possible for a unit to start-up and shutdown during the same period.

$$y_{i,t} - z_{i,t} = u_{i,t} - u_{i,(t-1)} \quad \forall i,t \tag{5.63}$$

$$y_{i,t} + z_{i,t} \leq 1 \quad \forall i,t \tag{5.64}$$

Changes in the output power are limited by the ramp-up and ramp-down constraints described by Equations 5.65 and 5.66. ΔT is the length of a period in minutes.

$$P_{i,t}^{S} - P_{i,(t-1)}^{S} \leq \Delta T \cdot \mathrm{RU}_i \quad \forall i,t \tag{5.65}$$

$$P_{i,(t-1)}^{S} - P_{i,t}^{S} \leq \Delta T \cdot \mathrm{RD}_i \quad \forall i,t \tag{5.66}$$

The procurement of spinning-up, nonspinning, and spinning-down reserves is limited by constraints in Equations 5.67 through 5.69. T^S and T^{NS} is the time in minutes the reserves should respond.

$$0 \leq R_{i,t}^{D} \leq T^S \cdot \mathrm{RD}_i \cdot u_{i,t} \quad \forall i,t \tag{5.67}$$

$$0 \leq R_{i,t}^{U} \leq T^S \cdot \mathrm{RU}_i \cdot u_{i,t} \quad \forall i,t \tag{5.68}$$

$$0 \leq R_{i,t}^{NS} \leq T^{NS} \cdot \mathrm{RU}_i \cdot \left(1 - u_{i,t}\right) \quad \forall i,t \tag{5.69}$$

5.4.2 Model of Wind Power Producers

The constraint in Equation 5.70 limits the scheduled wind power production. In [20], the minimum and maximum limits are considered as parameters that are submitted by the wind power producer together with its bidding and the maximum limit is considered infinite. Other limits that could be imposed are the maximum wind scenario values, the wind power forecast, and the installed capacity of the wind farm.

$$0 \leq P_{w,t}^{WP,S} \leq P_{w,t}^{WP,max} \quad \forall w,t \tag{5.70}$$

A portion of available wind production can be spilled if it is necessary to facilitate the operation of the power system. This is enforced by Equation 5.71.

$$0 \leq S_{w,s,t} \leq P_{w,s,t}^{WP} \quad \forall w,s,t \tag{5.71}$$

From an economical perspective, the wind producer is not considered as a price maker and is compensated by a feed in tariff scheme. Thus, for the day-ahead market clearing, the wind producer is considered to offer all its available energy (minus the portion spilled) without cost.

5.4.3 Model of Network

A linear representation of the network is considered, neglecting active power losses. A linear approximation of the active power losses may be included as explained in [20]. Constraints in Equations 5.72 through 5.76 stand for the general network representation adopted. It is enforced for scenario independent variables (first-stage), but it may be also enforced for the second-stage variables by substituting them with the corresponding second-stage variables.

$$A_{n,w}^{wf} \sum_{w \in W} P_{w,t}^{WP,S} + A_{n,i}^{un} \sum_{i \in I} P_{i,t}^{S} - \sum_{l \in L:n \equiv nn} f_{l,t} + \sum_{l \in L:n \equiv n} f_{l,t}$$
$$= A_{n,r}^{inel} \sum_{r \in R} L_{r,t} + A_{n,d}^{ind} \sum_{d \in D} P_{d,t}^{ind,S} \quad \forall n,t \tag{5.72}$$

$$f_{l,t} = B_{l,n} \cdot (\delta_{n,t} - \delta_{nn,t}) \quad \forall (n,nn) \equiv l,n,t \tag{5.73}$$

$$-\pi \leq \delta_{n,t} \leq \pi \quad \forall n,t \tag{5.74}$$

$$\delta_{n,t} = 0 \quad \forall t, n \equiv \text{reference} \tag{5.75}$$

$$-f_l^{max} \leq f_{l,t} \leq f_l^{max} \quad \forall l,t \tag{5.76}$$

Specifically, Equation 5.72 enforces the power balance for each node. $A_{n,x}$ is the node to resource x incidence matrix, where x can be unit, wind-farm, inelastic load, or industrial consumer. The constraint in Equation 5.73 determines the power flow through the transmission lines. Angles are constrained by Equation 5.74, whereas the reference bus angle is fixed to zero by (5.75). Transmission capacity constraints are imposed by Equation (5.76).

5.5 Formulation of the Two-Stage Stochastic Optimization Problems

5.5.1 Load-Following Reserve Procurement Considering Industrial Customers

To accommodate the uncertain nature of wind power production, a network-constrained day-ahead market-clearing model is proposed under a two-stage stochastic programming framework, considering energy and reserve services procured by industrial consumers.

The first stage of the model represents the day-ahead market, where energy and reserves are jointly scheduled to balance wind volatility. The variables of this stage do not depend on any specific scenario realization and constitute here-and-now decisions. The second stage of the model stands for several actual system operation possibilities. The variables of this stage are scenario-dependent and have different values for every single wind scenario. The second stage variables constitute wait-and-see decisions. A set of constraints that link the day-ahead decisions and the several possible outcomes of the random wind production are also developed.

5.5.1.1 Objective Function

The objective function (Equation 5.77) stands for the minimization of the total expected cost (EC) emerging from the system operation. The first line considers the cost of the committed energy to be produced by generators, the start-up and shutdown costs of the generating units as well as the reserve commitment cost. The second line stands for the cost of committing reserves from the industrial consumers as well as the utility benefit that emerges from the energy usage by the industrial costumer. The next two lines are scenario dependent. The third line accounts for the cost because of the commitment status of generating units and the extra cost because of deploying reserves. Finally, the fourth line considers the cost of deploying reserves from the industrial consumers. Also, two artificial costs are taken into account: wind spillage cost and cost of energy not supplied (because of involuntary load shedding).

$$
\begin{aligned}
\text{EC} = \sum_{t} \Bigg\{ & \sum_{i} \Bigg[\sum_{f \in F^i} \left(C_{i,f,t} \cdot b_{i,f,t} \right) + \text{SUC}_i \cdot y_{i,t} + \text{SDC}_i \cdot z_{i,t} \\
& + C_{i,t}^{R^D} \cdot R_{i,t}^D + C_{i,t}^{R^U} \cdot R_{i,t}^U + C_{i,t}^{R^{NS}} \cdot R_{i,t}^{NS} \Bigg] \\
& + \sum_{d} \left(C_{d,t}^{R^{D,In}} \cdot R_{d,t}^{D,ind} + C_{d,t}^{R^{U,In}} \cdot R_{d,t}^{U,ind} \right. \\
& \left. + C_{d,t}^{R^{NS,In}} \cdot R_{d,t}^{NS,ind} - \lambda_{d,t}^D \cdot P_{d,t}^{ind,S} \right) \Bigg\} \\
& + \sum_{s} \pi_s \sum_{t} \Bigg\{ \sum_{i} \Big[\text{SUC}_i \cdot \left(y_{i,t,s}^2 - y_{i,t} \right) + \text{SDC}_i \cdot \left(z_{i,t,s}^2 - z_{i,t} \right) \\
& + \sum_{f \in F^i} \left(C_{i,f,t} \cdot r_{i,f,t,s}^G \right) \Big] \\
& + \sum_{d} \lambda_{d,t}^D \sum_{g \in G} \sum_{p \in P} \left(r_{d,g,p,t,s}^{U,pro} - r_{d,g,p,t,s}^{D,pro} - r_{d,g,p,t,s}^{NS,pro} \right) \\
& + \sum_{w} \left(V^S \cdot S_{w,t,s} \right) + \sum_{r} \left(V^{LOL} \cdot L_{r,t,s}^{shed} \right) \Bigg\}
\end{aligned}
\tag{5.77}
$$

5.5.1.2 Constraints

- **First-Stage Constraints**

  In this stage, the relations constituting the industrial load model (Equations 5.7 through 5.28 are incorporated as constraints of the optimization problem. Detailed modeling of the generating units (Equations 5.57 through 5.69) and wind power producer (Equation 5.70) is also directly included in the first stage of the problem. It is to be noted that network constraints are not enforced in the first stage. A day-ahead market energy balance is imposed instead (Equation 5.78). Nonetheless, any other scheme may be implemented within the proposed structure.

$$
\sum_{i \in I} P_{i,t}^S + \sum_{w \in W} P_{w,t}^{WP,S} = \sum_{r \in R} L_{r,t} + \sum_{d \in D} P_{d,t}^{ind,S} \quad \forall t
\tag{5.78}
$$

- **Second-Stage Constraints**

  First of all, constraints that limit the output power of the conventional units should be enforced (Equations 5.79 and 5.80).

$$
P_{i,t,s}^G \geq P_i^{min} \cdot u_{i,t,s}^2 \quad \forall i,t,s
\tag{5.79}
$$

$$
P_{i,t,s}^G \leq P_i^{max} \cdot u_{i,t,s}^2 \quad \forall i,t,s
\tag{5.80}
$$

The minimum up- and down-time constraints of the conventional units are also enforced in the second stage by substituting the first-stage variables of constraints in Equations 5.61 and 5.62 by the appropriate second-stage variables. Similarly, the unit-commitment logic constraints are included in the second stage of the problem by appropriate substitutions of the first-stage variables in Equations 5.63 and 5.64. Ramp-up and ramp-down limits are rendered active in the second stage as well by substituting $P_{i,t}^S$ with $P_{i,t,s}^G$ in Equations 5.65 and 5.66.

Apart from the constraints that allow for a portion of available wind production to be spilled (Equation 5.71), as a last resort, the ISO may decide to shed a part of the inelastic demand in order to retain the consistency of the system. This is enforced by Equation 5.81.

$$0 \leq L_{r,t,s}^{\text{shed}} \leq L_{r,t} \quad \forall r,t,s \tag{5.81}$$

The scenario-dependent analogous of the relevant industrial load constraints presented and explained in the first stage of the problem is enforced by substituting in Equations 5.7 and 5.10 through 5.19 the first-stage variables with the appropriate scenario-dependent second-stage variables (as described in the nomenclature section).

Finally, in the second stage of the problem, network constraints are also taken into account (Equations 5.73 through 5.76). The nodal balance is expressed by Equation 5.82.

$$
\begin{aligned}
A_{n,w}^{\text{wf}} \cdot \sum_{w \in W}(P_{w,t,s}^{\text{WP}} - S_{w,t,s}) + A_{n,i}^{\text{un}} \cdot \sum_{i \in I}P_{i,t,s}^G - \sum_{l \in L:n \equiv nn}f_{l,t,s} \\
+ \sum_{l \in L:n \equiv n}f_{l,t,s} = A_{n,r}^{\text{inel}} \cdot \sum_{r \in R}(L_{r,t} - L_{r,t,s}^{\text{shed}}) \\
+ A_{n,d}^{\text{ind}} \cdot \sum_{d \in D}P_{d,t,s}^{\text{ind},C} \quad \forall n,t,s
\end{aligned}
\tag{5.82}
$$

- **Linking Constraints**

This set of constraints links the market and the actual operation of the power system. It enforces the fact that reserves in the actual operation of the power system are no longer a standby capacity, but are materialized as energy.

The constraint in Equation 5.83 involves the scheduled day-ahead unit outputs with the scenario-dependent deployed power. It is clear that up-spinning and nonspinning reserves stand for an increase of the output power and down-spinning stands for a decrease.

$$P_{i,t,s}^G = P_{i,t}^S + r_{i,t,s}^U + r_{i,t,s}^{\text{NS}} - r_{i,t,s}^D \quad \forall i,t,s \tag{5.83}$$

Constraints in Equations 5.84 through 5.86 stipulate that the deployed reserves cannot be greater than their respective scheduled values.

$$0 \leq r_{i,t,s}^{U} \leq R_{i,t,s}^{U} \quad \forall i,t,s \tag{5.84}$$

$$0 \leq r_{i,t,s}^{NS} \leq R_{i,t,s}^{NS} \quad \forall i,t,s \tag{5.85}$$

$$0 \leq r_{i,t,s}^{D} \leq R_{i,t,s}^{D} \quad \forall i,t,s \tag{5.86}$$

Constraints in Equations 5.87 through 5.89 decompose the deployed reserves into energy blocks.

$$r_{i,t,s}^{U} + r_{i,t,s}^{NS} - r_{i,t,s}^{D} = \sum_{f \in F^{i}} r_{i,f,t,s}^{G} \quad \forall i,t,s \tag{5.87}$$

$$r_{i,f,t,s}^{G} \leq B_{i,f,t} - b_{i,f,t} \quad \forall i,f,t,s \tag{5.88}$$

$$r_{i,f,t,s}^{G} \geq -b_{i,f,t} \quad \forall i,f,t,s \tag{5.89}$$

Constraints in Equations 5.90 and 5.91 determine the actual consumption of the industrial load. Especially, Equation 5.91 reallocates the power of every single process (through the determination of reserves), whereas Equation 5.90 sums all the consumptions of the single processes up to the actual consumption of the industry.

$$P_{d,t,s}^{ind,C} = D_{d,t}^{min} + \sum_{g \in G} \sum_{p \in P} P_{p,g,d,t,s}^{pro,C} \quad \forall d,t,s \tag{5.90}$$

$$P_{p,g,d,t,s}^{pro,C} = P_{p,g,d,t,s}^{pro,S} + r_{p,g,d,t,s}^{D,pro} - r_{p,g,d,t,s}^{U,pro} + r_{p,g,d,t,s}^{NS,pro} \quad \forall p,g,d,t,s \tag{5.91}$$

Constraints in Equation 5.92 through 5.100 determine the reserves provided by the reallocation of the energy needs of the processes.

$$r_{p,g,d,t,s}^{U,pro} = a_{p,g,d,t,s}^{up,rt} \cdot P_{p,g,d}^{line} \quad \forall p,g,d,t,s \tag{5.92}$$

$$0 \leq r_{p,g,d,t,s}^{U,pro} \leq R_{p,g,d,t}^{U,pro} \quad \forall p,g,d,t,s \tag{5.93}$$

$$0 \leq a_{p,g,d,t,s}^{up,rt} \leq a_{p,g,d,t}^{up} \quad \forall p,g,d,t,s \tag{5.94}$$

$$r_{p,g,d,t,s}^{D,pro} = a_{p,g,d,t,s}^{down,rt} \cdot P_{p,g,d}^{line} \quad \forall p,g,d,t,s \tag{5.95}$$

$$0 \leq r_{p,g,d,t,s}^{D,pro} \leq R_{p,g,d,t}^{D,pro} \quad \forall p,g,d,t,s \tag{5.96}$$

$$0 \leq a_{p,g,d,t,s}^{down,rt} \leq a_{p,g,d,t}^{down} \quad \forall p,g,d,t,s \tag{5.97}$$

$$r_{p,g,d,t,s}^{NS,pro} = a_{p,g,d,t,s}^{ns,rt} \cdot P_{p,g,d}^{line} \quad \forall p,g,d,t,s \tag{5.98}$$

$$0 \le r_{p,g,d,t,s}^{NS,pro} \le R_{p,g,d,t}^{NS,pro} \quad \forall p,g,d,t,s \tag{5.99}$$

$$0 \le a_{p,g,d,t,s}^{ns,rt} \le a_{p,g,d,t}^{ns} \quad \forall p,g,d,t,s \tag{5.100}$$

5.5.2 Load-Following Reserve Procurement Considering the Economic Model of Responsive Demand and Aggregation of PEVs

5.5.2.1 Objective Function

The required spinning reserve capacity of each power system is determined based on its required reliability. The methods to determine the required reserve is generally classified into two major categories: deterministic methods and probabilistic methods. The simplest method to determine the required spinning reserve capacity is the deterministic method. Due to simple and fast calculation, this method has very wide applications. In this method, the required spinning reserve capacity is considered as a percent of total load or the largest committed generation unit [53–55]. In this method, probabilistic nature of power system is not considered, whereas the components of power system are faced with several contingencies. Moreover, in this method, between different types of customers, with different importance, the distinction is not made. Therefore, the spinning reserve supply by the deterministic method may cause that ISO purchases capacity over requirements and increases the costs. Vice versa, this method may result that ISO purchases capacity less than needed and occurrence of an event causes significant damage to customers.

The probabilistic method is a method that considers probabilistic nature of various components of power system. This method is applied in some real-world markets (e.g., Pennsylvania New Jersey Maryland [PJM] market) in order to determine the required reserve capacity [56–59]. In [56–59], the unit commitment problem has been solved based on the forced outage rate (FOR) of generation units. One of the constraints of the unit commitment problem is the risk of system that should be less than a predetermined value.

In probabilistic methods, the probability of generation deficit is calculated, not the value of one. In other words, using this method, the amount of reserve is increased until the value of risk is less than a default value. Therefore, in this method, when the load is not supplied, the amount of customers' damage is ignored.

The reliability of power system is improved by increasing the amount of spinning reserve capacity. However, supplying this amount of capacity will increase the operation cost of the system. Therefore, a compromise between network reliability and the cost of its provision is necessary. To this end, in this chapter, a benefit–cost method is employed. The benefit–cost method is a more complete method to calculate the optimal amount of reserve. This method overcomes the disadvantages of above methods, such as not considering the amount of customers' damage. In the benefit–cost method, the reserve capacity is determined in a manner that the obtained benefits of the reserve supply to be greater than its costs. In these bases, first, ISO converts the amount of lost load (related to reliability of system) to the cost, using the value of lost load (VOLL) factor. Next, it schedules conventional units and DR resources such that the total operation costs of the system with large amount of wind power are minimized including cost of the spinning reserve supply. On this basis, an optimal wind-thermal generation scheduling is determined considering different DR programs with the aim of increasing system flexibility to facilitate wind power integration. The mathematic formulation of the objective function is expressed as follows:

$$\text{Min} \sum_s \omega_s \cdot \left\{ \begin{array}{l} \sum_i \sum_t \left[\left(C_{i,f,t} \cdot b_{i,f,t} \right) + SUC_i.y_{i,t} + SDC_i.z_{i,t} \right. \\ \left. + C_{i,t}^{R^D} \cdot R_{i,t}^D + C_{i,t}^{R^U} \cdot R_{i,t}^U \right] + \sum_t \left(C_t^{EDRP} + Cost_t^{Risk} \right) \end{array} \right\} \qquad (5.101)$$

In Equation 5.101, the first two terms are start-up and shutdown costs of unit i at hour t. The spinning reserve capacities are formulated by the next term. In the next two terms, the cost function of the generation unit is linearized by a set of piecewise blocks. Afterward, the cost of deploying spinning is represented. The two last terms are associated with the emergency DR program (EDRP) cost and the cost of system risk. The cost of system risk is expressed in Equation 5.102.

$$Cost_t^{Risk} = \sum_j \lambda_{j,t}^{outage} \cdot \sum_{L=1}^{N_L} LNS_{L,t}.VOLL_{L,t} \qquad (5.102)$$

$$\lambda_{j,t}^{outage} = \frac{FOR_j}{1 - FOR_j} \prod_j \left(1 - FOR_j \right) \qquad (5.103)$$

$$ELNS_t = \sum_{j=1}^{N_j} \lambda_{j,t}^{outage} \cdot \sum_{L=1}^{N_L} LNS_{L,t} \qquad (5.104)$$

Equation 5.103 expresses the outage probability of the element j at time t. Equation 5.104 denotes the value of system risk or in other words, the expected load not supplied at time t.

5.5.2.2 Constraints

In this section, the constraints of the optimization problem are presented. First of all, the model described in Section 5.5.2.1 for the conventional generating units (Equations 5.57 through 5.69) is included in the constraints of the problem. Furthermore, to ensure the power system security, hourly generation and load dispatch in each scenario must satisfy the power balance constraint at each bus. In this regard, DC load flow equation is applied, as it can be seen in Equation 5.105.

$$\sum_i \left[P_{i,t,s}^{tot} - P_{n,t}^{mod} \right] = \sum_l F_{l,t,s} \tag{5.105}$$

where:

$$P_{i,t,s}^{tot} = P_i^{min}.I_{i,t} + \sum_m P_{i,t,s}^e(m) + \mathrm{sr}_{i,t,s} \tag{5.106}$$

$$0 \le P_{i,t,s}^e(m) \le P_i^{max}(m) \tag{5.107}$$

Moreover, in Equation 5.105, $P_{n,t}^{mod}$ is the modified demand of bus n at hour t after implementing DR, which is allocated to appropriate buses, as represented in Equations 5.108 and 5.109 for price-based and incentive-based DR programs, respectively.

$$P_{n,t}^{mod} = LD_n.\left\{ (1-\eta_d).d_{0,t} + \eta_d.d_{0,t}.\left[\sum_{t'=1}^{24} E(t,t').\frac{[\lambda_{t'} - \lambda_{0,t'}]}{\lambda_{0,t'}} \right] \right\} \tag{5.108}$$

$$P_{n,t}^{mod} = LD_n.\left\{ (1-\eta_d).d_{0,t} + \eta_d.d_{0,t}.\left[\sum_{t'=1}^{24} E(t,t').\frac{A_{t'}}{\lambda_{0,t'}} \right] \right\} \tag{5.109}$$

5.6 Performance Indices

Strong growth of renewable resources increases the need for ramp-up/-down services by the conventional generation units. In order to provide the ramping requirements of the system, increasing system flexibility seems a crucial issue. In fact, the more flexibility means the less regulation services.

Due to technical restrictions of conventional generating units, such as ramp-rate constraints, and minimum up/down times, the need for more flexible resource is essential. In the 24th wind task of the International Energy Agency (IEA), which investigates issues, impacts, and economics of wind power grid integration, DR resources were introduced as the most flexible and a cost-effective option to facilitate the grid integration of wind power [60]. In this regard, since load changes are very important for regulated activities of wind power, some novel measures have been proposed in this section.

Based on this, in order to investigate the impact of different DR programs on facilitating grid integration of wind power, a novel measure is applied as in [61]. Average DR benefit (ADRB) represents the decrease in system operation cost as a result of an additional 1-MWh integration of wind power. In other words, this measure represents the impression of DR implementation on the average cost reduction of 1-MWh additional wind power injection to the power system. The measure is presented in Equation 5.110.

$$\text{ADRB} = \frac{1}{24} \sum_{t=1}^{24} \frac{\left[TCost_t^{\text{NoDR}} - TCost_t^{\text{DR}} \right]}{\sum_s \omega_s . W_{st}^{\text{int}}} \tag{5.110}$$

Moreover, in order to measure the impact of DR programs on the load curve and consequently facilitating grid integration of wind power, three other measures are proposed. Based on this, load turbulence index (LTI) is proposed to indicate the smoothness of the load curve. The lower LTI shows the smoother load curve and the easier regulation. The index is presented in Equation 5.111. Another feature of load curve that is very important for regulated activities is the rate of demand change. Bigger changes of demand cause more difficulties in following the load. On this basis, maximum load up and down indices (MLU and MLD) are utilized to measure the maximum rate of demand changes. The measures are presented in Equations 5.112 and 5.113, respectively.

$$\text{LTI} = \frac{1}{24} \sum_{t=1}^{24} \frac{|d_t - d_{t-1}|}{d_t} \tag{5.111}$$

$$\text{MLU} = \max\{d_t - d_{t-1}, t = 1, \ldots, 24\} \tag{5.112}$$

$$\text{MLD} = \max\{d_{t-1} - d_t, t = 1, \ldots, 24\} \tag{5.113}$$

5.7 Case Studies on the Insular Power System of Crete

In order to illustrate the potential effects of a flexible industrial consumer that offers energy and reserve services, an extensive test case is performed on the insular power system of Crete. Crete is the largest noninterconnected Greek island both in terms of area and population. Currently, it comprises 25 conventional units of different technologies with an installed capacity of 799 MW in three power stations across the island (Xania, Linoperamata, and Atherinolakkos). Furthermore, there are 29 wind parks with a total installed capacity of 184 MW. Additionally, the installed photovoltaic (PV) power is 94 MW. However, it is reported that the technically exploitable potential of producing power from solar and wind resources in the island is more than 1 GW. An overview of the generation mix of the power system of Crete at the end of 2013 is presented in [62], whereas a description of the transmission system characteristics may be found in [63].

The facts described above render evident that it is of interest to explore the potential effects of responsive demand on the operation and economics of this insular power system. It is to be noted that for the purposes of the following test cases, the installed PV generation is not taken into account.

5.7.1 Case 1: Flexible Industrial Consumer

5.7.1.1 Description of the Tests

In order to demonstrate the potential benefits that flexible industrial consumers may provide to the operation of the insular power system, a base case is first analyzed in which the industrial consumption is considered inelastic. For the sake of simplicity, we consider that the load that corresponds to industrial consumers, discussed later in this chapter, contributes to energy and reserve services is located at a single bus that is located in the center of the island and supplies approximately 15% of the total load. Furthermore, it should be stated that the aggregated industrial load assumed in this study does not necessarily correspond to a single industrial customer but may represent several industries. The total energy required by the industrial load is 280 MWh/day (2.3% of the total daily energy requirements of the whole system). Regarding the cost to procure energy and reserve services from the industrial consumer, a simplifying assumption is made: since the total energy required by the industrial customers to accomplish their purposes is guaranteed to be provided during the day, the utility of this type of load may be considered equal to zero since no economic losses occur. Nevertheless, industrial consumers are compensated with 0.5 €/MWh in order to be committed to offer load-following reserve services. This cost is significantly less than the price at which conventional units offer reserve and energy services. It is to be stated that in order to motivate industrial customers to dispose a part of their load to be controlled by the system operator, they may be offered attractive billing plans or other economic incentives that are out of the scope of the daily scheduling. Finally, as a last resort to maintain the consistency of the power system, the system operator may involuntarily curtail a part of a load under a high penalty (1000 €/MWh).The total system load together with the industrial load considered inelastic are depicted in Figure 5.7.

The operator of the power system of the island requires that the OCGT unit located at Xania is a must-run unit in order to support the voltage in the west side of the island and to provide AS. All the rest conventional units are primarily considered to be able to contribute to spinning reserves. The formulation proposed allows for nonspinning reserves; that is, units that are offline in the day-ahead scheduling may be committed to change their

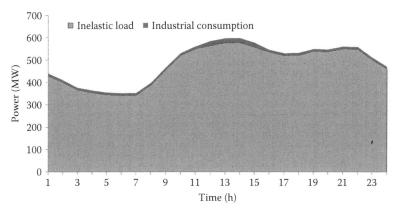

FIGURE 5.7
Total system load.

status if needed. However, since contingencies (sudden loss of generating units or transmission lines) or significant wind-ramping events are out of the scope of this study, nonspinning reserves are not considered as part of the day-ahead scheduling procedures of the power system of Crete. It is also considered that wind farms are compensated based on a feed-in-tariff system and as a result, they have zero cost as regards the operation of the day-ahead unit commitment and economic dispatch. Previously, it has been stated that wind spillage cost is an artificial cost that penalizes each MWh of available wind power generation that is curtailed. This cost is used in order to promote the utilization of the wind power generation but is difficult to determine its suitable value. For this reason, in this study, it is considered that the wind spillage cost is zero in order to obtain unbiased results concerning the operational impacts of the uncertain wind power generation.

To account for wind power generation stochasticity, five hundred equiprobable wind power scenarios were initially created for each wind park, based on a sampling approach. Specifically, an appropriate artificial neural network (ANN) forecasting model was first determined and trained with the available historical wind power data [62]. Then, the ANN model was used in order to formulate the wind power scenarios for the next day. In order to take into account the statistical correlation of the power output from neighboring wind plants, the scenario generation algorithm was appropriately modified to allow for the generation of spatial and temporal cross-correlated scenarios. To allow for computational tractability, a scenario reduction technique was then applied resulting in a reduced scenario set comprising 10 scenarios for the wind power generation. In the final scenario set, there is one dominant scenario (with a probability of 82%) and nine equiprobable scenarios (with a probability of 2%). The total available wind power per scenario is displayed in Figure 5.8.

After obtaining the results of the base case, different test cases are examined considering that the industrial consumption is flexible. The model presented

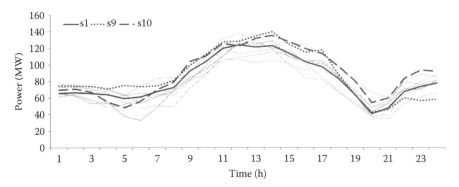

FIGURE 5.8
Available wind power production scenarios.

may cover a wide range of different processes that may be rendered available to control by the system operator in order to facilitate the grid operation. However, for the illustrative purposes of this chapter, only several characteristic types of processes and their parameters are examined.

Finally, comparisons are performed regarding the obtained results and relevant discussion is made.

5.7.1.2 Base Case

The optimal day-ahead scheduling that is employed to cover the needs of the load for the base case is presented in Figure 5.9. The daily wind power generation is 1945.88 MWh, which stands for the 16.56% of the total energy needs of the system. The cost of scheduling power from conventional units is 1113774.891 €. Despite the fact that wind power generation is free, its volatile nature imposes the need of scheduling reserves in order to cope with the

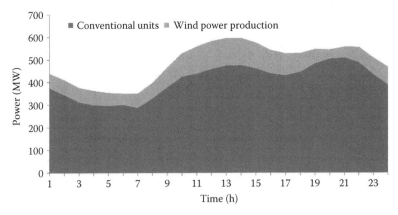

FIGURE 5.9
Day-ahead power scheduling (base case).

changes that occur during the actual operation of the power system regarding the available wind production (scenarios). The up-spinning reserve stands for an increase in the load in order to cover a deficit of wind production in real-time, whereas the down-spinning reserve stands for the opposite. It stated that the first-stage decision (day-ahead scheduling) regarding wind generation does not depend on any specific scenario. The cost of committing load-following reserves from the conventional generating units is 13153.184 €.

The fact that reserves are scheduled under a relatively high cost causes a percentage of wind production that is actually available (scenarios) to be spilled since the relative benefit of integrating this amount of free wind-farm energy (energy cost reduction) is less than the cost to compensate its uncertainty (reserve scheduling and deployment cost). Figure 5.10 presents the amount of wind energy spilled in each of the 10 scenarios. The basic observation based on Figure 5.10 is that scenario 1 bears the least wind energy spillage in contrast with scenarios 9 and 10. This may be justified in a straightforward way. First, scenario 1 is dominant and, therefore, has a major impact on the day-ahead scheduling of available wind power resulting in a reduced need of reserves to cover the deviations of this scenario. On the other hand, scenarios 9 and 10 are characterized by relatively higher amounts of available wind power generation. Since their probabilities of occurrence are insignificant compared to the dominant scenario, the reserves that would be needed to cover these relatively higher deviations would lead to an unjustified uneconomic operation of the power system.

The high probability of occurrence of scenario 1 affects the total expect cost of the daily operation of the insular power system. The total expected operational cost associated with the base case is 1151877,754 €. However, as it can be seen in Figure 5.11, in which the cumulative distribution function of the cost is presented, there is the possibility of 18% of incurring a cost up to 2.36% higher than the EC.

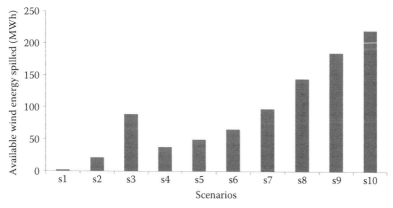

FIGURE 5.10
Available wind energy spilled in scenarios (base case).

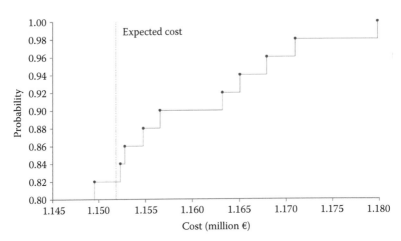

FIGURE 5.11
Cumulative distribution function of the cost (base case).

5.7.1.3 Flexible Industrial Consumer

It has been mentioned before that the flexibility of the industrial consumption by means of offering energy and reserve services in order to accommodate the uncertainty related to the wind power generation depends on the type of the processes that are subject to control of the system operator. Since the processes are considered to consume energy in discrete blocks, accordingly, the effect of their respective parameters on system operation should be examined in order to assess the potential benefits.

However, in order to illustrate the general functionality of the model, an illustrative test is first performed (Case 1–C1). In this respect, a total 280 MWh is considered available to be optimally allocated in 560 discrete blocks of 0.5 MW each during the 24 hours of the horizon. This stands for the case of a totally flexible process of type 2 (discontinuous) illustrated in Figure 5.6. with the following parameters: $T_{max} = 24h, a_{max} = 560, a_{max}^{h} = 560, P^{line} = 0.5$ MW.

Figure 5.12 presents the day-ahead power generation scheduling by both conventional and wind power generating units. The utilization of demand-side resources has led to a slight decrease in the power generation of the conventional units (~1.1%). As a result, the energy cost is also decreased. Furthermore, the optimal scheduling of the available flexible industrial consumer energy has led to a reduction of 3.34% as regards the peak power of the power system demand. In the same time, the scheduling of the load has a load valley filling effect since a portion of the consumption has been shifted to the relatively low-consumption periods of the day. Thus, the responsive industrial demand contributes toward mitigating the ramping requirements of the conventional power production units. However, the most significant contribution of the industrial load scheduling is the reduction in the cost of

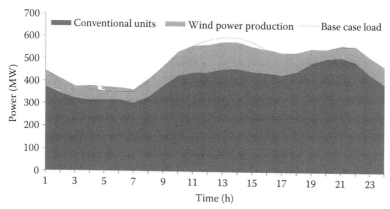

FIGURE 5.12
Day-ahead power scheduling (C1).

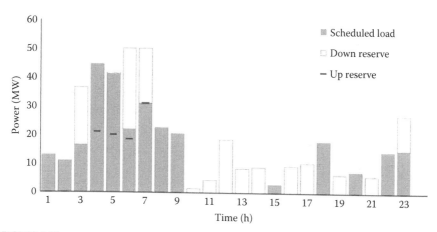

FIGURE 5.13
Scheduled industrial energy consumption and offered reserves (C1).

procuring reserve services by conventional units in order to accommodate the wind by 40.3%. The optimal scheduling of the flexible industrial consumer energy, together with the reserves scheduled by the demand side, is displayed in Figure 5.13.

Another profound effect of the introduction of the flexible industrial consumer into the operational assets of the insular power system is that the available wind production that is curtailed has been significantly reduced in comparison with the base case in all scenarios, as it can be seen in Figure 5.14. As a result, the total EC of the operation of the power system has dropped down to 1139192,039 € (a reduction of 1.1% in comparison with the base case). Besides, the probability of having a cost up to 2.27% greater than the expected has dropped down to 12% as can be seen in Figure 5.15. Considering this fact, it can be noticed that the introduction of controllable industrial consumption

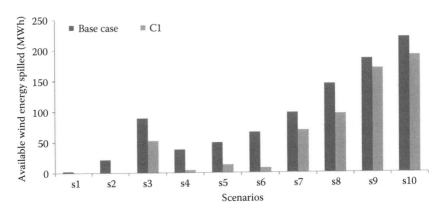

FIGURE 5.14
Available wind energy spilled in different scenarios (C1).

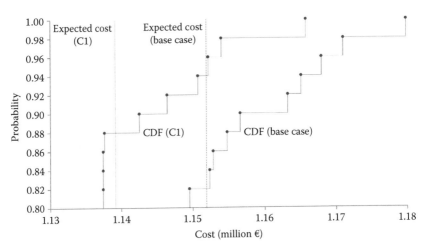

FIGURE 5.15
Cumulative distribution function of the cost (C1).

is a means of not only mitigating the total EC but also to control the dispersion of the cost in the different scenarios that are considered.

5.7.1.4 Investigation of the Flexible Industrial Demand Parameters Impact

In Sections 5.7.1.2 and 5.7.1.3, the base case in which demand-side resources were neglected as well as an illustrative case in which a significant amount of energy could be allocated by the system operator in a totally flexible fashion were presented. In order to evaluate the effect of the parameters of the flexible processes on the operation of the power system, several tests are performed and the results regarding their economic and operational impacts

TABLE 5.1

Parameters of the Flexible Industrial Processes

	Case	Energy (MWh)	Type	T_{max} (h)	a_{max}	a^h_{max}	P^{line} (MW)	Time
Totally flexible	C1	280	2	24	560	560	0.5	–
	C2-a	100	2	24	200	200	0.5	12–15: 80 MWh rest: 1 MWh
	C2-b	100	2	24	50	50	2	12–15: 80 MWh rest: 1 MWh
	C2-c	100	2	24	25	25	5	12–15: 80 MWh rest: 1 MWh
Flexible	C3-a	100	1	5	200	100	0.5	12–15: 80 MWh rest: 1 MWh
	C3-b	100	1	5	200	50	0.5	12–15: 80 MWh rest: 1 MWh
	C3-c	100	1	5	200	45	0.5	12–15: 80 MWh rest: 1 MWh
Inflexible	C4	80	1	4	1	1	20	12–15: 80 MWh

are discussed. The data of the cases examined are presented in Table 5.1 and the relevant results are gathered in Table 5.2.

First, the effects of totally flexible processes are analyzed. For cases C2-a to C2-c, an energy amount of 100 MWh (80 MWh originally allocated to periods 12–15 and 1 MWh from the rest periods) is rendered available at power blocks of different size. The results obtained suggest that the total EC, as well as the cost of committing conventional units to produce energy, increases with the increase of the size of the power blocks. In the same time, the cost of scheduling reserves from conventional units decreases. These results may be interpreted if it is noticed that the demand-side services are provided in discrete amounts of energy or reserve capacity. If a power block is scheduled to provide reserve services, then it should be deployed as a whole. As the size of the blocks increases, if a block of power is committed to contribute to reserves, then directly a larger amount of reserve is shifted from the generation-side and results into less reserve scheduling cost. Following the same rationale, the energy cost provision by generation side is increasing with the increase of the size of the power blocks because the ability of evenly allocating the power blocks to relatively low-demand periods is limited.

The total EC of cases C2-a to C2-c is less than the one of the base case. It may be also noticed that the probability of experiencing costs higher than the EC, together with the highest probable cost, significantly decrease. This is the direct result of allowing for more wind power generation penetration to the system.

Afterward, the relatively flexible processes are studied. In cases C3-a to C3-c, 100 MWh of industrial consumption are rendered available. It is

TABLE 5.2

Results for the Different Test Cases

Case	Conventional Unit Reserve Scheduling Cost (€)	Conventional Unit Energy Scheduling Cost (€)	Total Expected Cost (€)	Maximum Probable Cost (% greater than expected)	Probability of Bearing Costs Higher Than Expected (%)	Total Expected Wind Production Spilled (MWh)
Base case	13153.184	1113774.891	1151877.754	2.36	18	20.22
C1	7853.253	1104898.491	1139192.039	2.27	12	12.05
C2-a	10374.267	1108256.514	1144540.438	2.28	12	13.96
C2-b	10370.019	1108629.899	1144575.271	2.29	12	14.01
C2-c	9645.908	1111269.999	1147049.579	2.34	12	13.81
C3-a	10780.246	1108747.543	1145110.972	2.38	18	16.68
C3-b	10959.781	1108549.693	1144996.059	2.38	18	16.87
C3-c	11154.304	1108981.649	1145442.888	2.37	18	17.10
C4	12129.552	1111073.594	1147214.226	2.44	18	17.23

considered that they should be delivered within any five consecutive periods in 200 power blocks of 0.5 MW each. The flexibility of this type of processes is further limited by imposing a maximum number of power blocks that may be allocated at any of these periods. The results obtained clearly state that the cost of scheduling generation-side reserves is mitigated because of the use of the industrial consumption-offered services. However, due to limited flexibility, in comparison with the totally flexible processes, the energy and reserve services cost reduction is less. Furthermore, greater wind spillage is noticed since the re-allocation of the relatively flexible processes has a narrower temporal effect and as a result during several periods, the demand-side services to support wind integration are not applicable.

Another important remark is that even if the total EC is, of course (because of greater wind integration and less reserves procured by conventional units), less than the one of the base case, the risk of having larger deviations in the probable cost is greater in contrast with the cases C2-a to C2-c.

Finally, the case of an inflexible process (C4) is studied. It is considered that the 80 MWh of the periods 12–15 are rendered available to be shifted during the day as a whole. Naturally, the total EC is less than the one of the base case. Energy and reserve procurement cost from conventional units is also reduced since the consumption is shifted to periods of relatively low load. It is evident that this case offers the least flexibility that mitigates the potential benefits of the demand-side assets for the system.

Considering all the cases presented, it is evident that any type of flexible process regardless of its parameters provides economic benefits to the system and facilitates the integration of wind power production.

5.7.2 Case 2: Aggregation of PEVs and Economic Model of Responsive Demand

In order to indicate the effectiveness of implementing DR programs and the effect of PEVs on the operation of the insular power system, two cases are investigated. Case 1 is associated with studying the impact of PEVs' behavior, whereas in case 2, influences of different DRPs are analyzed.

5.7.2.1 Aggregation of PEVs

In this case, a PEV aggregator and 5000 EVs are considered. The details of PEV aggregator's technical and economic data have been presented in Tables 5.3 and 5.4, respectively. The market prices have been considered

TABLE 5.3

Technical Data for the PEV Aggregator

η^D (%)	η^C (%)	SOC^{min} (pu)	SOC^{max} (pu)	FOR^{Agg}	$Ramp^{C/D}$ (pu/h)	L_{ET} (kWh)
82	90	0.3	0.9	0.05	0.2	43840

TABLE 5.4

Economic Data for the PEV Aggregator

$C_{battery}$ (€)	Cost^{Wiring} (€)	Cost$^{On-board}$ (€)	N_y (year)	dr (%)
13170	481	296	10	10

to be stationary stochastic parameters. For the sake of simplicity, only one session has been considered for the intraday market. In addition, the imbalance ratios are calculated by taking the average of historical data and considered to be hourly variable throughout the day.

The hourly profits of the PEV aggregator are indicated in Figure 5.16. It is supposed that all the PEV owners agree with the operation in V2G mode. According to Figure 5.16, the hourly profit of the PEV aggregator in the off-peak period is higher than in the peak and base-load periods. It is because of higher prices in off-peak compared to those in the base-load period and higher availability of PEVs in comparison with that in the peak period.

In order to investigate the effect of intraday market on the behavior of PEV aggregator, two cases have been simulated with and without the intraday market. The results of these two cases have been compared in Table 5.5.

According to Table 5.5, the major part of the PEV aggregator's income is resulted from its participation in the spinning reserve market. However, by implementation of the intraday market, the aggregator can have more participation in the energy market. Indeed, with participating in the intraday market, the PEV aggregator prefers to designate a part of its offer to the market mentioned and modify its offers by participating in it. Therefore, the aggregator employs the intraday market to compensate its imbalance costs

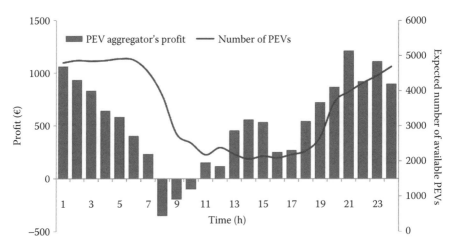

FIGURE 5.16
Hourly profit of the PEV aggregator.

TABLE 5.5

Different Terms of the PEV Aggregator's Profit

	Without Intraday Market	With Intraday Market
Income of day-ahead market [€]	6309	6511
Income of spinning reserve market [€]	165,997	162,843
Income of charging [€]	51,983	52,821
Income of intraday market [€]	0	165
Imbalance cost [€]	1052	535
Cost of intraday market [€]	0	303.0
Cost of battery degradation [€]	4904	4823
Cost of energy market to meet obligations [€]	528	507
Cost of charging [€]	50,733	48,775
Fixed costs [€]	154,374	154,374
Expected profit [€]	12,698	13,021

associated with the uncertain behavior of PEV owners. Therefore, the PEV aggregators can maximize their profits by participating in the market mentioned to cover the high risk of prices in the balancing market.

5.7.2.2 Economic Model of Responsive Demand

In this case, it is assumed that each generation unit offers at its marginal cost according to linearized cost curves of generation units in three blocks. Involuntary load-shedding cost is considered to be 2000 €/MWh in all buses. It is assumed that one DRP is found in all load buses. The load curve is divided into three intervals: low load period (1:00–9:00), off-peak period (16:00–24:00), and peak period (10:00–15:00). Moreover, time of use (TOU) tariff, real-time pricing (RTP), and EDRP are considered. The self- and cross-price elasticities of electricity demand are illustrated in Table 5.6. In addition, in order to indicate the effect of tariff on the consumers' behavior, three different types of TOU have been considered. The details of the tariff for the TOU programs are presented in Table 5.7.

TABLE 5.6

Self- and Cross-Elasticities

	Peak	Off-Peak	Low-Load
Peak	−0.10	0.016	0.012
Off-peak	0.016	−0.10	0.010
Low-load	0.012	0.010	−0.10

TABLE 5.7

Time of Use Tariffs

Time (h)	Type 1	Type 2	Type 3	Time (h)	Type 1	Type 2	Type 3
1	16	12	8	13	36	48	72
2	16	12	8	14	36	48	72
3	16	12	8	15	36	48	72
4	16	12	8	16	24	24	24
5	16	12	8	17	24	24	24
6	16	12	8	18	24	24	24
7	16	12	8	19	24	24	24
8	16	12	8	20	24	24	24
9	16	12	8	21	24	24	24
10	36	48	72	22	24	24	24
11	36	48	72	23	24	24	24
12	36	48	72	24	24	24	24

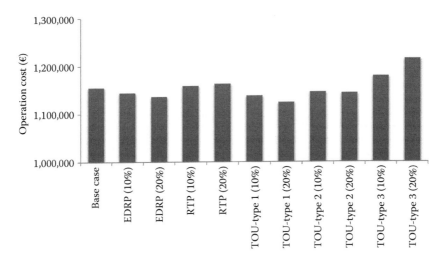

FIGURE 5.17
Impact of various DRPs on the expected operation cost.

The expected operation costs of the electricity system considering various DRPs have been compared in Figure 5.17. As it can be seen, for EDRP, TOU-type 1, and TOU-type 2, the increase of maximum participation level of DR decreases the total operation cost, whereas it increases the total operation cost for RTP and TOU-type 3. In addition, TOU-type 1 can cause the minimum operation cost among different DRPs. Behind the first type of TOU program, EDRP and TOU-type 2 are the most effective DRPs to decrease the operation cost. It can be observed that TOU-type3 has the worst effect to minimize the operation cost.

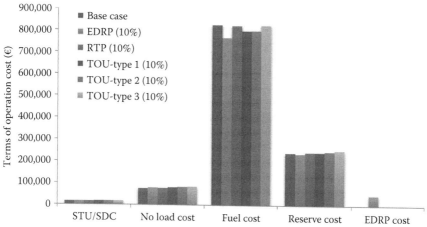

FIGURE 5.18
Impact of various DRPs on the terms of operation cost.

The impact of the DRPs mentioned on the different terms of operation cost has been presented in Figure 5.18. As it can be seen, although EDRP can significantly reduce the fuel cost, the market payments to responsive demands due to take part in the program causes the effect of EDRP on the operation cost to be less than the first type of TOU.

The changes of daily load profile because of various DRPs are illustrated in Figures 5.19 through 5.23. As can be seen in Figure 5.19, EDRP can decrease the amount of load peak and consequently, it can produce a flatter load curve compared to the base case, although it has no significant effect on demand in low-load and off-peak periods. According to Figure 5.20, RTP has an insignificant effect on the load curve compared to other DRPs, whereas, as can be

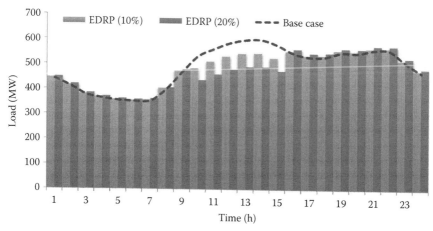

FIGURE 5.19
Impact of EDRP on the load profile.

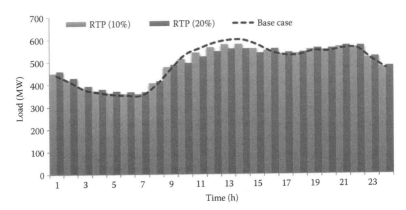

FIGURE 5.20
Impact of RTP on the load profile.

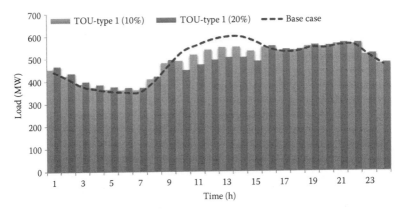

FIGURE 5.21
Impact of TOU-type 1 on the load profile.

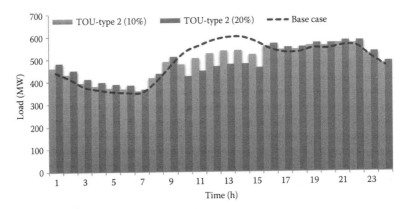

FIGURE 5.22
Impact of TOU-type 2 on the load profile.

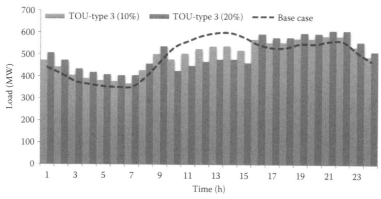

FIGURE 5.23
Impact of TOU-type 3 on the load profile.

observed in Figures 5.21 through 5.23, TOU programs can reduce the demand peak and also increase the low-load and off-peak demand. Therefore, the programs can cause the load curve to be smoother. It is noteworthy that the implementation of TOU programs results in a change in the hour of demand peak, which is shifted to about hour 21. As can be observed in Figures 5.21 through 5.23, if the system operator aims to have smoother load curves, using very different prices for different hours of a day (e.g., implementation of the third type of TOU program), it can cause negative impacts and even cause some higher demand peaks.

The effect of the DRPs mentioned on the hourly total operation cost has been illustrated in Figures 5.24 through 5.28. As it can be seen, the DRPs can reduce the operation costs in peak period because of decreasing the electricity loads and consequently electricity prices. The implementation of TOU programs causes an increase in the operation cost in off-peak and low-load periods because of the load shifting feature.

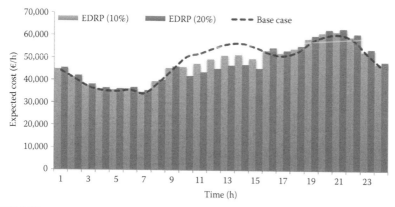

FIGURE 5.24
Impact of EDRP on hourly operation cost.

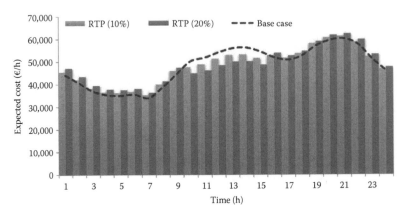

FIGURE 5.25
Impact of RTP on hourly operation cost.

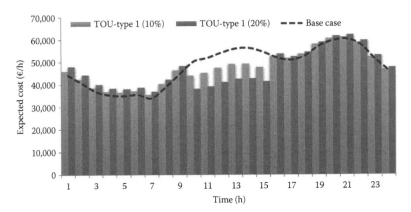

FIGURE 5.26
Impact of TOU-type 1 on hourly operation cost.

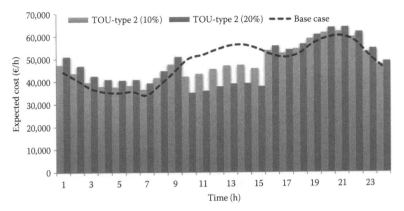

FIGURE 5.27
Impact of TOU-type 2 on hourly operation cost.

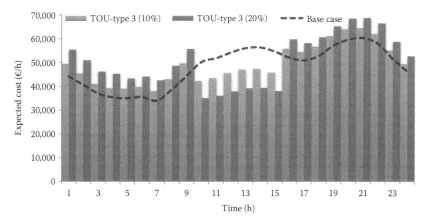

FIGURE 5.28
Impact of TOU-type 3 on hourly operation cost.

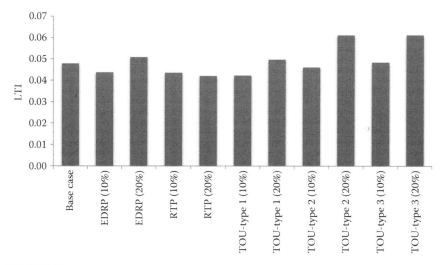

FIGURE 5.29
Impact of various DRPs on the LTI.

In Figures 5.29 through 5.32, the various DRPs have been compared using the proposed indices. As can be seen in Figure 5.32, the second type of TOU program has the highest ADRB, hence the program has the most effect on decreasing the operation cost due to wind power generation. However, the program mentioned can produce high load turbulence as reflected in all of LTI, MLU, and MLD indices. The second type of TOU program has been followed by the first type of TOU program in terms of the index of ADRB.

TOU-Type 1 causes acceptable load turbulence in comparison with the other DRPs. This program has been followed by TOU-Type 3 to have a higher

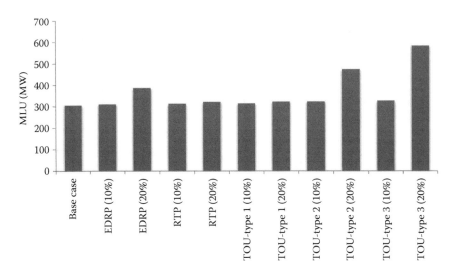

FIGURE 5.30
Impact of various DRPs on the MLU.

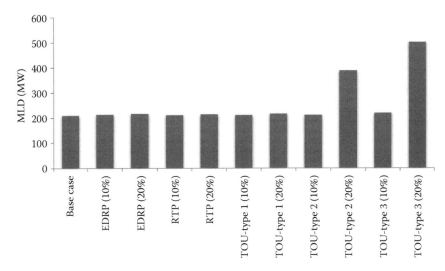

FIGURE 5.31
Impact of various DRPs on the MLD.

ADRB. However, according to LTI, MLU, and MLD indices, the TOU-Type 3 has caused the highest load turbulence among the studied programs. TOU-Type 3 program has been followed by the EDRP to have the higher ADRB index. According to LTI, MLU, and MLD indices, the EDRP has created less load turbulence than TOU-Type 1. Finally, the ADRB index shows that the RTP program has the least effect on reducing the operation cost because of wind power generation compared with other DRPs.

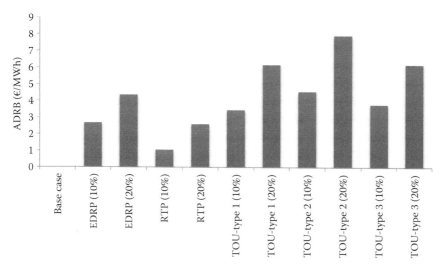

FIGURE 5.32
Impact of various DRPs on the ADRB.

5.8 Conclusions

In this chapter, several responsive demand models, namely, a model for a responsive industrial consumer, the bidding model of a PEVs aggregator, and an economic model for DR have been presented and analyzed. All the aforementioned models are developed around two-stage stochastic optimization and aim to provide energy and reserve services in order to constitute an exploitable resource to cope with power system uncertainties. Especially, the introduction of high levels of RES in power systems will bring upon the need for procuring larger amounts of reserves in order to maintain the generation and consumption balance at any instance.

Insular power systems that, by definition, are more fragile in terms of having limited generation capacity and are characterized by the absence of interconnections that may increase the security of the grid are more likely to be affected by the penetration of RES. The application of the proposed stochastic optimization models has proven that several types of demand-side resources are capable of providing energy and reserve services in order to immunize the system against contingencies and RES volatility. In all the test cases presented, the significant economic and operational benefits that emerge from the utilization of demand-side resources are evident.

Research on the field of integrating demand-side resources within the AS mechanisms has already proven to be fruitful, and many ISOs around the world are beginning the restructuring of their mechanisms in order to integrate as much responsive load as possible. Nevertheless, there are many issues that are likely to be addressed by researchers in the following years, mainly regarding the determination of the available amount of responsive demand, the design of innovative policies, and the exploitation of the full economic and operational potential of the demand-side resources.

Nomenclature

The main nomenclature used throughout the text is presented here to facilitate the reader. Several variables that may be needed (e.g., the scenario dependent equivalent variables) is considered to be deduced trivially. Other symbols and abbreviations are defined where they first appear.

Sets and indices

$d(D)$	index (set) of industrial loads
$f(F^i)$	index (set) of steps of the marginal cost function of unit i
$g(G^d)$	index (set) of groups of processes of industry d
$i(I)$	index (set) of conventional generating units
$j(J)$	index (set) of contingencies
$l(L)$	index (set) of transmission lines
$n(N)$	index (set) of nodes
$p(P^d)$	index (ordered set) of industrial processes
p_{type}^h	set of industrial process types: $h = 1$ for continuous, $h = 2$ for interruptible
$r(R)$	index (set) of inelastic loads
$s(S^W)$	index (set) of wind power scenarios of wind farm W
$t(T)$	index (set) of time periods
$v(V)$	index (set) of PEVs
$w(W)$	index (set) of wind farms
$\omega(\Omega)$	index (set) of PEV scenarios

Parameters

$a_{p,g,d}^{max}$	positive integer—maximum number of available production lines for process p of group g of industry d
$a_{p,g,d}^{max,h}$	positive integer—maximum number of production lines per hour for process p of group g of industry d
A_t	amount of incentive award in DR program (€/MWh)
$B_{i,f,t}$	size of step f of unit i marginal cost function in period t (MW)
$B_{l,n}$	susceptance of line l (per unit)
$C_{battery}$	battery capital cost (€)
C_d	battery degradation cost (€/cycle)

$C_{i,f,t}$	marginal cost of step f of unit i marginal cost function in period t (€/MWh)
$C_{i,t}^{R^D}$	offer cost of down reserve by unit i in period t (€/MWh)
$C_{i,t}^{R^{D,In}}$	offer cost of down reserve by industrial load d in period t (€/MWh)
$C_{i,t}^{R^{NS}}$	offer cost of nonspinning reserve by unit i in period t (€/MWh)
$C_{i,t}^{R^{NS,In}}$	offer cost of nonspinning reserve by industrial load d in period t (€/MWh)
$C_{i,t}^{R^U}$	offer cost of up reserve by unit i in period t (€/MWh)
$C_{i,t}^{R^{U,In}}$	offer cost of up reserve by industrial load d in period t (€/MWh)
$Cost_{\text{On-board}}$	on-board incremental cost (€)
$Cost_{\text{Wiring}}$	wiring upgrade cost (€)
$d_{0,t}$	initial electricity demand (MW)
$D_{d,t}^{\min}$	minimum power for industry d in period t (MW)
DT_i	minimum down-time of unit i (h)
$E(t,t)$	self-elasticity of demand
$E(t,t')$	cross-elasticity of demand
$f_l^{\max}$	maximum capacity of transmission line l (MW)
FOR^{Agg}	probability of inability of PEV aggregator to generate energy
FOR_j	forced outage rate [failures/day]
L_{ET}	battery lifetime (cycles)
$L_{r,t}$	demand of inelastic load r in period t (MW)
$P_i^{\max}$	maximum power output of unit i (MW)
$P_i^{\min}$	minimum power output of unit i (MW)
$P_{p,g,d}^{\text{line}}$	power of production line of process p of group g of industry d (MW)
$P_{w,s,t}^{\text{WP}}$	power output of wind farm w in scenario s in period t (MW)
$P_{w,t}^{\text{WP,max}}$	maximum wind power that can be scheduled from wind farm w in period t (MW)
pen_t	amount of penalty in DR program (€/MWh)
r_t^+, r_t^-	positive and negative imbalance ratios
RD_i	ramp-down rate of unit i (MW/min)
RU_i	ramp-up rate of unit i (MW/min)
SDC_i	shutdown cost of unit i (€)
SUC_i	start-up cost of unit i (€)
$T_{p,g,d}^{c,\max}$	maximum completion time of process p of group g of industry d (h)
$T_{p,g,d}^{g,\max}$	maximum time interval between process p and $p+1$ of group g of industry d (h)
$T_{p,g,d}^{g,\min}$	minimum time interval between process p and $p+1$ of group g of industry d (h)
UT_i	minimum up-time of unit i (h)
V^{LOL}	cost of involuntary load shedding for inelastic loads (€/MWh)
V^S	cost of wind energy spillage (€/MWh)
$VOLL_{L,t}$	value of lost load (€/MWh)

$\lambda_{d,t}^{D}$ utility of industrial d load in period t (€/MWh)
π_s probability of wind power scenario s
$\pi_{t,\omega}$ occurrence probability of PEV scenario ω
$\lambda_{0,t}$ electricity price before implementing DR programs (€/MWh)
$\lambda_t^{\text{TariffCharge}}$ tariff of buying energy of EV owner (€/MWh)
$\lambda_t^{\text{TariffRes}}$ tariff of spinning reserve of EV owner (€/MWh)

Variables

$a_{p,g,d,t}$ positive integer variable—number of production lines scheduled in period t for process p of group g of industry d
$a_{p,g,d,t,s}^{2}$ positive integer variable—actual number of production lines in period t of scenario s for process p of group g of industry d
$a_{p,g,d,t}^{\text{down}}$ positive integer variable—number of production lines that contribute in scheduled down reserve in period t from process p of group g of industry d
$a_{p,g,d,t,s}^{\text{down},rt}$ positive integer variable—actual number of production lines contributing in down reserve in period t of scenario s from process p of group g of industry d
$a_{p,g,d,t}^{\text{ns}}$ positive integer variable—number of production lines that contribute in scheduled nonspinning reserve in period t from process p of group g of industry d
$a_{p,g,d,t,s}^{\text{ns},rt}$ positive integer variable—actual number of production lines contributing in nonspinning reserve in period t of scenario s from process p of group g of industry d
$a_{p,g,d,t}^{\text{up}}$ positive integer variable—number of production lines that contribute in scheduled up reserve in period t from process p of group g of industry d
$a_{p,g,d,t,s}^{\text{up},rt}$ positive integer variable—actual number of production lines contributing in up reserve in period t of scenario s from process p of group g of industry d
$\text{Act}_{t,\omega}^{\text{Res}}$ quantity of reserve activated by ISO (MW)
$b_{i,f,t}$ power output scheduled from the fth block by unit i in period t (MW)
C_t^{EDRP} cost of customer's participation in EDRP (€)
$\text{Cost}_{\text{Infra}}$ annualized infrastructure cost (€)
$\text{Cost}_{t,\omega}^{\text{Charge}}$ purchase cost of electricity to charge EVs (€)
$\text{Cost}_{\omega}^{\text{Deg.}}$ degradation cost of EV battery due to operation in V2G mode (€/cycle)
$\text{Cost}_{t,\omega}^{\text{Intra}}$ cost of buying energy from intraday market (€)
$\text{Cost}_{t,\omega}^{\text{Imb}}$ cost of imbalance penalties (€)
$\text{Cost}_{t,\omega}^{\text{Obl}}$ purchase cost of electricity to meet the aggregator obligations (€)
$\text{Cost}_{\omega}^{\text{Res}}$ payment cost to EV owners regarding participation in spinning reserve market (€)

d_t	optimal curtailed load (MW)
E_Ω	expected value obtained from scenario set Ω
$f_{l,t}$	power flow through line l in period t (MW)
$\text{Income}_{t,\omega}^{\text{Call}}$	income resulting from being called to generate in the spinning reserve market (€)
$\text{Income}_{\omega}^{\text{Charge}}$	income resulting from receiving the charge cost from EV owners (€)
$\text{Income}_{t,\omega}^{\text{Energy}}$	income resulting from the participation in the electricity market (€)
$\text{Income}_{t,\omega}^{\text{Imb}}$	income resulting from imbalances (€)
$\text{Income}_{t,\omega}^{\text{Intra}}$	income resulting from selling energy to the intraday market (€)
$\text{Income}_{t,\omega}^{\text{Res}}$	income resulting from participation in the spinning reserve market (€)
$\text{LNS}_{L,t}$	amount of Lth load not supplied at time t (MW)
$L_{r,t,s}^{\text{shed}}$	portion of inelastic load r shed in period t in scenario s (MW)
$P_{d,t}^{\text{ind,S}}$	scheduled consumption in period t for industry d (MW)
$P_{d,t,s}^{\text{ind,C}}$	actual power consumption in period t for industry d in scenario s (MW)
$P_{i,t}^{\text{S}}$	power output scheduled for unit i in period t (MW)
$P_{i,t,s}^{\text{G}}$	actual power output for unit i in period t in scenario s (MW)
$P_{n,t}^{\text{mod}}$	modified demand of bus n (MW)
$P_{p,g,d,t}^{\text{pro,S}}$	scheduled consumption in period t for process p of industry d in period t (MW)
$P_{p,g,d,t,s}^{\text{pro,C}}$	actual consumption in period t for process p of industry d in period t in scenario s (MW)
$P_{t,\omega}$	actual generation of the PEV aggregator (MW)
$P_{t,\omega}^{\text{DA}}$	offer to participate in the electricity market (MW)
$P_{t,\omega}^{\text{del}}$	probability of being called to generate
$P_{t,\omega}^{\text{Intra,buy}}$	power bought in intraday market (MW)
$P_{t,\omega}^{\text{Intra,sell}}$	power sold in intraday market (MW)
$P_{t,\omega}^{\text{Sch}}$	scheduled generation of the PEV aggregator (MW)
$P_{t,\omega}^{\text{Res}}$	offer to participate in the reserve market (MW)
$P_{v,t,\omega}^{\text{G2V}}$	injected power from grid to PEV v (MW)
$P_{v,t,\omega}^{\text{V2G}}$	injected power from PEV v to grid (MW)
$P_{w,t}^{\text{WP,S}}$	scheduled wind power in period t by wind farm w (MW)
$R_{d,t}^{\text{D,ind}}$	down reserve scheduled in period t by industry d (MW)
$R_{d,t}^{\text{NS,ind}}$	nonspinning reserve scheduled in period t by industry d (MW)
$R_{d,t}^{\text{U,ind}}$	up reserve scheduled in period t by industry d (MW)
$R_{i,t}^{\text{D}}$	down reserve scheduled in period t by unit i (MW)
$R_{i,t}^{\text{NS}}$	nonspinning reserve scheduled in period t by unit i (MW)
$R_{i,t}^{\text{U}}$	up reserve scheduled in period t by unit i (MW)
$R_{p,g,d,t}^{\text{D,pro}}$	down reserve scheduled by process p of group g by industry d in period t (MW)
$R_{p,g,d,t}^{\text{NS,pro}}$	nonspinning reserve scheduled by process p of group g by industry d in period t (MW)

$R_{p,g,d,t}^{\mathrm{U,pro}}$ | up reserve scheduled by process p of group g by industry d in period t (MW)

$r_{v,t,\omega}^{\mathrm{charge}}$ | charging rate of PEV v

$r_{i,f,t,s}^{\mathrm{G}}$ | reserve deployed from the fth block of unit i in scenario s in period t (MW)

$r_{i,t,s}^{\mathrm{D}}$ | down reserve deployed by unit i in period t in scenario s (MW)

$r_{i,t,s}^{\mathrm{NS}}$ | nonspinning reserve deployed by unit i in period t in scenario s (MW)

$r_{i,t,s}^{\mathrm{U}}$ | up reserve deployed by unit i in period t in scenario s (MW)

$r_{v,t,\omega}^{\mathrm{discharge}}$ | discharging rate of PEV v

$r_{p,g,d,t,s}^{\mathrm{D,pro}}$ | down reserve deployed by process p of group g of industry d in period t in scenario s (MW)

$r_{p,g,d,t,s}^{\mathrm{NS,pro}}$ | nonspinning reserve deployed by process p of group g of industry d in period t in scenario s (MW)

$r_{p,g,d,t,s}^{\mathrm{U,pro}}$ | up reserve deployed by process p of group g of industry d in period t in scenario s (MW)

$S_{w,t,s}$ | wind spilled in scenario s from wind farm w in period t (MW)

$\mathrm{SOC}_{v,t,\omega}$ | state of charge of PEV v

$u_{i,t}$ | binary variable—1 if unit i is committed during period t, else 0

$u_{i,t,s}^{2}$ | binary variable—1 if unit i is committed during period t in scenario, else 0

$y_{i,t}$ | binary variable—1 if unit i is starting-up during period t, else 0

$y_{i,t,s}^{2}$ | binary variable—1 if unit i is committed during period t in scenario s, else 0

$z_{i,t}$ | binary variable—1 if unit i is shutting-down during period t, else 0

$z_{i,t,s}^{2}$ | binary variable—1 if unit i is committed during period t in scenario, else 0

$\Delta_{t,\omega}$ | total deviation (MW)

$\Delta_{t,\omega}^{+}$ | positive deviation (MW)

$\Delta_{t,\omega}^{-}$ | negative deviation (MW)

$\eta_{v}^{\mathrm{C}}, \eta_{v}^{\mathrm{D}}$ | charging and discharging efficiencies

$\lambda_{j,t}^{\mathrm{outage}}$ | price in contingency j (€/MWh)

$\lambda_{t,\omega}^{\mathrm{Bal}}$ | balancing market price (€/MWh)

$\lambda_{t,\omega}^{\mathrm{DA}}$ | day-ahead market price (€/MWh)

$\lambda_{t,\omega}^{\mathrm{Intra}}$ | intraday market price (€/MWh)

$\lambda_{t,\omega}^{\mathrm{Res}}$ | price of spinning reserve market (€/MWh)

$\upsilon_{p,g,d,t}$ | binary variable—1 if process p of group g of industry d is in progress in period t, else 0

$\zeta_{p,g,d,t}$ | binary variable—1 if process p of group g of industry d is terminated in period t, else 0

$\psi_{p,g,d,t}$ | binary variable—1 if process p of group g of industry d is beginning in period t, else 0

$\delta_{n,t}$ | voltage angle of node n in period t (rad)

References

1. European Union Webpage for the EU Climate and Energy Package. http://ec.europa.eu/clima/policies/package/ (accessed December 11, 2013).
2. Cappers P., Goldman C., and D. Kathan. 2010. Demand response in U. S. electricity evidence. *Energy* 35: 1526–35.
3. Kirby B.J. 2006. *Demand Response for Power System Reliability: FAQ.* TN, ORNL/TM-2006/565. Oak Ridge, TN: Oak Ridge National Laboratory.
4. Walawalkar R., Fernands S., Thakur N., and K.R. Chevva. 2010. Evolution and current status of demand response (DR) in electricity markets: Insights from PJM and NYISO. *Energy* 35: 1553–60.
5. Torriti J., Hassan M.G., and M. Leach. 2010. Demand response experience in Europe: Policies, programmes and implementation. *Energy* 35: 1575–83.
6. Kapetanovic T., Buchholz B.M., Buchholz B., and V. Buehner. 2008. Provision of ancillary services by dispersed generation and demand side response-needs, barriers and solutions. *Elektrotechnik und Informationstechnik* 125: 452–9.
7. Lawrence Berkeley National Laboratory. 2007. *Demand Response Spinning Reserve Demonstration.* Report no. LBNL-62761. http://certs.lbl.gov/pdf/62761.pdf (accessed February 17, 2015).
8. Rebours Y.G., Kirschen D.S., Trotignon M., and S. Rossignol. 2007. A survey of frequency and voltage control ancillary services- Part I: Technical Features. *IEEE Transactions on Power Systems* 22: 350-7.
9. Rebours Y.G., Kirschen D.S., Trotignon M., and S. Rossignol. 2007. A survey of frequency and voltage control ancillary services—Part II: Economic features. *IEEE Transactions on Power Systems* 22: 358–66.
10. Ma O., Alkadi N., Cappers P., Denholm P., Dudley J., Goli S., Humman M. et al. 2013. Demand response for ancillary service. *IEEE Transactions on Smart Grid* 4: 1988–95.
11. Navid N., Rosenweld G., and D. Chatterjee. 2012. *Ramp Capability for Load Following in the MISO Markets.* Midwest Independent System Operator. https://www.misoenergy.org (accessed February 17, 2015).
12. Xu L., and D. Tretheway. 2012. *Flexible Ramping Products.* California Independent System Operator. http://www.caiso.com/Documents/DraftFinalProposal-FlexibleRampingProduct.pdf (accessed February 17, 2015).
13. Jafari A.M., Zareipour H., Schellenberg A., and N. Amjady. 2014. The value of intra-day markets in power systems with high wind power penetration. *IEEE Transactions on Power Systems* 29: 1121–32.
14. Bouffard F., and F.D. Galiana. 2008. Stochastic security for operations planning with significant wind power generation. *IEEE Transactions on Power Systems* 23: 306–16.
15. Zongrui D., Yuanxiong G., Dapeng W., and F. Yuguang. 2013. A market based scheme to integrate distributed wind energy. *IEEE Transactions on Smart Grid* 4: 976–84.
16. Qianfan W., Yongpei G., and W. Jianhui. 2012. A chance-constrained two-stage stochastic program for unit commitment with uncertain wind power output. *IEEE Transactions on Power Systems* 27: 206–15.

17. Chaoyue Z., and G. Yongpei. Unified stochastic and robust unit commitment. *IEEE Transactions on Power Systems* 28: 3353–61.
18. Constantinescu E.M., Zavala V.M., Rocklin M., Sangmin L., and M. Anitescu. 2011. A computational framework for uncertainty quantification and stochastic optimization in unit commitment with wind power generation. *IEEE Transactions on Power Systems* 26: 431–41.
19. Ruiz P.A., Philbrick C.R., and P.W. Sauer. 2010. Modeling approaches for computational cost reduction in stochastic unit commitment formulations. *IEEE Transactions on Power Systems* 25: 588–9.
20. Morales J.M., Conejo A.J., and J. Perez-Ruiz. 2009. Economic valuation of reserves in power systems with high penetration of wind power. *IEEE Transactions on Power Systems* 24: 900–10.
21. Sahin C., Shahidehpour M., and I. Erkmen. 2013. Allocation of hourly reserve versus demand response for security-constrained scheduling of stochastic wind energy. *IEEE Transactions on Sustainable Energy* 4: 219–28.
22. Papavasiliou A., Oren S.S., and R.P. O'Neill. 2011. Reserve requirements for wind power integration: A scenario-based stochastic programming framework. *IEEE Transactions on Power Systems* 26: 2197–206.
23. Parvania M., and M. Fotuhi-Firuzabad. 2010. Demand response scheduling by stochastic SCUC. *IEEE Transactions on Smart Grid* 1: 89–98.
24. Karangelos E., and F. Bouffard. 2010. Towards full integration of demand-side resources in joint forward energy/reserve electricity markets. *IEEE Transactions on Power Systems* 27: 280–9.
25. Shan J., Botterud A., and S.M. Ryan. 2013. Impact of demand response on thermal generation investment with high wind penetration. *IEEE Transactions on Smart Grid* 4: 2374–83.
26. Peng X., and P. Jirutitijaroen. 2013. A stochastic optimization formulation of unit commitment with reliability constraints. *IEEE Transactions on Smart Grid* 4: 2200–8.
27. Guodong L., and K. Tomsovic. 2012. Quantifying spinning reserve in systems with significant wind power penetration. *IEEE Transactions on Power Systems* 27: 2385–93.
28. Vrakopoulou M., Margellos K., Lygeros J., and G. Andersson. 2013. A probabilistic framework for reserve scheduling and N-1 security assessment of systems with high wind power penetration. *IEEE Transactions on Power Systems* 28: 3885–96.
29. Meibom P., Barth R., Hasche B., Brand H., Weber C., and M. O'Malley. 2011. Stochastic optimization model to study the operational impacts of high wind penetrations in Ireland. *IEEE Transactions on Power Systems* 26: 1367–79.
30. Ortega-Vazquez M.A., and D.S. Kirschen. 2009. Estimating the spinning reserve requirements in systems with significant wind power generation penetration. *IEEE Transactions on Power Systems* 24: 114–24.
31. Conejo A.J., Carrión M., and J.M. Morales. 2010. Uncertainty characterization via scenarios. In *Decision Making Under Uncertainty in Electricity Markets*. New York: Springer.
32. Dupacova J., Growe-Kuska N., and W. Romisch. 2003. Scenario reduction in stochastic programming: An approach using probability metrics. *Mathematical Programming Series A* 95: 493–511.
33. Heitsch H., and W. Romisch. 2003. Scenario reduction algorithms in stochastic programming. *Computational Optimization and Applications* 24: 187–206.

34. Niknam T., Azizipanah-Abarghooee R., and M.R. Narimani. 2012. An efficient scenario-based stochastic programming framework for multi-objective optimal micro-grid operation. *Applied Energy* 99: 455–70.

35. Amjady N., Aghaei J., and H.A. Shayanfar. 2009. Stochastic multiobjective market clearing of joint energy and reserves auctions ensuring power system security. *IEEE Transactions on Power Systems* 24: 1841–54.

36. Kumamoto H., and E.J. Henley. 2001. *Probabilistic Risk Assessment and Management for Engineers and Scientists*. New York: John Wiley and Sons.

37. Meliopoulos S. 2009. *Power System Level Impacts of Plug-In Hybrid Vehicles*. Power Systems Engineering Research Center. http://www.pserc.wisc.edu/documents/publications/reports/2009_reports/meliopoulos_phev_pserc_report_t-34_2009.pdf (accessed February 17, 2015).

38. Domínguez-García A.D., Heydt G.T., and S. Suryanarayanan. 2011. *Implications of the Smart Grid Initiative on Distribution Engineering*. Power Systems Engineering Research Center. http://www.pserc.wisc.edu/documents/publications/reports/2011_reports/PSERC_T-41_Final_Project_Report_2011_ExecSum.pdf (accessed February 17, 2015).

39. Pasaoglu G., Fiorello D., Zani L., Martino A., Zubaryeva A., and C. Thiel. 2012. *Driving and Parking Patterns of European Car Drivers—A Mobility Survey*. Joint Research Centre, European Commission, European Union. http://setis.ec.europa.eu/system/files/Driving_and_parking_patterns_of_European_car_drivers-a_mobility_survey.pdf (accessed February 17, 2015).

40. Nemry F., Leduc G., and A. Muñoz. 2009. *Plug-In Hybrid and Battery-Electric Vehicles: State of the Research and Development and Comparative Analysis of Energy and Cost Efficiency*. European Communities, http://ftp.jrc.es/EURdoc/JRC54699_TN.pdf (accessed April 25, 2014).

41. Sutiene K., Makackas D., and H. Pranevicius. 2010. Multistage K-means clustering for scenario tree construction. *Informatica* 21: 123–38.

42. Conejo A.J., Nogales F.J., and J.M. Arroyo. 2002. Price-taker bidding strategy under price uncertainty. *IEEE Transactions on Power Systems* 17: 1081–88.

43. Paterakis N.G., Erdinc O., Bakirtzis A.G., and J.P.S. Catalão. 2015. Load-following reserves procurement considering flexible demand-side resources under high wind power penetration. *IEEE Transactions on Power Systems* 30: 1337–50.

44. Morales J.M., Conejo A.J., and J. Pérez-Ruiz. 2010. Short-term trading for a wind power producer. *IEEE Transactions on Power Systems* 25: 554–64.

45. Kempton W., and J. Tomic. 2005. Vehicle-to-grid power fundamentals: Calculating capacity and net revenue. *Journal of Power Sources* 144: 268–79.

46. Gan L., Topcu U., and S.H. Low. 2013. Optimal decentralized protocol for electric vehicle charging. *IEEE Transactions on Power Systems* 28: 940–51.

47. Kirschen D., and G. Strbac. 2004. *Fundamentals of Power System Economics*. Chichester: John Wiley and Sons.

48. Aalami H.A., Moghaddam M.P., and G. Yousefi. 2010. Demand response modeling considering interruptible/curtailable loads and capacity market programs. *Applied Energy* 87: 243–50.

49. Aalami H.A., Moghaddam M.P., and G. Yousefi. 2010. Modeling and prioritizing demand response programs in power markets. *Electric Power Systems Research* 80: 426–35.

50. Baboli P.T., Eghbal M., Moghaddam M.P., and H. Aalami. 2012. Customer behavior based demand response model. *IEEE PES GM*, San Diego, CA, July 22–26.

51. Moghaddam M.P., Abdollahi A., and M. Rashidinejad. 2011. Flexible demand response programs modeling in competitive electricity markets. *Applied Energy* 88: 3257–69.

52. Kwag H.G., Kim J.O., Shin D.J., and C.H. Rhee. 2011. Modeling demand response by using registration and participation information of demand resources. *Transactions of the Korean Institute of Electrical Engineers* 60: 1097–102.

53. Li Z., and M. Shahidehpour. 2005. Security-constrained unit commitment for simultaneous clearing of energy and ancillary services markets. *IEEE Transactions on Power Systems* 20: 1079–88.

54. Cheung K.W., Shamsollahi P., Sun D., Milligan J., and M. Potishnak. 2000. Energy and ancillary service dispatch for the interim ISO new England electricity market. *IEEE Transactions on Power Systems* 15: 968–74.

55. Fu Y., Shahidehpour M., and Z. Li. 2005. Security-constrained unit commitment with AC constraints. *IEEE Transactions on Power Systems* 20: 1538–50.

56. Simopoulos D.N., Kavatza S.D., and C.D. Vournas. 2006. Reliability constrained unit commitment using simulated annealing. *IEEE Transactions on Power Systems* 21: 1699–706.

57. Cai Y.P., Huang G.H., Yang Z.F., and Q. Tan. 2009. Identification of optimal strategies for energy management systems planning under multiple uncertainties. *Applied Energy* 86: 480–95.

58. Xia L.M., Gooi H.B., and J. Bai. 2005. A probabilistic reserve with zero-sum settlement scheme. *IEEE Transactions on Power Systems* 20: 993–1000.

59. Bouffard F., and F.D. Galiana. 2004. An electricity market with a probabilistic spinning reserve criterion. *IEEE Transactions on Power Systems* 19: 300–7.

60. IEA Wind Task 24 Final Report. 2011. *Integration of Wind and Hydropower Systems.* Final Technical Report. www.nrel.gov/docs/fy12osti/50182.pdf (accessed February 17, 2015).

61. Heydarian-Forushani E., Moghaddam M.P., Sheikh-El-Eslami M.K., Shafiekhah M., and J.P.S. Catalão. 2014. A stochastic framework for the grid integration of wind power using flexible load approach. *Energy Conversion and Management* 88: 985–998.

62. Simoglou C.K., Kardakos E.G., Bakirtzis E.A., Chatzigiannis D.I., Vagropoulos S.I., Ntomaris A.V., Biskas P.N. et al. 2014. An advanced model for the efficient and reliable short-term operation of insular electricity networks with high renewable energy sources penetration. *Renewable and Sustainable Energy Reviews* 38: 415–27.

63. Karapidakis E.S., Katsigiannis Y.A., Georgilakis P.S., and E. Thalassinakis. 2011. Generation expansion planning of Crete power system for high penetration of renewable energy sources. *Materials Science Forum* 670: 407–14.

6

Electric Price Signals, Economic Operation, and Risk Analysis

Miguel Asensio, Pilar Meneses de Quevedo, Javier Contreras,
Cláudio Monteiro, Radu Porumb, Ion Trişţiu, and George Seriţan

CONTENTS

ABSTRACT This chapter addresses the integration of a pool of renewable energy sources (RESs) into traditional electricity power systems using conventional generation backed by generic storage for grid stability applications. It gives special emphasis on methodologies and tools for the analysis of the operation of a selected mix of resources in insular networks. It also contributes for power system stakeholders to become economically benefited from the intended savings in operation costs. The key aspects considered are the availability of energy sources and reserves, the complementary exploitation of different energy sources, the role of storage, the economics of the connection to the insular network, and the effects of the possible connection to external systems. In the first section, a description of most relevant terms used in power systems and distribution systems operation is given. In the second section more detail about weakly meshed distribution systems and their challenges is provided. Next, a brief description of concepts about the operation of distribution grids introducing different modeling is shown. Finally, a case study proposed for the analysis of future competitive operation in distribution systems is presented. In the fourth section, a scenario analysis for an insular electricity grid is developed using various case studies. The fifth section is dedicated to model electricity price signals in small insular systems starting with an assessment over the methodology used and the risk.

6.1 Introduction

One of the important parts of a power system is the electricity distribution network. Recently, due to the increase of distributed energy sources (DES) penetration, distribution grids have become active networks. The distribution system operator (DSO) is responsible for operating the network in a reliable manner, maintaining the required level of quality of supply. Furthermore, the DSO is responsible for a coordinated operation with the transmission system operator (TSO), in order to maintain the level of security.

This chapter is organized in the following way. In the second section, a description of most relevant terms used in distribution power systems operation, structures, and challenges are given. Also, different modeling approaches and an overview of their features are introduced. Next, the mathematical formulation of the considered problem, as well as a description of the different types of generation is presented. In the fourth section, a case study is considered and an in-depth analysis of the results obtained for future competitive operation of distribution systems with high DES penetration are provided, followed by conclusions in the last section. Finally, the modeling of electricity price signals in small insular systems is addressed in the fifth section.

6.2 Distribution System Operations

6.2.1 Introduction

The primary and the most imperative assignment for decision makers from the power systems operation point of view is to determine the load characteristics and to decide upon which units and what capacities should be committed to satisfy the demands for a certain period. This study is commonly referred to as *unit commitment*.

Usually, the timescale for such operations is one week (see Figure 6.1), so a detailed list of loads for all hours is generated to cover the week. However, when the exact dispatch hour arrives, the actual load is rather different from the predicted one. Hence, in order to cover such mismatches, it is necessary to make an additional reallocation exercise within the shortest time possible, not to say instantaneously. Such exercise to cover the offset may be based on some technical and/or economic considerations and is commonly referred to as optimal power flow (OPF). Further, there are additional solutions given for those mismatches, even for a narrower time span: the so-called automatic generation control. This control scheme should

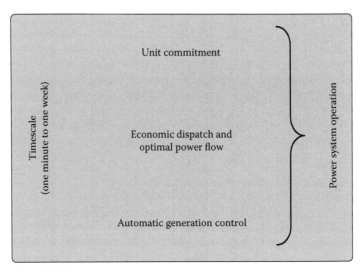

FIGURE 6.1
Power system operation framework applied to distribution networks.

be performed periodically (in minutes) otherwise the system frequency may change undesirably.

Therefore, formerly debatable technical decisions are of primary importance whenever the power generated is dispatched to any class of consumers. This endeavor related to a power delivery, with reduction of losses, voltage profile enhancement, and increased reliability is within the scope of power system operation assignments. The word *operation* is a common electric power term used when referring to the short term, whereas power system experts use the term *planning* to denote actions required in the future.

For the reasons given above, it is crucial for the system operator acting as a decision maker to ensure an acceptable performance level in distribution systems in order to deliver the energy generated as per a planned schedule.

6.2.2 Operational Challenges in Distribution Systems

Local distribution grids are part of the grid structure connected to primary distribution grids covering a neighborhood, which deliver a small amount of energy to the customers. Normally, the distribution grid is operated at medium and low voltages (LVs). Different configurations exist in distribution grids with radial, loop, and meshed grid structure, as illustrated in Figure 6.2, the radial structure being the simplest one.

Service interruption time during any grid component failure can be considered a limiting factor, especially in radial topology. Nevertheless, these drawbacks of radial systems can be solved using a loop structure, since all

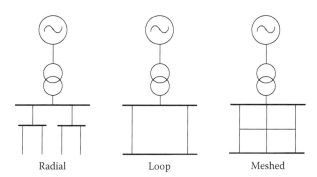

FIGURE 6.2
Illustration of the three structures of electrical distribution grids.

customers can be connected at least via two lines when the faulty section is split-up into two parts. Furthermore, in meshed structures, customers are connected by more than two lines. This potential benefit results in improved security, reduced grid losses, and improved voltage profiles along the feeders.

From the structures given above, the last two of them are more reliable compared to the radial grid, but they are more expensive due to large amount of wiring, switching, and protection system requirements to obtain selective fault clearing. Even though the electric distribution system (EDS) has a looped or meshed structure, the great majority of practical power systems are operated radially due to the simplicity of handling and creating openings at certain locations using switch disconnectors or tie switches in the grid [1].

In addition, the power flow in a radial operation of any distribution grid is unidirectional, and the radial operation also offers the possibility to apply a simple protective system, which is cost reducing. During disturbances, for a number of substations, the radial operation mode leads to interruption of supply, which can be restored by isolating the faulted line or cable and closing one or more grid openings.

The increasing penetration of intermittent and decentralized renewable generation capacities presents new challenges to ensuring the reliability and quality of power supply. A large share of recent RES installed capacity has already taken place in insular grids, since these regions are preferable due to their high RES potential and high prices imposed by already in place diesel generators. Most of these new energy resources (both in number and capacity) are being installed at the distribution network level, directly affecting its management and operation.

The integration of distributed generation (DG) has had a significant impact on distribution grids. Since no direct control is at hand for this type of generation, due to its special features, they are directly connected to the distribution network without being centrally planned or dispatched. Generally,

DG units can be divided into DG units based on a static energy conversion (such as photovoltaic and fuel cells) and DG units based on rotational energy conversion (wind turbine or combined heat and power).

Traditionally, distribution networks were designed and operated through a top-down approach, considering predictable flows in the electricity network. Unpredictable network flows, greater variations in voltage, and different network reactive power characteristics are expected as a consequence of the increasing integration of DES. Furthermore, grid constraints in the distribution network will occur more frequently, adversely affecting the quality of supply. Nevertheless, DSOs are in charge of operating their networks in a secure way, providing the required quality of supply to their customers.

A higher penetration of DG units in a local distribution grid may lead to voltage violation. It disturbs the classical way of voltage control affecting the feeder voltage or deteriorates power quality. However, the voltage profile is not violated when the injected power by DG is near the load of the feeder and the power factor of the DG is in line with the power factor of the load. In this case, the energy supplied by the grid is increasing, as well as the current through the feeder. This results in a reduced voltage drop. However, when the generated power exceeds the load of the feeder or the power factor is extreme, a rise in voltage occurs. This is due to the reversed power flow and is a function of the DG size, the power factor, and the overall impedance of the grid. The effect of reverse power flows gets worse when DG injects reactive power as well.

The impact of DG on voltage control is dependent on the power flow in the network; besides, it may also affect the performance of the protective system in overhead lines. Particularly, in conventional distribution grids with an entirely passive setup, where unidirectional power flow is considered, protective system's response to clear any fault is faster. However, for DG with synchronous generators (SGs), the time needed for the protective system to detect and clear the fault can exceed the stability limits. Hence DG has to be disconnected before stability limits are exceeded, triggering unnecessary disconnection of DG, which no longer is desirable since it reduces the expected benefits.

In addition, DG integration alters the power flow changing the fault currents, affecting the proper operation of the protection system in the distribution grid. The effect of DG on grid losses strongly depends on the injected power and location on the grid. Moreover, intermittent generation sources with a weak correlation to the load, such as wind turbines, can have a negative impact on grid losses. Especially at night, there is low demand and, in a high wind situation, the distribution grid can start to export power, which can increase grid losses. In this case, nearby storage systems can locally balance the power flow and prevent the export of power. It is also demonstrated that wind turbines connected sufficiently close to the load have a positive effect on grid losses.

There are several operational characteristics describing distribution systems since their operation is rather different from transmission systems. Hence, the primary objective of a reliability-driven design of a distribution system is to reduce the frequency and duration of power interruptions to customers [2]. According to the EPRI (Electric Power Research Institute, Palo Alto, California), the current worldwide agenda for the road map of future smart grid aims at ensuring that electricity networks continue to function in a way that optimizes cost and environmental performance without jeopardizing traditional high security and quality of supply and presiding over increasing penetration of renewable energy sources. A more active distribution network management would allow a more efficient integration of DES by leveraging the inherent characteristics of this type of generation.

To pursue the aforementioned goals, there is a large list of existing professional distribution system analysis tools typically used to study the effect of distributed energy resources. Some of these represent state-of-the-art commercial tools such as CYMDIST by CYME International; meanwhile, others represent a set of open-source tools such as OpenDSS by EPRI, GridLAB-D by PNNL, and Power System Toolbox and MATLAB by Cherry Tree Scientific Software. However, to use a commercial solver, it is desirable to obtain a linear equivalent for the nonlinear terms, including the objective function [3,4].

Consequently, the approach proposed here is distinctive in addressing the modeling challenges in radial or weakly meshed distribution networks in a simplified manner, using a linearized power flow (one-line diagram) grid representation. In addition, the modeling task focuses on issues related to distribution system operation integrating DG, distributed loads, and storage devices, as well as substations and intermittent renewable wind energy units located at certain buses.

6.2.3 Key Challenges in Distribution Systems

In theory, the increasing penetration of DG in distribution networks can result in several benefits. These benefits can be summarized as the overall energy efficiency increase, reduction of environmental impacts, line loss reduction, as peak shaving, voltage support, and deferred investments to upgrade existing generation and distribution systems.

The integration of DES in distribution networks represents a challenge for DSOs due to the particular characteristics of this type of generation. DES location is rarely related to distribution network needs, but to the availability of the resource. Furthermore, DES are mostly nondispatchable and their production profiles do not always match the demand. Grid operators and planners should recognize that this supply must be treated as any other supply source, being fully integrated in transmission and distribution network.

6.2.3.1 Distribution Network Operation

RES represents a challenge not only for system balancing but also for local network operation. Secure operation and capacity of the distribution system are determined by voltage and the physical current limits on the feeders. These limits are the voltage upper and lower constraints and the maximum allowable branch current capacity.

The connection of generation affects voltage just like consumption does. Overvoltage is the most frequent problem at the connection point for DG units and the relevant grid area. When DG production exceeds local consumption, power flows into the network impacting voltage profiles. Since the network is dimensioned so that the minimum voltage is sufficient even under periods with maximum load, voltage control has not yet been needed, choosing adequately the voltage at the substation (1.05 pu). Under this paradigm, voltage at the substation bus is chosen higher than the nominal value in order to account for voltage decrease toward the load connection. Since active voltage control is not in place, DSOs may have difficulties in maintaining the voltage profile at the customer connection points, in particular at the LV level, endangering security of supply.

When generation units are connected to the distribution network, the power flow may be reversed under certain periods, where the power injected from DG is larger than the local demand. Therefore, assuming decreasing voltages from the substation along the distribution network is no longer valid under all conditions.

Another relevant task for distribution network operation is congestion management. Congestion occurs whenever the network is not able to accommodate all the desired transactions due to the violation of one or more network constraints, leading to necessary emergency actions to curtail generation or demand.

Generation curtailment is used in cases of system-security-related events (i.e., congestion or voltage rise). However, the regulatory basis for generation curtailment in emergency situations differs across Europe. If available, DG control is mostly in the hands of the TSO. Since the TSOs do not required tools to monitor distribution network conditions (voltage and flows), DSOs can only react to DG actions. As a consequence, distribution networks with high DES penetration are already facing challenges in meeting some of their responsibilities. These challenges are expected to become more frequent in the future. Although telemeasuring and telecontrol devices in transmission networks are quite common, there is still not much telecontrol logic used in distribution systems, resulting in an inefficient generation profile. To allow a larger penetration of DES maintaining an adequate voltage profile along the distribution feeders at the same time, advanced control systems have been proposed in the literature, adopting centralized or decentralized structures. The present study incorporates generation curtailment and reactive power management as an additional DSO task, which is called *distributed generation active management* (DGAM). Results of the application of DGAM will be compared with those obtained connecting DG with no regulation in order to outline the benefits of DGAM.

6.2.3.2 Network Reinforcement

Additionally, the possibility of installing DG to produce electricity close to the demand reduces the need to use network capacity for transport over longer distances during certain hours. However, distribution networks must be designed for peak load, potentially increasing the overall investment cost. For example, solar production does not follow consumption generally, and peak demand usually accounts for evening hours.

Distribution networks have been traditionally designed to be prepared for all possible combinations of generation and demand situations. Short-term (ST) constraints trigger grids reinforcements, what is traditionally called the *fit-and-forget approach*. Even constraints happening for a few hours a year trigger grid adaptations. With the presence of DG, the utilization rate of network assets would decline even more, calling for a review on traditional reinforcement practices. The stochastic nature of DES motivates a probabilistic approach, but this has not yet completely considered in standard procedure for the planning of DG connection, based on the worst-case scenario.

6.2.3.3 Insular Distribution Systems Operation Context

One recent major concern in insular grids is related to the high level of penetration of renewable energy, affecting grid security, since conventional generation has become more volatile, due to renewable sources and the intermittence of other nondispatchable energy sources. On the other hand, despite experience in this subject, suitable techniques for insular grid systems are still insufficient. According to Eurelectric [5] questionnaire for European insular systems, insular economies are very fragile compared to the mainland ones due to their small size and declining employment rates and gross domestic product. In addition, the electricity production mix and technology in insular systems has been largely exempted from the large combustion plant directive, which they are currently expected to comply with. At present, insular generation sources are predominantly diesel technologies or heavy-fuel oil engines, characterized by a high economic cost and greenhouse gas emissions, although they are highly flexible in meeting daily and seasonal variations in energy demand. Currently, an energy paradigm shift is taking place in the European insular energy policy. Therefore, a stable and secure power grid management for insular applications requires more robust diagnosis tools in order to satisfy the existing grid code requirements.

Therefore, the security of the electricity grid remains a vital issue for European insular communities, healing their fragile economies and encouraging investment opportunities. For instance, in order to neutralize systemic risks, most insular systems operate with generation margins around 30%–40%, implying higher cost scenarios compared to 15%–20% on the mainland, which is a highly interconnected grid system [2].

6.2.4 Distribution Systems Modeling Approaches

Electrical distribution system operation planning has as an objective to determine the optimal operation conditions of existing grid parameters, in order to minimize their negative impacts. Statistically, the majority of service interruptions to customers come from distribution systems. Therefore, a detailed reliability evaluation of the distribution system has, therefore, become very important in the planning and operating stages of a power system [6].

The mainstream ways of solving operational and planning tasks are modeling them as optimization problems. Consequently, they require the following three basic steps to be taken into account. The first step is the problem definition; in this step, all optimization objective functions, system constraints as well as decision variables, both independent and dependent, should be clearly defined. The second step involves modeling the scope of the system problems. Here, a proper modeling approach should be followed so as to adequately control the operating variables of the system, as well as to achieve the necessary accuracy of the output variables using valid sets of data inputs. Last, but not least, is the identification of the solution algorithms. At this juncture, the adequacy of all previous steps should be measured and integrated to pursue the optimal solution performance.

The mathematical formulation of an optimization problem is dependent on the mathematical terms contained in the objective function and constraints. Nonlinear programming (NLP) requires at least a quadratic programming; meanwhile, linear functions can be treated using linear programming (LP) and other formulations and may require integer or mixed integer linear programming (IP or MILP) [7]. On the other hand, distribution systems are multifaceted problems. Distribution reconfiguration is essentially a combinatorial optimization problem, where the best possible combination of status (open or close) of switch disconnectors and tie switches has to be found, so that the objective function is minimized [1–3].

The most important aspect for the operation of a distribution grid is the topology. The complexity of such combinatorial issue is even more complex for large systems using conventional optimization approaches [8]. More recently, the introduction of smart grids has added further complexity to the problem requiring new methodologies and procedures for planning and operating electrical networks.

Some of these desired features in a smart distribution grid include low operation and maintenance costs and the ability for self-healing and self-reconfiguration. Network reconfiguration can be cast as an optimization problem resulting from switching decisions to reconfigure optimally the network topology. This subject requires fast and efficient algorithms to achieve the ideal topology among all possible configurations. Additionally, the number of possible configurations grows exponentially giving birth to the so-called combinatorial explosion due to the extensive number of switching operations acting as a limiting factor in all possible configurations.

For this reason, new models and approaches are required to deal with such complexities in order to provide system operators with a useful instrument to appropriately cope with distribution system constraints.

6.3 Problem Formulation of the Operation of Distribution Networks

6.3.1 Description of a Generic Electric Energy Storage Model

The model for generic storage is assumed to have a capability of transforming, storing, and reversing the process to release the stored energy when the demand is increased, including multiple applications to stabilize grid operations [9,10]. Defining generic storage has some mathematical advantages: primarily, it can be integrated within complex optimization problems. In addition, it can be easily modeled using LP suitable for practical large-scale cases. Further, simplifications are made to describe the characteristics of this ideal storage device as follows [11]:

1. There are no up or down ramps. A unit can go from not producing anything to full power instantly.
2. It is assumed that there no energy losses in storage.
3. There is no hysteresis during loading or discharging the storage device.
4. Storage devices have conversion losses expressed in terms of charge and discharge efficiency rates.

There are also two aspects that should be mentioned in the design of an energy storage system. Since wind power fluctuations exist, energy storage can provide support in different time periods. The second aspect is that the determination of the storage capacity is also an important factor in reducing the total cost as much as possible [12].

The use of a generic storage device is assumed in the model formulation. The storage system may be located at any feeder or bus, just like any generator or load. Then, for the sequential power flow studies, it is assumed that the storage element behaves either as a power source or as a sink depending on the time of period and the operational costs. For this purpose, active constraints are set for the storage model.

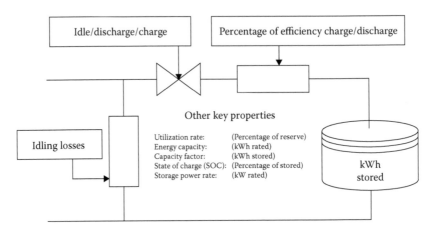

FIGURE 6.3
Generic ESS model. (Data from P. Medina et al., Electrical energy storage systems: Technologies' state-of-the-art. Techno-economic benefits and applications analysis. *Proceedings of the 47th Hawaii International Conference on System Sciences*, pp. 2295–2304, 2014.)

The electrical demand overtime by the storage or the charge profile is defined by storage size, storage type, charging rate, charging efficiency, and the storage response when it is connected. In the literature [13], there are three well-defined states of storage: charging, discharging, and idling, as illustrated in the generic storage model in Figure 6.3.

Thus, a generic energy storage system (ESS) is any device with the capability of transforming and storing energy, reverting the process by injecting the stored energy to the system [11]. This is represented by Equations 6.1 through 6.6:

$$x_{it\omega\xi} = x_{it-1\omega\xi} + \left[\eta_i^{ch} P_{it\omega\xi}^{ch} - \left(\frac{1}{\eta_i^{dis}} \right) P_{it\omega\xi}^{dis} \right] \tag{6.1}$$

$$\underline{x}_i \le x_{it\omega\xi} \le \overline{x}_i \tag{6.2}$$

$$\underline{s}_i^{ch} \le P_{it\omega\xi}^{ch} \le \overline{s}_i^{ch} \, b_{it}^{s} \tag{6.3}$$

$$\underline{s}_i^{dis} \le P_{it\omega\xi}^{dis} \le \overline{s}_i^{dis} \, b_{it}^{p} \tag{6.4}$$

$$b_{it\omega\xi}^{p} + b_{it\omega\xi}^{s} \le 1 \tag{6.5}$$

$$x_{it=0\omega} = x_0^{s} \tag{6.6}$$

$$\forall i \in \Omega_{st}; \forall t \in \Omega_t; \forall \omega \in \Omega_\omega; \text{ and } \forall \xi \in \Omega_\xi$$

Equation 6.1 refers to the storage transition function. The state of charge ($\hat{x}_{it\omega\xi}$) at the end of the first time period depends on the previous state $\hat{x}_{it-1\omega\xi}$. There are efficient production (η_i^p) and energy storage transformation (η_i^s) rates. Equation 6.2 refers to the maximum and minimum energy capacities. The minimum and maximum storage limits (charge), and productions (injection into the network) are defined in Equations 6.3 and 6.4. Equation 6.5 means that the storage can only produce or store within a period. Finally, the initial status of the storage device is considered in Equation 6.6.

6.3.2 Description of Conventional DG

Within the problem formulated in this study, the conventional generation cost has been expressed as a quadratic function of the power output. This quadratic fuel production cost function typically used in scheduling problems [14] can be formulated as follows:

$$\text{Cost}_{it\omega\xi}^{\text{conv}} = a^{\text{conv}} * v_{it\omega\xi} + b^{\text{conv}} * P_{it\omega\xi}^{\text{conv}} + c^{\text{conv}} * P_{it\omega\xi}^{\text{conv}} * P_{it\omega\xi}^{\text{conv}} \tag{6.7}$$

The cost function in Equation 6.7 can be accurately approximated by a set of piecewise blocks. For practical purposes, the piecewise linear function is an adequate approximation of the nonlinear model if enough segments are used.

The analytic representation of the linear approximation of the quadratic production cost function is the following one:

$$\text{Cost}_{it\omega\xi}^{\text{conv}} = A^{\text{conv}} * v_{it\omega\xi} + \sum_{l} \gamma_{ilt\omega\xi} * \delta_{ilt\omega\xi} \tag{6.8}$$

$$P_{it\omega\xi}^{\text{conv}} = \sum_{l} \delta_{ilt\omega\xi} + P_{\min}^{\text{conv}} * v_{it\omega\xi} \tag{6.9}$$

$$\delta_{i1t\omega\xi} \leq P_{i1t\omega\xi}^{\text{conv}} - P_{\min}^{\text{conv}} \tag{6.10}$$

$$\delta_{ilt\omega\xi} \leq P_{ilt\omega\xi}^{\text{conv}} - P_{il-1t\omega\xi}^{\text{conv}} \tag{6.11}$$

$$\delta_{il=L\omega\xi} \geq P_{\max}^{\text{conv}} - P_{iL-1t\omega\xi}^{\text{conv}} \tag{6.12}$$

$$\delta_{ilt\omega\xi} \geq 0 \tag{6.13}$$

$$\forall i \in \Omega_{st}; \forall l \in \Omega_l; \forall t \in \Omega_t; \forall \omega \in \Omega_\omega; \text{ and } \forall \xi \in \Omega_\xi$$

where:

a^{conv}, b^{conv}, and c^{conv} are constants of fuel cost function of a conventional generator (CONV)

$P_{\min}^{\text{conv}}$ and $P_{\max}^{\text{conv}}$ are the minimum and maximum power generation limits, respectively

$$A^{\text{conv}} = a^{\text{conv}} + b^{\text{conv}} * P_{\min\ it\omega}^{\text{conv}} + c^{\text{conv}} * P_{\min\ it\omega}^{\text{conv}} * P_{\min\ it\omega}^{\text{conv}} \tag{6.14}$$

6.3.3 Description of Renewable DG

The integration of renewable DG units typically installed close to load centers improves the performance of distribution systems through active and reactive power injections. This is a cost-effective solution to achieve voltage support minimizing the power loss and enhancing reliability.

In this chapter, we are considering wind turbines as the main source of RES in distribution systems. As many as 15 wind turbines have been considered in buses 16, 21, and 26, consistent with the aim of placing DES close to demand.

DG is usually used as a source in small-to-medium-scale low and medium voltage networks. Although telemeasuring and telecontrol devices in transmission networks are quite common, there is not much telecontrol logic used in distribution systems, resulting in an inefficient generation profile. Directly coupled induction generators cannot provide reactive power control without external equipment. Reactive power control is only possible in this case by external equipment (capacitor banks). This assumption and the lack of telecontrol devices have traditionally led to a direct connection of DG in distribution networks.

$$Q_{it\omega\xi}^{ren} = 0 \tag{6.15}$$

All other grid-coupling technologies such as doubly fed induction generators, directly coupled SGs, and inverters can provide reactive power control by itself [15].

To allow a larger penetration of DES, maintaining an adequate voltage profile along the distribution feeders at the same time, advanced control systems have been proposed in the literature, adopting centralized or decentralized structures. A DG control algorithm for coordinated voltage control in distribution systems with high DG penetration is developed in the present study, incorporating generation curtailment and reactive power management (Equations 6.16 and 6.17) as an additional DSO task, which is called DGAM. The voltage control is based on controlling the active and reactive power output from the DG units (see Figure 6.4).

$$Q_{it\omega\xi}^{ren} \leq P_{it\omega\xi}^{ren} * \tan\left[\cos^{-1}\left(FP^{ren}\right)\right] \tag{6.16}$$

$$Q_{it\omega\xi}^{ren} \geq P_{it\omega\xi}^{ren} * \tan\left[\cos^{-1}\left(-FP^{ren}\right)\right] \tag{6.17}$$

Monte Carlo-generated scenarios

Wind and demand forecast scenarios have been created using Monte Carlo simulation. The representative branching trees consisting of probability-weighed scenarios account for both forecasts and their forecast errors. Each forecast is defined for 24 hours.

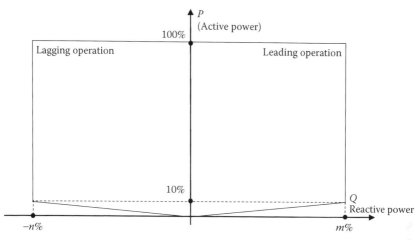

FIGURE 6.4
Conventional distributed cost function.

The number of considered wind forecasts has been limited to eight to minimize dimensionality in the problem formulation, maintaining accuracy in the provided information. These scenarios shown in Figures 6.9 and 6.10 are represented accounting for the required uncertainty. Additionally, reactive power demand scenarios have been included in the model, being correlated to the active demand scenarios. Each single scenario in the tree consists of time series for wind, and demand and is defined with an equiprobable probability.

OPF model description

The model presented describes a linearized OPF model approach for the minimization of active power losses and operation costs in a real distribution network using MILP. We focus on the operating cost optimization for 24 hours where the net power is supplied by a mix of sources: substation, renewable generation, and thermal-type conventional generation, supported by generic storage devices to satisfy a given demand at each bus for a 24-hour horizon. In addition to that, the model for renewable generation comprises different wind and demand scenarios with hourly profiles. Meanwhile, for the other types of generation, we merely consider power boundary constraints for the upper and lower power limits. Besides, in order to represent the steady-state operation of an EDS, the following basic assumptions are made:

- The EDS is balanced and represented by its single-phase (monophasic) equivalent model.
- In the ij branch, bus i is closer to the substation than bus j.

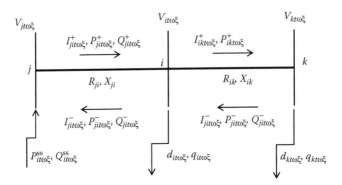

FIGURE 6.5
Illustrative radial distribution system.

- Active and reactive power losses in branch *ij* are concentrated at bus *i*.
- In order to guarantee a radial configuration of the grid, enforcement of constraints is incorporated in the model for the 24 looped connections of branches.

The theoretical background for the current model is based on the assumption that the basic EDS operation problem can be modeled as a mixed-integer nonlinear programming problem (MINLP), where the components of the current active and reactive power flows can be represented as illustrated in Figure 6.5.

Objective function

The objective function can take different forms. Three groups of case studies have been considered in this chapter, in which each group is computed with and without DGAM. As seen in Section 6.4, cases are analyzed there.

6.3.3.1 Power Losses Minimization

In Equation 6.18, the objective function is related to the minimization of the active power losses in a distribution system for 24 hours taking into account the probability of each of the wind and demand scenarios. It is assumed that there is not energy non-supplied (ENS).

$$\text{Min} \sum_{\omega\xi} \pi_\omega \pi_\xi \left(\sum_t \sum_{ij} R_{ij} I^{sqr}_{ijt\omega\xi} \right) \tag{6.18}$$

$$\forall ij \in \Omega_b; \forall t \in \Omega_t; \forall \omega \in \Omega_\omega; \text{ and } \forall \xi \in \Omega_\xi$$

6.3.3.2 Minimization of Generation Cost

The generation cost is defined as the sum of the renewable (REN) operation cost, fuel cost of the CONV, charging and discharging operation cost of
the ESS, active power cost coming upstream from the substation (SUB), and
also ENS and wind curtailment costs. The objective function is defined in
Equation 6.19.

$$\text{Min} \sum_{\omega} \pi_{\omega} \pi_{\xi} \left(\begin{array}{c} \sum_{t} \sum_{ij} C^{\text{ren}} P_{it\omega\xi}^{\text{ren}} + \text{Cost}_{ijt\omega\xi}^{\text{conv}} + C^{\text{ch}} P_{it\omega\xi}^{\text{ch}} + C^{\text{dis}} P_{it\omega\xi}^{\text{dis}} \\ + C^{\text{su}} P_{it\omega\xi}^{\text{su}} + C^{\text{ens}} P_{it\omega\xi}^{\text{ens}} + C^{\text{ren\_curt}} P_{it\omega\xi}^{\text{ren\_curt}} \end{array} \right) \tag{6.19}$$

$$\forall ij \in \Omega_b; \forall t \in \Omega_t; \forall \omega \in \Omega_\omega; \text{ and } \forall \xi \in \Omega_\xi$$

6.3.3.3 Simultaneous Minimization of Generation Costs and Power Losses

As seen in Equation 6.20, the objective function simultaneously considers
both the system power losses and the generation cost described in 6.3.2.

$$\text{Min} \sum_{\omega} \pi_{\omega} \pi_{\xi} \left(\begin{array}{c} \sum_{t} \sum_{ij} \rho R_{ij} I_{ijt\omega\xi}^{\text{sqr}} + C^{\text{ren}} P_{it\omega\xi}^{\text{ren}} + C^{\text{su}} P_{it\omega\xi}^{\text{su}} + \text{Cost}_{ijt\omega\xi}^{\text{conv}} + C^{\text{ch}} P_{it\omega\xi}^{\text{ch}} \\ + C^{\text{dis}} P_{it\omega\xi}^{\text{dis}} + C^{\text{ens}} P_{it\omega\xi}^{\text{ens}} + C^{\text{ren\_curt}} P_{it\omega\xi}^{\text{ren\_curt}} \end{array} \right) \tag{6.20}$$

$$\forall ij \in \Omega_b; \forall t \in \Omega_t; \forall \omega \in \Omega_\omega; \text{ and } \forall \xi \in \Omega_\xi$$

The constraints of the model are described in Sections 6.3.3.4 to 6.3.3.13.

6.3.3.4 Active Power Balance

The real-time operation of an EDS requires that the power generated (SU,
REN, CONV) must match all the demand at each bus for all renewable scenarios ω and demand scenarios ξ at any period t discounting line losses.
In addition to that, the ENS and wind curtailment are considered as a production deficit of the EDS. The active power balance equation is a power flow
balance obtained considering positive or downstream (forward) and negative or upstream (backward) power flow senses.

$$P_{it\omega\xi}^{\mathrm{ren}} + P_{it\omega\xi}^{\mathrm{conv}} + \sum_{j}\left(P_{jit\omega\xi}^{+} - P_{jit\omega\xi}^{-}\right) - \sum_{k}\left[\left(P_{ikt\omega\xi}^{+} - P_{ikt\omega\xi}^{-}\right) + R_{ij}I_{ijt\omega\xi}^{\mathrm{sqr}}\right]$$

$$+ P_{it\omega\xi}^{\mathrm{su}} + \left(P_{it\omega\xi}^{\mathrm{dis}} - P_{it\omega\xi}^{\mathrm{ch}}\right) = d_{it\xi} - P_{it\omega\xi}^{\mathrm{ens}} - P_{it\omega\xi}^{\mathrm{ren\_curt}}$$

(6.21)

$$\forall i \in \Omega_n; \forall ijk \in \Omega_b; \forall t \in \Omega_t; \forall \omega \in \Omega_\omega; \text{ and } \forall \xi \in \Omega_\xi$$

6.3.3.5 Reactive Power Balance

The reactive power balance equation is expressed in the same way as previously explained for the active power; accounting for every EDS component of the reactive power balance, with the exception of storage.

$$Q_{it\omega\xi}^{\mathrm{ren}} + Q_{it\omega\xi}^{\mathrm{conv}} + \sum_{j}\left(Q_{jit\omega\xi}^{+} - Q_{jit\omega\xi}^{-}\right) - \sum_{k}\left[\left(Q_{ikt\omega\xi}^{+} - Q_{ikt\omega\xi}^{-}\right) + X_{ij}I_{ijt\omega\xi}^{\mathrm{sqr}}\right] + Q_{it\omega\xi}^{\mathrm{su}} = q_{it\xi} \quad (6.22)$$

$$\forall i \in \Omega_n; \forall ijk \in \Omega_b; \forall t \in \Omega_t; \forall \omega \in \Omega_\omega; \text{ and } \forall \xi \in \Omega_\xi$$

6.3.3.6 Nominal Voltage Balance

The nominal voltage balance for the whole network is expressed by Equation 6.23. This equation is explained in detail in [16].

$$V_{it\omega\xi}^{\mathrm{sqr}} - 2\left[R_{ij}\left(P_{ijt\omega\xi}^{+} - P_{ijt\omega\xi}^{-}\right) + X_{ij}\left(Q_{ijt\omega\xi}^{+} - Q_{ijt\omega\xi}^{-}\right)\right] - Z_{ij}^{2}\,I_{ijt\omega\xi}^{\mathrm{sqr}} - V_{jt\omega\xi}^{\mathrm{sqr}} = 0 \quad (6.23)$$

$$\forall i \in \Omega_n; \forall ijk \in \Omega_b; \forall t \in \Omega_t; \forall \omega \in \Omega_\omega; \text{ and } \forall \xi \in \Omega_\xi$$

6.3.3.7 Nonlinear Apparent Power Expression of the Network

The apparent power expression defined in Equation 6.24 is nonlinear due to the product of quadratic voltage and current magnitudes and the squares of the active and reactive power flows, which are linearized in the following subsection.

$$V_{it\omega\xi}^{\mathrm{sqr}}\,I_{ijt\omega\xi}^{\mathrm{sqr}} = P_{ijt\omega\xi}^{2} + Q_{ijt\omega\xi}^{2}$$

(6.24)

$$\forall i \in \Omega_n; \forall ijk \in \Omega_b; \forall t \in \Omega_t; \forall \omega \in \Omega_\omega; \text{ and } \forall \xi \in \Omega_\xi$$

6.3.3.8 Nonlinearization of the Nonlinear Apparent Power

This set of equations represents a linearization of the nonlinear terms of the model formulation with the objective of expressing a distribution system using MILP. Note that $V_{it\omega\xi}^{sqr}$ and $I_{ijt\omega\xi}^{sqr}$ are variables that represent the square magnitude values of voltages and currents, respectively.

In the nonlinear apparent power expression for the network, there are two nonlinear terms:

1. $V_{it\omega\xi}^{sqr} \cdot I_{ijt\omega\xi}^{sqr}$: The linearization of a product of two variables can be done by discretizing $V_{it\omega\xi}^{sqr}$ within a small interval, though it implies that there are many binary variables that increase the computation time. Since the voltage magnitude is within a small range in distribution networks, the approximation used considers V_{nom}^2 as $V_{it\omega\xi}^{sqr}$. This can be proven because, if the model is run in a second iteration considering the previous result of $V_{it\omega\xi}^{sqr}$, it hardly changes.

2. $P_{ijt\omega\xi}^2 + Q_{ijt\omega\xi}^2$: The linearization of both terms can be done by a piecewise linear approximation [7], as shown below:

$$P_{ijt\omega\xi}^2 + Q_{ijt\omega\xi}^2 = \sum_f \left(m_{ijft\omega\xi} \, \Delta P_{ijft\omega\xi} \right) + \sum_f \left(m_{ijft\omega\xi} \, \Delta Q_{ijft\omega\xi} \right) \tag{6.25}$$

$$P_{jit\omega\xi}^+ + P_{jit\omega\xi}^- = \sum_f \Delta P_{ijft\omega\xi} \tag{6.26}$$

$$Q_{jit\omega\xi}^+ + Q_{jit\omega\xi}^- = \sum_f \Delta Q_{ijft\omega\xi} \tag{6.27}$$

$$0 \le \Delta P_{ijft\omega\xi} \le \Delta S_{ijft\omega\xi} \tag{6.28}$$

$$0 \le \Delta Q_{ijft\omega\xi} \le \Delta S_{ijft\omega}\xi \tag{6.29}$$

where the constant parameters $m_{ijft\omega\xi}$, the slope of the fth power block, and $\Delta S_{ijft\omega\xi}$ are defined as follows:

$$m_{ijft\omega\xi} = \left(2f - 1 \right) \Delta S_{ijft\omega\xi} \tag{6.30}$$

$$\Delta S_{ijft\omega\xi} = \frac{\left(V_{nom} \, \overline{I}_{ij} \right)}{F} \tag{6.31}$$

$$\forall i \in \Omega_n; \forall ijk \in \Omega_b; \forall t \in \Omega_t; \forall \omega \in \Omega_\omega; \text{ and } \forall \xi \in \Omega_\xi$$

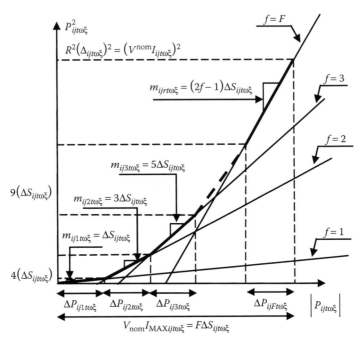

FIGURE 6.6
Modeling the piecewise linear function $P^2_{ijt\omega\xi}$.

The expressions on the right side of Equation 6.25 are linear approximations of $P^2_{ijt\omega\xi}$ and $Q^2_{ijt\omega\xi}$, respectively. Equations 6.26 and 6.27 ensure that $P_{ijt\omega\xi}$ and $Q_{ijt\omega\xi}$ are equal to the sum of all the values in each discrete block, respectively. The upper and lower bounds of the contribution of each block, $\Delta P_{ijft\omega\xi}$ and $\Delta PQ_{ijft\omega\xi}$, are defined in Equations 6.28 and 6.29, respectively.

The linearization of the active power flow is shown in Figure 6.6. It is important to note that the values of the blocks from the discretization must be filled sequentially. This condition can be ensured since the curve is quadratic; thus, it is convex according to the optimization problem that minimizes losses.

Therefore, finally in this model, the final linear equation is:

$$V^2_{nom}\, I^{sqr}_{ijt\omega\xi} = \sum_{f}\left(m_{ijft\omega\xi}\,\Delta P_{ijft\omega\xi}\right) + \sum_{f}\left(m_{ijft\omega\xi}\,\Delta Q_{ijft\omega\xi}\right) \qquad (6.32)$$

$$\forall i \in \Omega_n\,; \forall ijk \in \Omega_b\,; \forall t \in \Omega_t\,; \forall \omega \in \Omega_\omega\,; \text{ and } \forall \xi \in \Omega_\xi$$

6.3.3.9 Network Radiality Constraint

It is a known fact that EDS commonly operates with radial topology. Therefore, all models must incorporate it [17].

$$\sum_{ij} y_{ijt} \le N_{\text{loop}} - 1 \tag{6.33}$$

$$\forall i \in \Omega_n; \forall ijk \in \Omega_b; \forall t \in \Omega_t; \forall \omega \in \Omega_\omega; \text{ and } \forall \xi \in \Omega_\xi$$

6.3.3.10 Voltage Limits

The maximum voltage value must be checked for each bus, time and wind power, and demand scenarios, as explained in Equation 6.34.

$$\underline{V}^2 \le V^{\text{sqr}}_{it\omega\xi} \le \overline{V}^2 \tag{6.34}$$

$$\forall i \in \Omega_n; \forall ijk \in \Omega_b; \forall t \in \Omega_t; \forall \omega \in \Omega_\omega; \text{ and } \forall \xi \in \Omega_\xi$$

6.3.3.11 Maximum Branch Current Limit

This limit is the maximum allowable branch current capacity (Equation 6.35):

$$0 \le I^{\text{sqr}}_{ijt\omega\xi} \le \overline{I}^2_{ij} \tag{6.35}$$

$$\forall i \in \Omega_n; \forall ijk \in \Omega_b; \forall t \in \Omega_t; \forall \omega \in \Omega_\omega; \text{ and } \forall \xi \in \Omega_\xi$$

6.3.3.12 Active and Reactive Power Limits

These equations are related to the maximum power capacities of the branches. During any operation, the positive and negative flows for both active and reactive powers should be within the preestablished active and reactive limits for the whole network. Moreover, the active and reactive power flow capacities are represented in Equations 6.36 through 6.39. In order to distinguish the power flow directions, forward (downstream) and backward (upstream), two separate terms for both are used, where active and reactive flows are considered.

$$P^+_{ijt\omega\xi} \le V_{\text{nom}} \, \overline{I}_{ij} \, v^{P+}_{ijt\omega\xi} \tag{6.36}$$

$$P^-_{ijt\omega\xi} \le V_{\text{nom}} \, \overline{I}_{ij} \, v^{P-}_{ijt\omega\xi} \tag{6.37}$$

$$Q^+_{ijt\omega} \le V_{\text{nom}} \, \overline{I}_{ij} \, v^{Q+}_{ijt\omega\xi} \tag{6.38}$$

$$Q^-_{ijt\omega} \le V_{\text{nom}} \overline{I}_{ij} v^{Q-}_{ijt\omega\xi} \tag{6.39}$$

Equations 6.40 and 6.41 guarantee that the flow is only in one sense, which means it can be either upstream or downstream, but not both.

$$v^{P+}_{ijt\omega\xi} + v^{P-}_{ijt\omega\xi} \leq 1 \tag{6.40}$$

$$v^{Q+}_{ijt\omega\xi} + v^{Q-}_{ijt\omega\xi} \leq 1 \tag{6.41}$$

$$\forall i \in \Omega_n; \forall ijk \in \Omega_b; \forall t \in \Omega_t; \forall \omega \in \Omega_\omega; \text{ and } \forall \xi \in \Omega_\xi$$

6.3.3.13 Power Factor Constraints

The power factor conditions are set in such a way that they can have positive and negative values, in relation to the load condition. Therefore, there are power factor constraints relating to the active and reactive powers separately to account for positive and negative rates of the power factor. This applies to the REN (Equations 6.16 and 6.17), CONV (Equations 6.42 and 6.43), and SUB (Equations 6.44 and 6.45).

$$Q^{conv}_{it\omega\xi} \leq P^{conv}_{it\omega\xi} * \tan\left[\cos^{-1}\left(FP^{conv}\right)\right] \tag{6.42}$$

$$Q^{conv}_{it\omega\xi} \geq P^{conv}_{it\omega\xi} * \tan\left[\cos^{-1}\left(-FP^{conv}\right)\right] \tag{6.43}$$

$$Q^{su}_{it\omega\xi} \leq P^{su}_{it\omega\xi} * \tan\left[\cos^{-1}\left(FP^{su}\right)\right] \tag{6.44}$$

$$Q^{su}_{it\omega\xi} \geq P^{su}_{it\omega\xi} * \tan\left[\cos^{-1}\left(-FP^{su}\right)\right] \tag{6.45}$$

$$\forall i \in \Omega_n; \forall ijk \in \Omega_b; \forall t \in \Omega_t; \forall \omega \in \Omega_\omega; \text{ and } \forall \xi \in \Omega_\xi$$

6.4 Case Studies for the Operation of Distribution Grids

6.4.1 Network Case Overview

The context of the proposed model encompasses a possible island distribution system operation case including conventional thermal generation. According to the new European initiatives, intermittent renewable, nondispatchable distributed energy resources, and energy storage systems (ESS) located at different buses of the network have been included in the analyzed network. Therefore, the model considers a complex interaction between deterministic and intermittent generation sources, as well as storage systems, to cover the demand at any bus during 24 hours under several wind power production and demand scenarios. Concerning the power system economic analysis, the model proposed does not consider investment costs, only considers the optimal operation costs, since this does not involve planning decisions.

In the case study, a modified IEEE 34-bus distribution system (see Figure 6.7) supplied by different generators such as CONV, RN, SUB, and ESS are considered. The system has an optimal behavior in order to minimize (1) power system losses in the first case study, (2) generation costs in the second one, and (3) total operation cost, which includes (1) and (2). In addition, the system integrates intermittent wind generation and generic storage devices.

The distribution network analyzed consists of 26 buses, 25 branches, and 1 substation. Base power and voltage of the system are 1 MVA and 20 kV, respectively. In the study presented, the considered distribution network is connected through the substation to an existing island transmission system. Therefore, price signals considered in substation bus are those referred to the transmission system. Generation costs for the substation are presented in Figure 6.8. Substation costs are higher during valley hours for the considered system, due to high renewable penetration in the transmission network. Line data for the case study are presented in Table 6.1. Also, in Table 6.2, we include costs and other relevant data for the considered network.

The high valley-peak demand ratio is consistent with small island systems, where little industry is present and consumption is mostly driven by residential and services sectors. Conventional generation has been placed in bus 24, accounting for 250 kW installed capacity. Additionally, 15 wind turbines have been placed in each of the bus 16, 21, and 26. To compensate

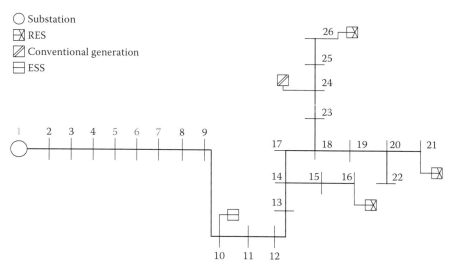

FIGURE 6.7
Modified IEEE-34 bus distribution network case study.

TABLE 6.1

Line Data for the Modified IEEE 34-Bus System

Sending Bus	Receiving Bus	Section Parameters	
		R (pu)	X (pu)
1	2	0.0013682	0.0010185
2	3	0.0009175	0.0006825
3	4	0.0170932	0.0012716
4	5	0.0198882	0.0147952
5	6	0.0157672	0.0117297
6	7	0.0000087	0.0000045
7	8	0.0002486	0.0001235
8	9	0.0081702	0.0040662
9	10	0.0006722	0.0003345
10	11	0.0163565	0.0081405
11	12	0.0004165	0.0002075
12	13	0.0294722	0.0146685
13	14	0.0000085	0.0000045
14	15	0.0294555	0.0632165
14	17	0.0039215	0.0019515
15	16	0.0054735	0.0040725
17	18	0.0046652	0.0023225
18	19	0.0016165	0.0008045
18	23	0.0002245	0.0000111
19	20	0.0021445	0.0010672
20	21	0.0006882	0.0003425
20	22	0.0002245	0.0001115
23	24	0.0010802	0.00053775
24	25	0.0029127	0.00144975
25	26	0.000424	0.00021155

TABLE 6.2

Case Study Input Data

C^{ens} (€/MWh)	C^{ren} (€/MWh)	C^{ch} (€/MWh)	ρ	FP^{conv}	FP^{su}	FP^{ren}
250	17	0.5	2	0.9	0.9	0.9

wind generation and stabilize the distribution system, a storage system has been considered in bus 10. In order to test the model, the input variables were specified (Table 6.2). It is expected that the system conceive profitable results for the company and the consumer.

Demand and wind scenarios generated using Monte Carlo simulation in this case study are represented in Figures 6.9 and 6.10.

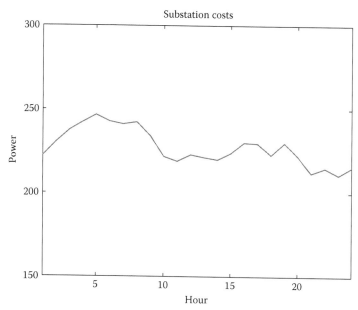

FIGURE 6.8
Energy costs for substation (€/MWh).

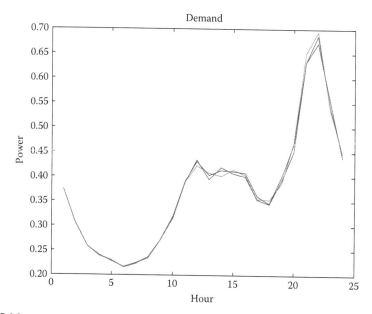

FIGURE 6.9
Demand scenarios.

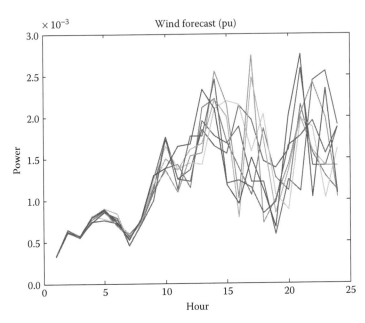

FIGURE 6.10
Wind power scenarios.

Wind generation has not been presented in the results since active power curtailment was not detected in the simulation outcomes. Additionally, the aim of the present study is to present the effect of DGAM on system management, especially on battery, conventional generation, and substation injection results.

It is important to note that in this chapter due to the larger number of graphical representations that are self-explanatory, only a short explanation has been given for some cases of interest.

6.4.2 Power Losses Minimization Without DGAM

This first case study presents the simulation run for the EDS minimizing power losses, considering the inclusion and noninclusion of direct control over distributed generation technologies (DGAM).

This case study presents the simulation results obtained for the cost minimization problem not considering DGAM. These results will serve as base to analyze the impact of the proposed direct control over DG technologies on the overall performance of the distribution system. Since conventional generation is closer to the demand than the substation and the objective function minimizes the power losses in the distribution network, the generator is dispatched close to its maximum output disregarding the increase in generation costs associated with the quadratic formulation of its costs function. Conventional generation is reduced during the valley hours, where wind highly contributes to cope with the considered demand (Figure 6.11).

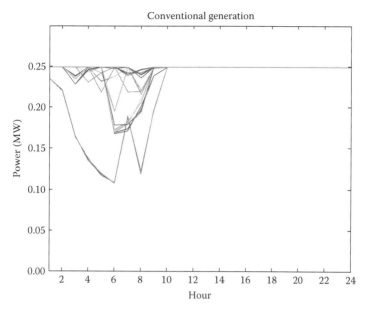

FIGURE 6.11
Conventional generation profile for power losses minimization without DGAM.

The optimal management of the storage system when considering the minimization of power losses as objective function of the system takes advantage of the valley hours to fully charge the storage device, reducing the power injection from the substation during peak hours (hour 21) (Figure 6.12).

The reduction of power inflow from the substation results in lower power losses, since load centers are placed close to the end of distribution lines (Figure 6.13).

6.4.3 Power Losses Minimization with DGAM

This case study presents the simulation results obtained for the cost minimization problem considering DGAM. DGAM substantially modifies the generation profile (including reactive power) for the considered technologies (conventional, storage, wind, and substation) as can be observed in Figures 6.14 through 6.17.

The inclusion of DGAM improves voltage profile and, therefore, reduces power losses in the distribution network and results in a higher injection from conventional generation during the valley hours, as can be observed comparing Figures 6.11 and 6.14. Especially during hours 5, 6, and 7, it can be observed how dispatch decisions for all scenarios result in a power output closer to the maximum power allowed.

The optimal management of the storage system when considering DGAM results in important variations when comparing the results with

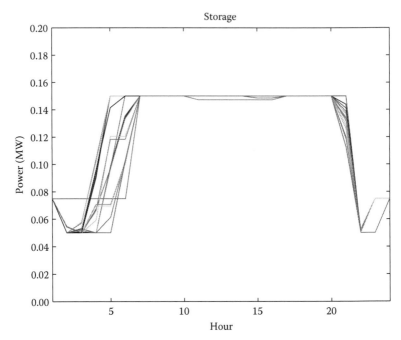

FIGURE 6.12
Storage profile for power losses minimization without DGAM.

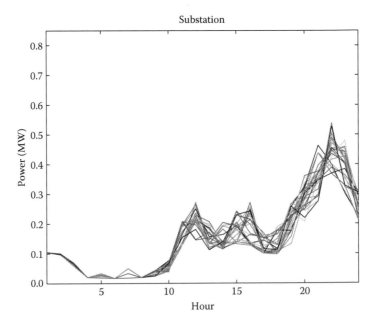

FIGURE 6.13
Substation power injection profile for power losses minimization without DGAM.

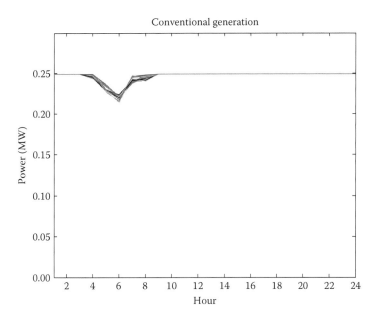

FIGURE 6.14
Conventional generation profile for power losses minimization with DGAM.

those obtained without DGAM (Figure 6.15). The fact that wind turbines are now providing reactive (Figure 6.16) power highly influences optimal storage behavior, since the objective function is the minimization of power losses, which are dependent on current flowing over the network and, therefore, is inversely dependent on the network voltage profile. Storage management in both cases results in a charging behavior during off-peak hours and discharge during system's peak hours contributing to the voltage stability of the distribution system over the considered time horizon (Figure 6.15).

Figure 6.18 shows the effect of DGAM on the distribution network voltage profile in bus 5 during a particular valley hour (hour 6), comparing the results with those obtained without DGAM.

Controlling the reactive power flows in the network can help to reduce system power losses, and, thus, improve the stability and performance of the system without generation rescheduling or topological changes in the network. Placing DG close to demand, as it is usually the case, may either increase or decrease losses. This effect depends of the location and the amount of the DG capacity as well as the correlations with the load.

In the analyzed modified IEEE 34-bus distribution network, increasing the voltage profile reduces the circulating current and therefore power losses. Inclusion of DGAM results in a reduction of 20.07% over the previous case.

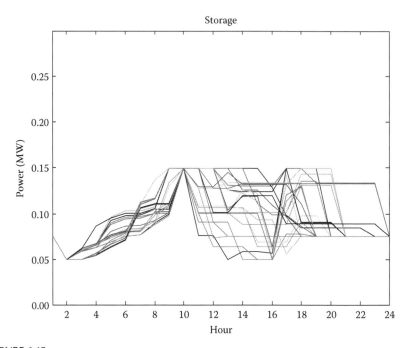

FIGURE 6.15
Storage profile for power losses minimization with DGAM.

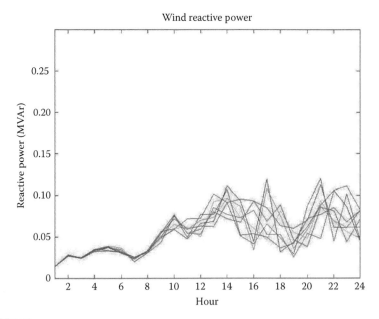

FIGURE 6.16
Reactive power profile from wind for power losses minimization with DGAM.

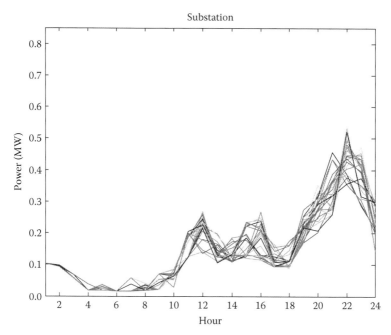

FIGURE 6.17
Substation power injection profile for power losses minimization with DGAM.

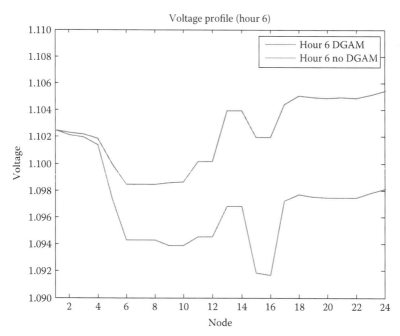

FIGURE 6.18
(See color insert.) Impact of DGAM on voltage profile during valley hours.

6.4.4 Generation Cost Minimization Without DGAM

This case study presents the model simulation run for the IEEE-34 bus distribution network minimizing operation costs without considering DGAM. Simulation results for conventional generation, storage, and substation are presented in Figures 6.19 through 6.21. These results will serve as base to analyze the impact of the proposed direct control over DG technologies on the overall performance of the distribution system.

Comparing these results with those obtained for case 6.4.2, it may be noted how conventional generation is substantially reduced, since its cost function has been expressed as a quadratic function of the power output. Storage and substation management are consequently influenced by conventional generation management through the power balance equation.

Additionally, it can be observed how conventional output increases during valley hours, where substation costs are higher. Hours 16, 17, and 19 present higher substation costs, increasing conventional generation for these hours.

The storage system management is fully used to reduce generation costs, discharging during high price hours (5, 6, and 15). The storage system also helps to mitigate voltage variations in distribution networks with high renewable penetration.

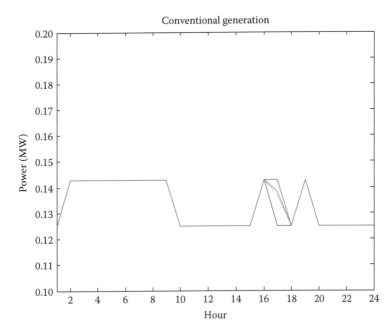

FIGURE 6.19
Conventional generation profile for generation costs minimization without DGAM.

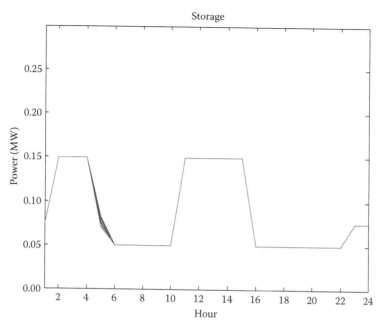

FIGURE 6.20
Storage profile for generation costs minimization without DGAM.

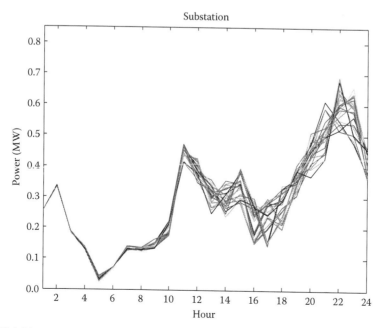

FIGURE 6.21
Substation power injection profile for generation costs minimization without DGAM.

6.4.5 Generation Cost Minimization with DGAM

This case study presents the model simulation run for the modified IEEE 34-bus distribution network minimizing operation costs without considering DGAM. Simulation results for conventional generation, storage, and substation are presented in Figures 6.22 through 6.24. It can be observed how DGAM substantially modifies the generation profile for the considered technologies (conventional, storage, and substation) (Figure 6.22).

DGAM allows an improved management of conventional generation, slightly reducing its output during hour 17, resulting in a generation costs reduction.

Almost no variation on storage management can be detected when comparing Figures 6.27 and 6.30. The storage charge and discharge takes place during the same hours.

The overall system costs are reduced by 1.69% when DGAM is used in the dispatch procedure when considering a fuel cost minimization problem as objective function (Table 6.3).

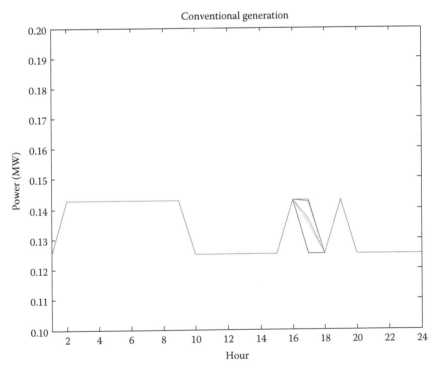

FIGURE 6.22
Conventional generation profile for generation costs minimization with DGAM.

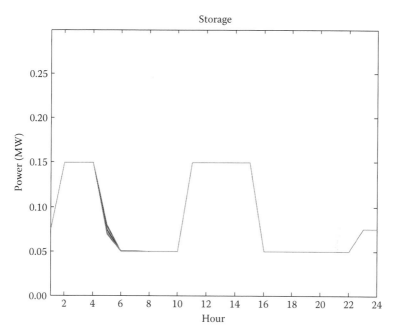

FIGURE 6.23
Storage profile for generation costs minimization with DGAM.

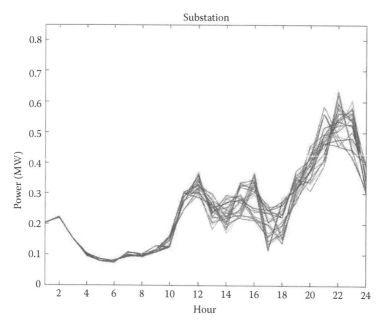

FIGURE 6.24
Substation power injection profile for generation costs minimization with DGAM.

TABLE 6.3

Objective Function Value (€)

No DGAM	DGAM
2267.91	2229.65

6.4.6 Simultaneous Minimization of Generation Cost and Power Losses Without DGAM

This case study presents the model simulation run for the modified IEEE 34-bus distribution network simultaneously minimizing fuel costs and power losses without considering DGAM. Simulation results for conventional generation, storage, and substation are presented in Figures 6.25 through 6.27.

The inclusion of network losses in the objective function, even with a small factor, varies significantly the conventional generation output profile. Conventional generation is higher in every hour, since the diesel generator considered is closer to the demand nodes than the substation, thus reducing overall network losses in the system.

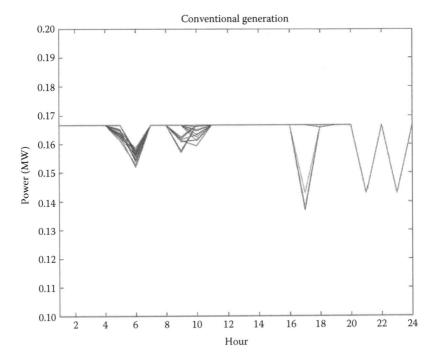

FIGURE 6.25
Conventional generation profile for simultaneous generation costs and power losses minimization without DGAM.

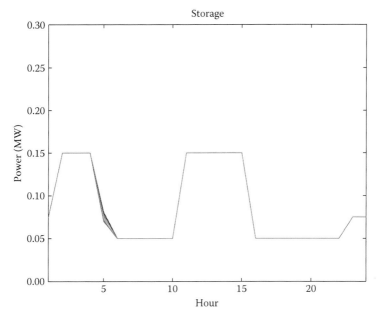

FIGURE 6.26
Storage profile for simultaneous generation costs and power losses minimization without DGAM.

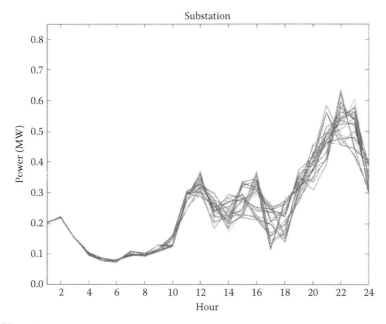

FIGURE 6.27
Substation power injection profile for simultaneous generation costs and power losses minimization without DGAM.

The storage system management is fully used to reduce generation costs, discharging during high price hours (5, 6, and 15). The storage system helps as well to mitigate voltage variations in distribution networks with high renewable penetration.

6.4.7 Simultaneous Minimization of Generation Cost and Power Losses with DGAM

This case study presents the model simulation run for the modified IEEE 34-bus distribution network simultaneously minimizing fuel costs and power losses considering DGAM. Simulation results for conventional generation, storage, and substation are presented in Figures 6.28 through 6.30.

The inclusion of DGAM does not affect the optimal generation schedule for conventional generation and storage. Increasing the losses weight factor modifies both profiles. The overall system costs are reduced by 1.75% when DGAM is considered in the dispatch procedure (Table 6.4). Additionally, the DGAM stabilizes the voltage profile along the distribution network. A higher penetration of DG units in a local distribution grid may lead to a violation of voltage rise, disturb the classical way of voltage control affecting the feeder voltage, or deteriorate power quality. Therefore, DGAM can help distribution networks to accommodate a higher renewable

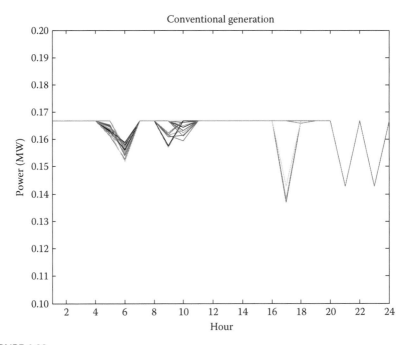

FIGURE 6.28
Conventional generation profile for simultaneous generation costs and power losses minimization with DGAM.

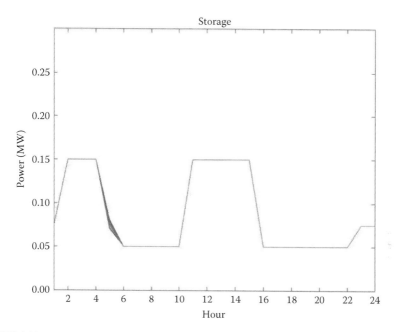

FIGURE 6.29
Storage profile for simultaneous generation costs and power losses minimization without DGAM.

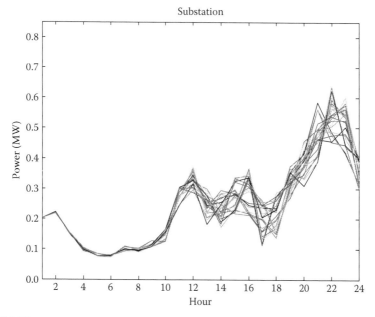

FIGURE 6.30
Substation power injection profile for simultaneous generation costs and power losses minimization with DGAM.

TABLE 6.4

Objective Function Value (€)

No DGAM	DGAM
2280.43	2240.83

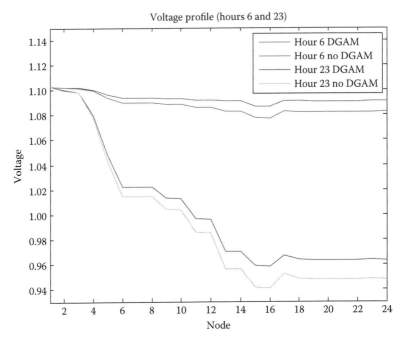

FIGURE 6.31
(See color insert.) Impact of DGAM on voltage profile during peak and valley hours.

penetration without endangering quality of supply. The impact of DGAM on the distribution network profile for peak and valley hours is represented in Figure 6.31.

6.5 Electric Price Signals in Insular Systems

6.5.1 Model Description

The following section of the document intends to explain the methodology used to determine an electricity price signal in small electrical insular systems. This process is based on average and marginal costs of production, and it becomes relevant not only for the opportunity to monitor the electric consumption but also because this control allows the production company to encourage consumer behavior changes according to consumption levels.

The model is developed from the earnings of the production company, and it is not based on the law of supply and demand. Since there is no electrical market in operation, the company works in monopoly and is willing to share its profit with the consumers in order to maintain the financial balance between the production and the consumption. Having said that, the company will share an agreed percentage of its marginal profit (Equation 6.7), which will be applied to the price to pay for electric power. Knowing that it is a dynamic model, the discount (Equation 6.11) will vary with the energy produced and the time of day. During peak hours, there is more consumption and the elasticity considered in these moments is higher than the off-peak hours; therefore, it is more probable that the consumer is prepared to increase or decrease the electrical consumption according to the consumption signal issued by the producing company.

6.5.2 Description of the Model with Risk-Embedded Assessment

The cost curve of a firm, represented in Figure 6.34, depends on the prices of inputs and the production function (the least costly combination of inputs that the firm can choose and that can give the desired level of production). The methodology considered also takes into account the probability of the generators operating near their minimum, which is treated as a risk and, therefore, is charged at a higher value.

In addition to the variable costs, which represent the variable expense with the level of production, the method also considers the fixed costs, the monetary expense that is supported even if there is no production. These costs are not subject to any variation with the increase or decrease of the production.

In order to understand the evolution of the profitability of the model, marginal cost is considered (Equation 6.48). It represents the additional or extra cost due to increased production of one unit, and the difference between its value and the market price represents the accuracy of the process. For the profit to be positive, the company will have to produce at a lower marginal cost compared to the market price. In this case, the market price in the end of the process will be the specific price (Equation 6.56). In order to break even, the specific price must equal the marginal cost. If, for any reason, the marginal cost is higher than the specific price, the company is not profitable.

Regarding the revenue, a fixed revenue is considered that is based on the operating and production costs. The revenue (Equation 6.49) will be defined by the fixed revenue and by the production load at a given price.

The system tested is the island of São Miguel, in Azores. It consists of the following:

- One wind farm
- One geothermal plant

- Seven hydropower plants
- One thermal power plant
  - Small generators [3, 8 and 7, 2 MW]
  - Big generators [8, 4, 10 MW]

The variables considered in the development of this methodology are as follows:

- Fixed costs (€/h)—Costs associated with electricity generation and with the transmission and distribution systems
- Fixed revenue (€/h)—Based on the operating costs
- Flat tariff (€/MWh)—Fixed tariff for all hours applied to the sale of energy to the consumer
- Profit share (%)—Percentage of the profit that is shared with the consumer
- Dynamic tariff (€/MWh)—Price charged to the consumer per MWh consumed assuming the marginal deviation
- Elasticity (%)—The price elasticity of demand is the percentage change that occurs in quantity demanded of a good resulting from a 1% change in the price of the asset.

$$\varepsilon_P^D = \left| \frac{dQ^D}{dP} * \frac{P}{Q^D} \right| \tag{6.46}$$

where:
ε_P^D is the elasticity of the price of demand
P is the market price for a certain product
Q^D is the demand for the respective product for a determined amount at a given price

For this study, the values are arbitrated according to the time of day and its consumption. Considering that at peak times there is a greater consumption, the ability of consumers to change their behavior will be higher. This variable is defined as a function of time.

In the estimation of the elasticity, four periods of consumption for 24 hours a day are considered:

- Peak hours: (09:00–11:00), (17:30–20:00)
- Near-peak hours: (08:00–09:00), (11:00–17:30), (20:00–22:00)
- Off-peak hours: (05:30–08:00), (22:00–01:30)
- Super off-peak hours: (1:30–05:30)

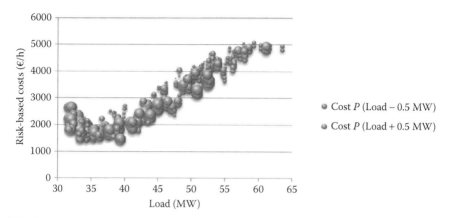

FIGURE 6.32
(See color insert.) Cost P (Load − 0.5 MW) and Cost P (Load + 0.5 MW) charts (the higher sized points represent higher wind generation).

Once the variables are set, the process begins. The cost P values for the load plus and minus 0.5 MW will be obtained using a set of simulations of scheduling and dispatches (Figure 6.32).

The variability of the costs when the system is producing low loads (Figure 6.32) is due to the fact that at that operational level, the generators may be working outside their normal operation range. That is harmful to the machine, since it will be running well below the optimum yield, which will reduce the average lifetime of the generator. As for the normal operation range, the risk-base cost of producing an extra portion of load for a certain load is higher than if it produced a smaller amount of load, since it is linear.

The costs calculated before will be used to determinate the cost P based on the following expression:

$$\text{Cost}\,P(\text{Load}) = \frac{\text{Cost}\,P(\text{Load}+0.5\,\text{MW})+\text{Cost}\,P(\text{Load}-0.5\,\text{MW})}{2} \qquad (6.47)$$
$$+ \text{Fixed costs}$$

where:
Cost P (Load) (€/h) is the cost of production for a determined amount of load
Cost P (Load + 0.5 MW) (€/h) is the value obtained from the simulations carried out for a small increase of the load considered
Cost P (Load − 0.5 MW) (€/h) is the value obtained from the simulations carried out for a small decrease of the load considered
Fixed costs (€/h) is the costs associated with electricity generation and with the transmission and distribution systems

The average of both costs and the fixed costs stipulated previously is considered.

The marginal cost can be calculated using the following equation:

$$\text{Marginal cost} = \frac{\text{Cost } P(\text{Load} + 0.5\,\text{MW}) + \text{Cost } P(\text{Load} - 0.5\,\text{MW})}{1\,\text{MW}} \quad (6.48)$$

where:
Marginal cost (€/MWh) is the additional or extra cost due to increased production of one unit
Cost P (Load + 0.5 MW) (€/h) is the value obtained from the simulations carried out for a small increase of the load considered
Cost P (Load − 0.5 MW) (€/h) is the value obtained from the simulations carried out for a small decrease of the load considered

In order to determine the profit, it is required to calculate the revenue.

$$\text{Revenue} = \text{Fixed revenue} + \text{Dynamic tariff} * \text{Load} \quad (6.49)$$

where:
Revenue (€/h) is the revenue earned by the company resulting from the sale of a load quantity at the price P = Dynamic tariff
Fixed revenue (€/h) is based on the operating costs
Dynamic tariff (€/MWh) is the price charged to the consumer per MWh consumed assuming the marginal deviation
Load (MW) is the amount of load produced

The same concept is applied for the calculation of the respective revenue for the load plus and minus 0.5 MW (Figure 6.33).
The income of the production company is linear, since the market price is set according to MW produced. However, the production cost is not, and it is

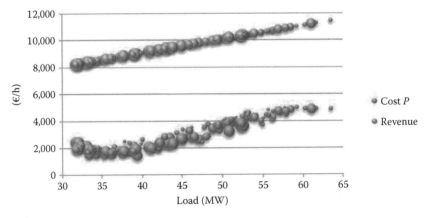

FIGURE 6.33
(See color insert.) Cost and revenue per load (the large-sized points represent higher wind generation).

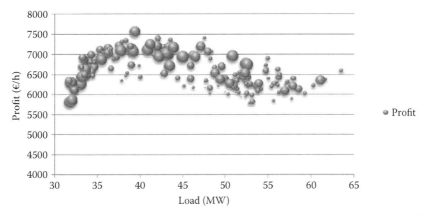

FIGURE 6.34
Profit per load chart (the higher sized points represent higher wind generation).

based on risk (the operating status of the generators and the penetration of renewable energy into the dispatch).

The profit value is determined by the following equation:

$$\text{Profit} = \text{Revenue} - \text{Cost }P \tag{6.50}$$

where:

Profit (€/h) is the gains of the company for the production of a certain amount of load

Revenue (€/h) is the revenue earned by the company resulting from the sale of a load quantity at the price $P = \text{Dynamic tariff}$

Cost P (€/h) is the cost of production for a determined amount of load

All the variables are in order of load (Figure 6.34).

In order to determine the marginal profit, first, it is required to obtain the values of the profit for the load plus and minus 0.5 MW.

That is obtained using the same expression that was used for the calculation of the profit, taking into account the cost and revenue associated with the increase and decrease of the load in 0.5 MW.

$$\text{Profit}(\text{Load} \pm 0.5\,\text{MW}) = \text{Revenue}(\text{Load} \pm 0.5\,\text{MW})$$
$$- \text{Cost }P(\text{Load} \pm 0.5\,\text{MW}) \tag{6.51}$$

$$\text{Marginal Profit} = \frac{\text{Profit}(\text{Load} \pm 0.5\,\text{MW}) - \text{Profit}(\text{Load} \pm 0.5\,\text{MW})}{1\,\text{MW}} \tag{6.52}$$

where:

Marginal profit (€/MWh) is the variation of the profit that is derived from the sale of an additional unit of the product

Profit (Load + 0.5 MW) (€/h) is the gains of the company for the production of a certain amount of load + 0.5 MW

Profit (Load – 0.5 MW) (€/h) is the gains of the company for the production of a certain amount of load – 0.5 MW

In the best-case scenario, the marginal profit is equal to zero.

If, at a given time, the marginal cost is equal to the marginal revenue, the marginal profit is zero. This is the most profitable rate of sale because all opportunities to make marginal profit have been exhausted. If the marginal revenue is lower than the marginal cost, there will be marginal loss and total profit will be reduced.

To obtain a relationship between the revenue (Equation 6.49) and the respective load, a tax named *specific profit* was created:

$$\text{Specific profit} = \frac{\text{Revenue}}{\text{Load}} \tag{6.53}$$

where:

Specific profit (€/MWh) is the ratio that expresses the variation of revenue in order to the load

Revenue (€/h) is the revenue earned by the company resulting from the sale of a load quantity at the price P = Dynamic tariff

Load (MW) is the amount of load produced (Figure 6.35)

It can be concluded that when the load increases, the specific profit decreases. This fact can be proved by simplifying the expression:

$$\text{Specific profit} = \frac{\text{Fixed revenue} + \text{Dynamic tariff} * \text{Load}}{\text{Load}} \tag{6.54}$$

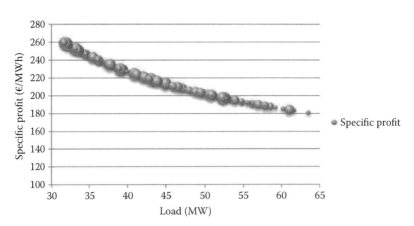

FIGURE 6.35
Specific profit per load chart (the higher sized points represent higher wind generation).

$$\text{Specific profit} = \frac{\text{Fixed revenue}}{\text{Load}} + \text{Dynamic tariff} \qquad (6.55)$$

where:

Specific profit (€/MWh) is the ratio that expresses the variation of revenue in order to the load

Fixed revenue (€/h) is based on the operating costs

Dynamic tariff (€/MWh) is the price charged to the consumer per MWh consumed assuming the marginal deviation

Load (MW) is the amount of load produced

Therefore, the fixed revenue does not vary, so the rate will be inversely proportional to the load. Its descent corresponds to a dilution of fixed revenue.

The methodology applied is based on the principle that dictates the sharing of profit with the consumer. For that, a discount (Equation 6.56) to be applied on the cost of purchasing electric energy is calculated (Equation 6.57). It is variable and it depends on the profit share percentage agreed and the marginal profit.

$$\text{Discount} = \text{Marginal profit} * \text{Profit share} \qquad (6.56)$$

where:

Discount (€/MWh) is a certain value to be applied on the cost of purchasing electric energy (dynamic tariff)

Marginal profit (€/MWh) is the variation of the profit that is derived from the sale of an additional unit of the product

Profit share (%) is the percentage of the profit that is shared with the consumer

This discount (Equation 6.56) will be subtracted from the original rate originating the specific price.

$$\text{Specific price} = \text{Dynamic tariff} - \text{Discount} \qquad (6.57)$$

where:

Specific price (€/MWh) is the price applied to the consumer for the production of the electric energy

Dynamic tariff (€/MWh) is the price charged to the consumer per MWh consumed assuming the marginal deviation

Discount (€/MWh) is a certain value to be applied on the cost of purchasing electric energy (dynamic tariff) (Figure 6.36)

This will be the price charged to the consumer for a specific load. Please note that the specific price (Equation 6.57) increases with the load since it is inversely proportional to the marginal profit. The marginal profit is very high when the load is low (the increase in production will originate a higher profit for the production company) which represents a bigger discount (Equation 6.56) for the consumer. That fact justifies the low specific price set for smaller loads and its evolution with the increase of load.

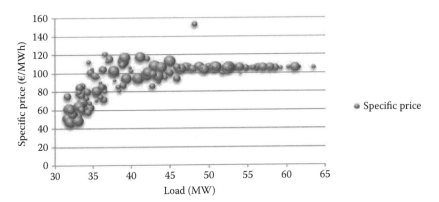

FIGURE 6.36
Specific price per load chart (the higher sized points represent higher wind generation).

According to Figure 6.37, it is possible to determine a relationship between daily consumption and the specific price. Specific price follows a constant pattern throughout the day with only a few spikes during peak hours. Over the off-peak and super off-peak hours, it will decrease drastically. This is due to the fact that at these times of the day, the production is low, which results in high values of marginal profit.

The quotient between the discount (Equation 6.56) and the dynamic tariff will create a ratio that will influence the load. Supposing that the dynamic tariff is charged for the total amount of the load, the following indicator (Equation 6.58) is used to estimate the value of the extra load created by the corresponding relation Discount/Dynamic tariff. It is also considered the elasticity that corresponds to the consumer's response to this share. Meaning, that the consumer is willing to change his or her behavior according to the consumption scenario. For example, only a small percentage of the consumer will react to the price reduction applied to all consumers.

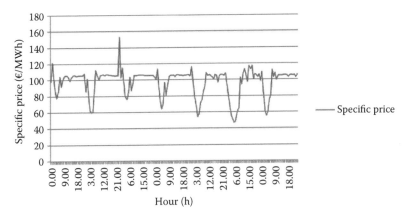

FIGURE 6.37
Specific price per hour chart.

$$\text{Extra load} = \frac{\text{Discount}}{\text{Dynamic tariff}} * \text{Elasticity} * \text{Load} \tag{6.58}$$

where:

Extra load (MW) is the load created by the rate Discount/Dynamic tariff, which reflects the percentage of the discount in relation to the dynamic tariff, applied to the entire load

Discount (€/MWh) is a certain value to be applied on the cost of purchasing electric energy (dynamic tariff)

Dynamic tariff (€/MWh) is the price charged to the consumer per MWh consumed assuming the marginal deviation

Elasticity (%) is the price elasticity of demand is the percentage change that occurs in quantity demanded of a good resulting from a 1% change in the price of the asset

Load (MW) is the amount of load produced

In order to determine the company earnings (Equation 6.59) and the customer gains (Equation 6.60), a basic tariff for the usage of the production system is taken into consideration. For the first calculation, it is assumed that the company receives the difference between the total load including the load variation [Load + Extra load (Equation 6.58)] sold to the customer for the specific price (Equation 6.57) and the estimated income for the production of the load excluding the variation sold at the basic tariff, adding the corresponding profit for the production of the extra load. According to the methodology applied in this system, the extra load (Equation 6.13) is created through marginal changes in production, so it is necessary to include the rate of income that comes from the marginal profit (Equation 6.52).

$$\text{Company earnings} = \frac{\begin{bmatrix} (\text{Load} + \text{Extra load}) * (\text{Specific price}) \\ - (\text{Load} * \text{Flat tariff}) + (\text{Marginal profit} \\ * \text{Extra load}) \end{bmatrix}}{1\text{h}} \tag{6.59}$$

where:

Company earnings (€) is the company's earnings after the application of the method for the sale of electric energy

Extra load (MW) is the load created by the rate Discount/Dynamic tariff, which reflects the percentage of the discount in relation to the dynamic tariff, applied to the entire load

Specific price (€/MWh) is the price applied to the consumer for the production of the electric energy

Marginal profit (€/MWh) is the variation of the profit that is derived from the sale of an additional unit of the product

Load (MW) is the amount of load produced

As for the customer gains (Equation 6.60), the calculation is similar but it is considered that profit (Equation 6.50) is determined by the difference between what the customer was supposed to pay and what he or she is actually paying. The first parcel is represented by load * flat tariff and the second one includes the extra load determined by the marginal overview of the problem: Specific price * (Load + Extra load).

$$\text{Customer gains} = \frac{\left[\begin{array}{c}(\text{Load} * \text{Flat tariff}) \\ - \text{Specific price} * (\text{Load} + \text{Extra load})\end{array}\right]}{1\,\text{hour}} \tag{6.60}$$

where:

Customer gains (€) is the gains of the consumer after the application of the method to purchase electric energy

Flat tariff (€/MWh) is the fixed tariff for all hours applied to the sale of energy to the consumer

Specific price (€/MWh) is the price applied to the consumer for the production of the electric energy

Extra load (MW) is the load created by the rate Discount/Dynamic tariff, which reflects the percentage of the discount in relation to the dynamic tariff, applied to the entire load

Load (MW) is the amount of load produced (Figure 6.38)

Even though the risk is shared between producer and consumer, when lower loads are produced, the production cost increases (Figure 6.8). The smaller the load, the higher the cost (due to the risk of generators operating at its minimum and penetration of renewable energy in dispatch).

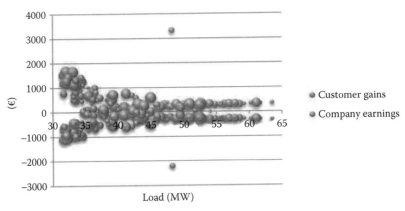

FIGURE 6.38
(See color insert.) Company earnings and customer gains per load chart (the higher sized points represent higher wind generation).

Regarding the marginal cost for a small load, it will be negative and it will increase as the load increases. This means that it will be less expensive to produce over that load value until it reaches an optimal value (Optimal load zone → Marginal cost = 0). The implemention of load over the optimal load zone will require an increase in cost of production, which has as a consequence a positive marginal cost.

As for the marginal profit, this will have an opposite pattern. It will be decreasingly positive as it approaches its optimum load zone.

Given that the marginal profit directly sets the discount applied to the consumer (Equation 6.56) that will be reflected in the specific price (Equation 6.57) to pay for the electrical energy, it is understandable that the consumer will have a high benefit when the load is low, whereas when the load exceeds the optimal zone load, the system will be stable.

The results are detailed in Tables 6.5 through 6.7.

First, the average marginal cost in lower than the average specific price. This means that the company is profitable over time test.

TABLE 6.5

Elasticity

Hour	%
0	9.00
1	7.00
2	5.00
3	5.00
4	5.00
5	7.00
6	9.00
7	9.00
8	13.00
9	17.00
10	17.00
11	13.00
12	13.00
13	13.00
14	13.00
15	13.00
16	13.00
17	15.00
18	17.00
19	17.00
20	13.00

(Continued)

TABLE 6.5 (*Continued*)

Elasticity

Hour	%
21	13.00
22	9.00
23	9.00

TABLE 6.6

Input Parameters for the Presented
Example, One Week Simulation

Fixed costs (€/h)	2000
Fixed revenue (€/h)	5000
Flat tariff (€/MWh)	100
Profit share (%)	10
Dynamic tariff (€/MWh)	102

TABLE 6.7

Output Parameters for the Presented Example, One
Week Simulation

Average marginal cost (€/h)	55.04
Average profit (€/h)	4574.72
Average specific profit (€/MWh)	215.93
Total extra load (MW)	1.74
Average extra load (MW)	0.01
Average specific price (€/MWh)	97.30
Total company earnings (€)	6284.67
Total customer gains (€)	6072.29
Average company earnings (€)	37.19
Average customer gains (€)	35.93

$$\text{If Specific price} > \text{Marginal cost} \rightarrow \text{Profit} > 0$$

In order to prove that fact, the average profit, determined by the model, is positive.

Second, the average specific price is 97.30 €/MWh, which means that starting from a higher price, it is possible to reduce the average market price (dynamic tariff) in 4.7 €/MWh. It varies between 46,600 and 152,808 €/MWh. When the value of the specific price is at its maximum, the company earnings will be also be maximum. However, when the value of the specific price is at its minimum, the customers gains will be maximal. When the marginal profit is at its maximum, the specific price is minimal. The opposite also occurs.

The total extra load produced was 1739 MW. On average, it produced 0.010 MW/hour, but the process is not linear, so it is not possible to say that an increase of production of 0.010 MW will allow a discount of 4.70 € in the market price.

The specific profit (Equation 6.53) is a balance between the revenue (Equation 6.49) obtained for a certain load produced. Thus, on average, the company gets a specific profit of 215.93 €/MWh for the production during the test period.

Regarding the overall gains of producers and consumers, the methodology enables profit for both. The profit share percentage and the other variables defined permit to obtain equilibrated profits for both parties: 6284.67 € for the company and 6072.29 € for the consumers. As for the average gains, they have very similar values being only separated by 1257 €.

6.6 Conclusions

The stochastic nature of DES motivates a probabilistic approach. DES installed close to consumption has, theoretically, a positive impact on network performance, reducing the need to use the network capacity. Since stochastic generation does not usually follow the consumption pattern, distribution networks are designed under the worst-case scenario considering peak load and reduced DES generation, potentially increasing the overall investment cost. Even constraints taking place for a few hours a year trigger grid adaptations. With the presence of DG the utilization rate of network assets would decline even more, calling for a review of traditional reinforcement practices.

This report addressed the methodologies and tools for analysis of the operation of a selected mix of resources in the insular network application. A multidimensional analysis was performed, in order to assess different alternatives, each of which containing set of resources exploiting various energy vectors. An array of technical solutions was analyzed, in order to establish the most convenient solution in terms of economic operation of the insular system and sustainability under uncertainty.

The key aspects considered were the availability of energy sources and reserves, the complementary exploitation of different energy sources, the role of storage, the economics of the connection to the insular network, and the effects of the possible connection to external systems.

This chapter addressed the electricity price signal in small electrical insular systems starting with an assessment over the methodology used. The methodology discussed considered the probability of the generators that are operating near their minimum and the situation that was treated as a risk and therefore charged at a higher value. Subsequently, the cost curve of a company depends on the prices of inputs and the production function (least costly combination of inputs which the company can choose and that can give the desired level of production).

In addition to the variable costs, which represent the variable expense with the level of production, the analysis method also considered the fixed costs, the monetary expense that is supported, even if there is no production. These costs are not subject to any variation with the increase or decrease of the production. However, the production cost is based on risk (the operating status of the generators and the penetration of renewable energy into the dispatch).

The variability of the costs when the system is producing low loads is harmful to the machine, since it runs well below the optimum yield, which will reduce the average lifetime of the generator. As for the normal operation range, the risk-base cost of producing an extra portion of load for a certain load is higher than if it is produced a smaller amount of load, since it is linear. The assessment process was based on average and marginal costs of production and it becomes relevant not only for the opportunity to monitor the electric consumption but also because this control allows the production company to encourage consumer behavior changes according to consumption levels.

Regarding the scenario analysis for the insular electricity grids, various case studies have been developed. The general case scenario encompasses a modified IEEE 34-bus case study of a distribution network for ST competitive operation using linearized MILP. The model considers a single substation, wind generation, conventional generation, and a storage system. Essentially three groups of case studies have been formulated for the analysis: for monitoring the system power losses, for operation costs, and for the simultaneous minimization of fuel costs and power losses. Additionally, DGAM has been included in each case study.

DGAM contributes to an efficient integration of an increasing DG penetration. Additionally, it may postpone the need for network reinforcements. The operation of the distribution network in an effective manner is the responsibility of the DSO, assuring adequate quality and security of supply, as well as coordinating the operation of the coordinated high-voltage grid in cooperation with the TSO. The increasing level of DES/DG stress network operation, endangering the quality of service for end customers. In order to correctly satisfy its duties, DSOs need new advanced services and tools. Such services, as described in the present study, include generation curtailment and reactive power management as an additional DSO task and its benefits are presented in the case studies. Additionally, a higher coordination and information exchange between TSOs, DSOs, and DES is necessary to optimize stochastic generation.

Different simulations have been run relocating generators and storage devices at different buses to see the grid parameters response to those changes, in order to analyze the distribution grid basic operating parameters, A pattern about the changes taking place on the grid parameters cannot be clearly defined by a rule of thumb, due to a complexity of combinatorial issues. However, the simulation yields some important hints about the electrical distribution system operational behavior, as depicted in different figures presented for the analysis of power losses and operation costs. The storage location, the penetration of CONV and RN, as well as the substation cost have serious impacts on the competitive operation of the system. The reconfiguration switches follow a regular pattern during all simulation tasks, showing that the model is robust enough, and a radial configuration of the EDS is guaranteed. The case studies presented are important to compare between different alternative solutions for the optimal and competitive operation strategies of island systems. Further modeling improvements and tests should be conducted in order to prove their effectiveness.

Another vision developed for the purpose of scenario analysis assumes that retailers and consumers receive price signals from insular system although in terms of annual price they pay the average of peninsular market signals. The aim of this approach is to achieve greater consumer awareness in order to increase the efficiency of the insular electrical system through demand modulation.

It set up a unit commitment depending on the actual costs of consumption of existing groups, adding then all stakeholders as wind producers, photovoltaic producers, energy storage, and so on. The results showed that the energy sector stakeholders can receive a price signal to decide how to get into the market. Also, the dispatch was set to operate at variable costs, thus the project that maintains and improves office, and among others, set performance standards, the new facility incorporates allotted through tenders, and so on. The existence of scalable efficient technologies (such as modern diesel units, cogeneration, or small combined cycle) offers the possibility of introducing some competition in manageable nonpeninsular generation systems.

A different view was offered by the simulation analysis that illustrates that certain electric power system (Crete, for example) can withstand higher RES penetration levels up to a certain threshold (e.g., +20% wind +20% PV with respect to the current levels), with notable operational savings and without endangering the system security and reliability. The operation of a pumped storage plant (PSP) or other energy storage system (e.g., battery energy storage system) is deemed necessary to efficiently accommodate higher levels of RES integration.

Different scenarios regarding the daily simulation of the power system were considered, which are differentiated only in terms of the installed capacity of the wind plants, PV, and PSP systems that are supposed to operate. The scenarios ranged from the ones with highly intermittent large-scale RES production to the ones with technologies embedded in order to alleviate the RES production intermittence (such as PSP). Also, a close investigation was performed in order to assess each RES technology contribution to the power system daily operation.

The results showed that a more detailed cost–benefit analysis that takes into account other economic parameters (e.g., units' fixed costs, RES feed-in-tariffs, interest rates, and payback periods) is required to yield more accurate results and investigate the economic viability of the various power system in the long term, as well as the total electricity cost to be undertaken by the final electricity consumers. This is due to the extraordinarily high variability of scenarios for each islanded power system in itself, as well as for the entire array of islands in the European Union.

The results also indicated that the best solution for a competitive operation of the power systems in islands is a local analysis performed on multiple layers (geographical position, energy sources availability, consumption profile, demand elasticity, risk-aversion, and social acceptance for market operation), in order to generate a locally tailored energy framework, suited for a specific island.

Nomenclature

Sets

Ω_b	set of branches of the network
Ω_{conv}	set of buses with conventional generation
Ω_f	set of blocks used for the linearization of the quadratic power
Ω_l	set of blocks used for the linearization of the quadratic cost function
Ω_{loop}	set of network loops
Ω_n	set of buses of the network
Ω_{ren}	set of the bus location of renewable generators
Ω_{st}	set of the bus location of storage units
Ω_{su}	set of the bus location of substations
Ω_t	set of time periods
Ω_ω	set including every wind scenario
Ω_ξ	set including every demand scenario

Indexes

f	partition segment of the blocks used for the linearization $(f = 1, 2 \ldots F)$
i,j,k	index for bus
t	time index for 24 hours
ω	wind scenario index
ξ	demand scenario index

Scalars

a^{conv}	coefficient of cost function
A^{conv}	coefficient of cost function
b^{conv}	coefficient of cost function

c^{conv}	coefficient of cost function
C^{ch}	cost of storage charge (€/MWh)
C^{ens}	cost of not served energy (€/MWh)
C^{ren}	cost of renewable generation (€/MWh)
C^{su}	cost of substation (€/MWh)
F	number of blocks used in the linearization segments
FP^{conv}	power factor of conventional power generation
FP^{ren}	power factor of renewable generation
FP^{su}	power factor of substation generation
L	number of segments of the piecewise linear production cost function
l_{ij}	maximum current flow of branch ij (A)
N_{loop}	total number of buses in each possible loop of the network
$P_{\text{max}}^{\text{conv}}$	maximum output power of conventional power generation (MW)
$P_{\text{min}}^{\text{conv}}$	minimum output power of conventional power generation (MW)
$\underline{s}_i^{\text{ch}}$	minimum charge storage power capacity (MWh)
$\bar{s}_i^{\text{ch}}$	maximum charge storage power capacity
$\underline{s}_i^{\text{dis}}$	minimum discharge storage power capacity (MWh)
$\bar{s}_i^{\text{dis}}$	maximum discharge storage power capacity (MWh)
$\underline{x}_i$	minimum energy storage capacity (MWh)
$\bar{x}_i$	minimum energy storage capacity (MWh)
$\underline{V}^2$	minimum voltage magnitude of the network (kV)
$\bar{V}^2$	maximum voltage magnitude of the network
V_{nom}	nominal voltage magnitude of the network (kV)
η_i^{ch}	storage charge efficiency
η_i^{dis}	storage discharge efficiency
ρ	weighting factor for power losses

Parameters

$C_{t\omega\xi}^{\text{conv}}$	cost of conventional generation at time t and scenarios ω and ξ (€/MWh)
$d_{it\xi}$	active power demand at bus i, time t, and scenarios ω and ξ (MW)
$m_{ijft\omega\xi}$	slope of block f of the piecewise linearization of branch ij, at time t and scenarios ω and ξ
$P_{it\omega\xi}^{\text{ren}}$	renewable active power generation at bus i, time t, and scenarios ω and ξ (MW)
$q_{it\xi}$	reactive power demand at bus i and time t, and scenarios ω and ξ (MVAr)
R_{ij}	resistance of branch ij (Ω)
$\Delta S_{ijrt\omega\xi}$	upper limits of the discretization of the apparent power of branch ij, at time t and scenarios ω and ξ (MVA)
X_{ij}	reactance of branch ij (Ω)
π_ω	probability of wind scenarios ω
π_ξ	probability of wind scenarios ξ

Variables

$P_{it\omega\xi}^{su}$ active power of a substation located at bus i, in period t and scenarios ω and ξ (MW)

$Q_{it\omega\xi}^{conv}$ active convectional power generation at bus i, time t and scenarios ω and ξ (MVAr)

$Q_{it\omega\xi}^{ens}$ reactive power demand curtailment at bus i, in period t and scenarios ω and ξ (MVAr)

$Q_{it\omega\xi}^{ren}$ renewable reactive power generation at bus i, time t and scenarios ω and ξ (MVAr)

$Q_{it\omega\xi}^{su}$ reactive power of a substation located at bus i, in period t and scenarios ω and ξ (MVAr)

Positive Variables

$Cost_{it\omega\xi}^{conv}$ distributed production cost of conventional generators located at bus i, in period t and scenarios ω and ξ (€/MWh)

$I_{it\omega\xi}^{sqr}$ square of the current flow of branch ij, in period t and scenarios ω and ξ (A^2)

$P_{jit\omega\xi}^{+}$ active power of branch ij, in period t and scenarios ω and ξ, when going downstream (MW)

$P_{jit\omega\xi}^{-}$ active power of branch ij, in period t and scenarios ω and ξ, when going upstream (MW)

$\Delta P_{ijft\omega\xi}$ value of the rth block associated with the active power through branch ij, in period t and scenarios ω and ξ (MWh)

$P_{ijt\omega\xi}^{2}$ value of the rth block associated with the active power through branch ij, in period t and scenarios ω and ξ (MWh)

$P_{it\omega\xi}^{ch}$ real-time storage power for a unit at bus i, in period t and scenarios ω and ξ (MW)

$P_{it\omega\xi}^{conv}$ active power of conventional generation at bus i, in period t and scenarios ω and ξ (MW)

$P_{it\omega\xi}^{dis}$ real-time production power for a unit at bus i, in period t and scenarios ω and ξ (MW)

$P_{it\omega\xi}^{ens}$ active demand curtailment at bus i, in period t and scenarios ω and ξ (MW)

$P_{it\omega\xi}^{ren}$ active wind power generation of one turbine at bus i, in period t and scenarios ω and ξ (MW)

$P_{it\omega\xi}^{ren\_cur}$ active power wind curtailment generation of one turbine at bus i, in period t and scenarios ω and ξ

$\Delta Q_{ijft\omega\xi}$ value of the rth block associated with the reactive power through branch ij, in period t and scenarios ω and ξ (MVAr)

$Q_{ijt\omega\xi}^{2}$ value of the rth block associated with the reactive power through branch ij, in period t and scenarios ω and ξ (MVAr)

$Q_{jit\omega\xi}^{+}$ active power of branch ij, in period t and scenario ω and ξ, when going downstream (MW)

$Q_{\overline{jit\omega\xi}}$ active power of branch ij, in period t and scenario ω and ξ, when going upstream (MW)

$V_{it\omega\xi}^{sqr}$ square of the voltage magnitude of branch ij, in period t and scenarios ω and ξ (kV²)

$\delta_{ilt\omega\xi}$ segment of the active power of the piecewise linearization of the conventional production generation cost at bus at period t and scenarios ω and ξ (MW)

$\gamma_{ilt\omega}$ slope of segment l of the piecewise linearization of the conventional production generation cost at bus at period t and scenarios ω and ξ

Binary Variables

$b_{it\omega\xi}^{p}$ binary variables used for production, in period t and scenarios ω and ξ

$b_{it\omega\xi}^{s}$ binary variables used for storage, in period t and scenarios ω and ξ

$v_{ijt\omega\xi}$ binary variable to indicate of the plant is running for branch ij, in period t and scenarios ω and ξ

$v_{ijt\omega\xi}^{P+}$ binary variable used for active power (upstream) for branch ij, in period t and scenarios ω and ξ

$v_{ijt\omega\xi}^{P-}$ binary variable used for active power (downstream) for branch ij, in period t and scenarios ω and ξ

$v_{ijt\omega\xi}^{Q+}$ binary variable used for reactive power (upstream) for branch ij, in period t and scenarios ω and ξ

$v_{ijt\omega\xi}^{Q-}$ binary variable used for reactive power (downstream) for branch ij, in period t and scenarios ω and ξ

References

1. E. J. Coster, *Distribution Grid Operation Including Distributed Generation: Impact on Grid Protection and Consequences of Fault Ride-Through Behavior*. PhD dissertation, Technische Universiteit Eindhoven, Eindhoven, the Netherlands, 2010.

2. W. H. Kersting, *Distribution System Modeling and Analysis*. CRC Press, Boca Raton, 2002.

3. A. W. Bizuayehu, E. M. G. Rodrigues, S. F. Santos and J. P. S. Catalão, Assessment on baseline and higher order grid security criteria: Prospects for insular grid applications. *Proceedings of the 2014 IEEE Power and Energy Society General Meeting*, July 27–31, 2014, Washington, DC.

4. R. Ayyanar and A. Nagarajan, *Distribution System Analysis Tools for Studying High Penetration of PV with Grid Support Features*. Power Systems Engineering Research Center, September 2013. Available at: http://www.PSERC.org.

5. T. Woodford, *EU Islands: Towards a Sustainable Energy Future*, June 2012. Available at: http://www.eurelectric.org.

6. B. Amanulla, S. Chakrabarti and S. N. Singh, Reconfiguration of power distribution systems considering reliability and power loss. *IEEE Transactions on Power Delivery*, vol. 27, pp. 918–926, April 2012.
7. J. F. Franco, M. J. Rider, M. Lavorato and R. Romero, A mixed-integer LP model for the reconfiguration of radial electric distribution systems considering distributed generation, *Electric Power Systems Research*, vol. 97, pp. 51–60, April 2013.
8. D. P. Bernardon, L. L. Pfitscher, L. N. Canha, A. R. Abaide, V. J. Garcia, V. F. Montagner, L. Comassetto and M. Ramos, Automatic reconfiguration of distribution networks using smart grid concepts. *Proceedings of the 10th IEEE/IAS International Conference on Industry Applications*, Fortaleza, CE, Brazil, pp. 1–6, November 5–7, 2012.
9. T. Senjyu, T. Miyagi, S. Ahmed Yousuf, N. Urasaki and T. Funabashi, A technique for unit commitment with energy storage system. *International Journal of Electrical Power and Energy Systems*, vol. 29. pp. 91–98, 2007.
10. A. W. Bizuayehu, P. Medina, J. Catalão, E. M. G. Rodrigues and J. Contreras, Analysis of electrical energy storage technologies state-of-the-art and applications on islanded grid systems. *Proceedings of the IEEE PES Transmission and Distribution Conference and Exposition*, Chicago, IL, April 14–17, 2014.
11. D. Pozo and J. Contreras, Unit commitment with ideal and generic energy storage units. *IEEE Transactions on Power Systems*, vol. 29 no. 6, November 2014, pp. 2974–2984, 2014.
12. P. Medina, A. W. Bizuayehu, J. P. S. Catalão, E. M. G. Rodrigues and J. Contreras, Electrical energy storage systems: Technologies' state-of-the-art. Techno-economic benefits and applications analysis. *Proceedings of the 47th Hawaii International Conference on System Sciences*, Hawaii, pp. 2295–2304, January 6–9, 2014.
13. X. Xiao, H. Yi, Q. Kang and J. Nie, A two-level energy storage system for wind energy systems. *Procedia Environmental Sciences*, vol. 12, Part A, pp. 130–136, 2011.
14. A. J. Wood and B. F. Wollenberg, *Power Generation, Operation, and Control*, 2nd edn., Wiley, New York.
15. T. Tran-Quoc, M. Braun, J. Marti, Ch. Kieny, N. Hadjsaid and S. Bacha, Using control capabilities of DER to participate in distribution system operation, *Paper presented at IEEE Power Tech 2007*, Paper ID: 567, Lausanne, Switzerland, July 1–5, 2007.
16. R. Piacente, Alocação Ótima de Reguladores de Tensão em Sistemas de Distribuição de Energia Elétrica Radiais Usando uma Formulação Linear Inteira Mista, Dissertation, Universidade Estadual Paulista. Facultade de Engenharia de Ilha Solteira, Ilha Solteira, SP, Brazil, 2012.
17. M. Lavorato, J. F. Franco and M. J. Rider, Imposing radiality constraints in distribution system optimization problems, *IEEE Transactions on Power Systems*, vol. 27, no. 1, pp. 172–180, February 2012.

7

Renewable Generation and Distribution Grid Expansion Planning

Gregorio Muñoz-Delgado, Sergio Montoya-Bueno, Miguel
Asensio, Javier Contreras, José I. Muñoz, and José M. Arroyo

CONTENTS

7.1 Introduction

A power system usually consists of generation units, transmission networks, distribution networks, consumption centers, system protections, and control equipment [1]. Distribution networks are an important part of the electric energy system since they supply energy from the distribution substations to the end users. Distribution networks are typically three-phase and the standard operating voltages are 30, 20, 15, and 10 kV. Regardless of their topology, which may be either meshed or radial, most distribution networks are operated in a radial way due to the economy and simplicity of such operation from the viewpoints of planning, design, and system protection. Distribution networks have been designed with wide operating margins, thereby allowing for a passive operation that results in a more economical management.

Distribution substations are fed through one or several subtransmission networks, although sometimes they can be directly connected to the transmission network. Each distribution substation supplies the energy demand by means of one or several feeders. Generally, a distribution substation consists of the following components: (1) protection devices, (2) metering devices, (3) voltage regulators, and (4) transformers [2].

From a centralized standpoint, distribution companies are responsible for the operation and planning of distribution networks. Distribution companies must satisfy the growing demand within quality standards and in a secure fashion. Therefore, planning models are used to obtain the least-cost investment, while meeting the security and quality requirements. These planning models determine the optimal expansion of the distribution networks that comprises (1) replacement and addition of feeders, (2) reinforcement of existing substations and construction of new substations, and (3) installation of new transformers [3].

This chapter addresses the expansion planning of a distribution system accounting for two aspects that remain open areas of research, namely, the presence of distributed generation (DG) and the incorporation of reliability.

The widespread growth of DG, mainly due to the penetration of renewable energy, inevitably requires the inclusion of this kind of generation in planning models [4]. DG refers to small-scale power units located near consumption centers. Manifold technologies are currently used for DG including wind turbines, photovoltaic (PV) plants, mini hydro plants, fuel cells, cogeneration plants, micro gas turbines, internal combustion engines, and energy storage devices such as batteries [5].

DG availability in distribution systems requires operating the network in an active way rather than passively since DG installation can have a significant impact on power flows, voltage profiles, system efficiency, and protections. The operational impact of DG depends on many factors such as the type, size, and location of generation units; the types of control equipment; and the characteristics of feeders and loads, among others.

The use of DG has numerous advantages related to system planning and operation [6]:

- Reduction in energy losses
- Control of the voltage profile
- Improvement of power quality
- Increase in system reliability
- Reduction or deferral of the network expansion
- Decrease in the emissions of carbon dioxide
- Short lead time
- Low investment risk
- Small-capacity modules
- Reduced physical size
- Availability of a wide range of DG technologies

However, it is worth mentioning that the increasing penetration of nondispatchable renewable-based technologies in the overall energy mix calls for the consideration of the associated uncertainty.

Although many tools have been successfully developed to accurately forecast load demands, wind and PV energy production uncertainties constitute challenging issues due to the high variability of these energy sources, which is strongly dependent on the meteorology.

The second aspect dealt with in the distribution planning model addressed in this chapter is reliability. It should be noted that most service interruptions experienced by end users take place at the distribution level [3]. Therefore, expansion plans should be driven not only by cost minimization but also by reliability standards. Distribution system reliability evaluation consists in assessing the adequacy of the various system components to perform their intended function. Rigorous analytical treatment of distribution reliability requires well-defined metrics. Unfortunately, reliability terminology has not been used consistently across the industry, and standard definitions are just beginning to be adopted [7].

Distribution systems should be operated and planned in order to guarantee that the demand will be supplied in a reliable way and at a minimum cost. An acceptable reliability level is defined either by the regulator on behalf of the customers, or by the customers themselves through customer damage functions. Once these indices are defined, a predictive reliability analysis allows the distribution company to optimize its investment in quality improvement.

In this chapter, a novel algorithm is presented to jointly plan the expansion of DG and the distribution network, while considering uncertainty and network reliability. The algorithm is built on the models described in [3,8–10] and comprises two steps.

First, a cost minimization model explicitly characterizing uncertainty is used to obtain a pool of cheap expansion plans with different topologies. The goal of such optimization is the minimization of the expected costs of investment, maintenance, production, losses, and unserved energy. Network expansion comprises several alternatives for feeders and transformers. Analogously, the installation of DG takes into account several alternatives for generators based on renewable energy sources (RESs) such as wind turbines and PV cells. Unlike what is customarily done, a set of candidate nodes for generator installation is considered. Thus, the optimal expansion plan identifies the best alternative, location, and installation time for the candidate assets. Uncertainty related to RESs and demand is accounted by generating a set of scenarios representing different realizations of the uncertain parameters.

Subsequently, several reliability indices and their associated costs are calculated for each solution of the pool. This information allows the decision maker to analyze the impact of reliability on the distribution expansion planning problem, so that the most convenient investment is selected.

The main contributions of this chapter are as follows:

1. A novel stochastic optimization model is presented for the dynamic expansion planning of both the distribution network and DG, considering the uncertainty related to demand and renewable generation.
2. The resulting multistage planning problem is formulated as a mixed-integer linear program suitable for efficient off-the-shelf software.
3. The system planner is provided with a new algorithm to make informed decisions on distribution planning based on costs and reliability.

The rest of this chapter is organized as follows. Section 7.2 provides an overview of the state of the art. Section 7.3 describes the production models for PV and wind power generation. In Section 7.4, uncertainty modeling is characterized. Section 7.5 presents the formulation for the proposed optimization problem. Reliability indices and the associated costs are described in Section 7.6. Section 7.7 outlines the proposed algorithm for distribution system planning. In Section 7.8, a case study from La Graciosa in the Canary Islands, Spain, illustrates the application of the approach to an insular system. Finally, some relevant conclusions are drawn in Section 7.9.

7.2 Literature Review

Traditionally, the distribution expansion planning problem has been addressed considering the expansion and reinforcement of feeders and substations as the only decision variables [3,4,8,11–25]. Since the relatively recent emergence of DG, several works have taken into account the decision making associated with the investment in distributed generators disregarding the typical decision variables modeling network investments [5,6,26–35]. In [8,16,19,22], the conventional distribution expansion planning problem modeled DG as existing generators in order to analyze their impact on the expansion plan, but DG investment decisions were neglected. In [36–40], investment in DG was considered within the framework of traditional distribution system planning. However, those works were focused on the distribution investment deferral as a consequence of the installation of DG units.

The joint expansion planning problem including investment decisions in both distribution network and DG has been addressed by mixed-integer nonlinear programming [9,41], metaheuristics [42–46], and mixed-integer linear programming [10]. Those approaches also differ from the modeling perspective. Thus, investment decisions related to the distribution network were exclusively associated with the reinforcement of feeders and substations in [9,41,43–46]. As a consequence, the radial topology of the network was not subject to modification and constraints imposing the radial operation were not required. In contrast, in [10,42], the construction of new feeders connecting new load nodes was also considered, thereby requiring specific constraints to impose radial operation. In addition, generation investment was limited to conventional plants in [9,10,43–46], whereas investment in renewable generation was also accounted for in [10,41,42].

Some of the aforementioned works on distribution system planning have included the uncertainty associated with demand and renewable generation [16,27,31,33,40,42]. In [16], a probabilistic load flow was developed taking into account the probability density functions (pdfs) of both the loads and the annual power production of each generating unit. In [27], a set of plausible scenarios associated with different wind power generation levels was specified through a pdf. Then, a genetic algorithm was applied to find the sizing and location of DG for each scenario. Finally, decision theory was applied to choose the best expansion plan among the candidate solutions. In [31], Atwa et al. developed a deterministic equivalent planning model relying on the consideration of all possible operating conditions of the renewable DG units with their probabilities. In [33], probabilistic models were used to generate synthetic data and a clustering method was proposed to aggregate similar system states in order to generate different scenarios. The scenario-based expansion planning was tackled by integrating particle swarm optimization and ordinal optimization. In [40], both wind power output and load were modeled as multistate variables. Furthermore, the optimal location of different types of DG units in the distribution system was addressed by a multiobjective optimization approach based on a genetic algorithm. Finally, in [42], uncertainties in demand values and production levels of DG units were represented through the analysis of multiple scenarios using two different methodologies. The first approach examined each scenario individually, whereas the second technique analyzed all the scenarios simultaneously. Both methods combined the multiscenario approach with a genetic algorithm as a way of introducing a probabilistic assessment into the process.

From distribution companies' viewpoint, reliability in distribution system planning has been addressed through different methodologies [47]. Relevant works on distribution system planning involving reliability are (1) [3,13,15], where DG units were disregarded; (2) [28,33,40], where investment in DG was considered while neglecting the construction of new feeders; and (3) [42,44], where expansion decisions on both network assets and DG were modeled. In [3], reliability indexes and associated costs were computed ex post for a set of solutions provided by a cost-minimization expansion planning model. In [13], distribution system reliability was integrated in the optimization via outage costs and costs of switching devices. The solution approach was based on an iterative algorithm wherein reliability costs were computed after knowing the network topology. In [15], a three-stage algorithm was described, where system reliability was evaluated through outage cost optimization. In [28], a value-based method was proposed to enhance reliability, which was calculated analytically. In [33], the cost of system reliability was evaluated based on the cost associated with energy not supplied (ENS). In [40], the interruption cost at each load node was calculated using Monte Carlo simulation. In [42], a genetic-algorithm-based approach was described wherein reliability indices were calculated by an analytical method considering all protection devices installed in the distribution network as well as DG units. Finally, in [44], the reliability cost was characterized by the cost of ENS using the unmet demand after an outage.

7.3 Renewable Power Generation Models

This section provides a brief overview of the models available for renewable generating units. For each generation technology, the corresponding production model relates the renewable energy resource to the power output. The focus is placed on PV and wind technologies, which are the most widely used types of renewable-based generation.

7.3.1 PV Energy

The photoelectric effect that takes place in PV cells consists in converting the solar radiation into direct current electricity. This process is achieved by the property of some materials to absorb photons and emit electrons [48]. The materials used in PV cells are semiconductors. Figure 7.1 shows the diagram of a PV cell.

The equivalent electric circuit of a PV cell consists of an ideal current source, a real diode, a series resistance, and a parallel resistance, as shown in Figure 7.2. The ideal current source supplies current in proportion to the solar irradiance.

The application of Kirchhoff's current law to the equivalent circuit for a PV cell yields:

$$I_{out} = I_{SC} - I_0 \left(e^{qV_d/(k_b T_p)} - 1 \right) - \frac{V_d}{R_p} \tag{7.1}$$

7.3.1.1 Power Output and I–V Curves

The power curve is the most important diagram of a PV panel comprising an array of PV cells. Figure 7.3 represents a generic I–V curve for a PV panel. Several parameters can be identified such as the short-circuit current, I_{SC}, and the open-circuit voltage, V_{OC}. Additionally, Figure 7.3 depicts the power supplied by the panel as the product of current and voltage. As can

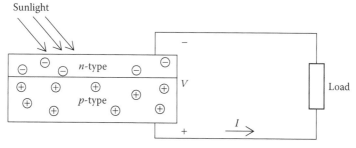

FIGURE 7.1
Photovoltaic cell.

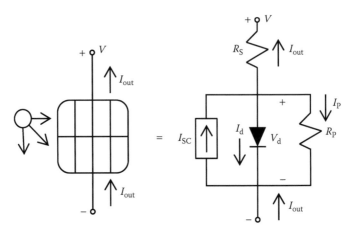

FIGURE 7.2
Equivalent electric circuit of a photovoltaic cell.

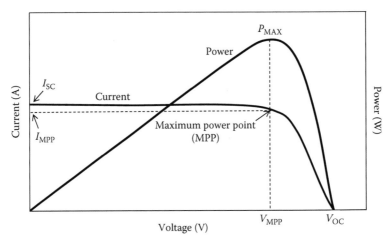

FIGURE 7.3
I–V curve and power output curve for a photovoltaic panel.

be observed, the power output is not constant. The maximum power output, P_{MAX}, is reached at current and voltage levels I_{MPP} and V_{MPP}, and is the usual operating point for PV panels.

Both curves are affected by temperature and irradiation. The relationship between irradiation and current is expressed as follows:

$$I_{SC} = G\frac{I_{SC}(STC)}{1000} \tag{7.2}$$

where:

STC denotes standard test conditions, which are characterized by a solar irradiance equal to 1000 W/m², a cell temperature equal to 25 ± 2°C, a spectral distribution of the irradiance according to the air mass factor equal to AM 1.5, and no wind

The temperature of a PV cell increases with ambient temperature and with the reheating from irradiance. The equation that considers both effects is:

$$T_{cell} = T_{amb} + \left(\frac{T_{NOC} - 20}{800} \right) G \qquad (7.3)$$

where:

T_{NOC} is the cell temperature in a panel at nominal operating cell temperature conditions (NOCT) which are (1) the ambient temperature at 20°C, (2) the solar irradiation equal to 800 W/m², and (3) the wind speed at 1 m/s

Figure 7.4 shows the effect of irradiation and temperature on the I–V curves.

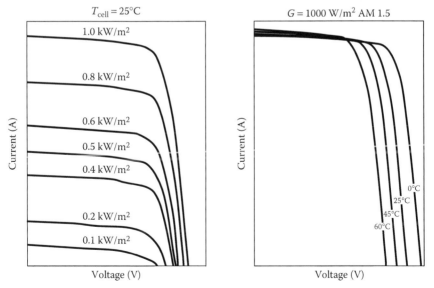

FIGURE 7.4
I–V curves under different irradiances and temperatures.

7.3.1.2 PV Generation Model

The PV generation model is based on the description provided in [49]. This model reflects the effect of temperature and relies on the power–temperature coefficient, δ. The power–temperature coefficient indicates the variation of the maximum power with increasing temperature. This coefficient can be expressed either as an absolute value (W/°C) or as a relative value (%/°C). The power output and the efficiency of a PV panel are determined as follows [49]:

$$P_{out} = A_p G \epsilon_{out} \tag{7.4}$$

$$\epsilon_{out} = \epsilon_{STC} \left[1 + \delta \left(T_{cell} - 25 \right) \right] \tag{7.5}$$

Combining Equations 7.4 and 7.5, the power output, P_{out}, of a PV panel can be expressed in terms of the power obtained under standard test conditions, P_{STC}:

$$P_{out} = P_{STC} \left\{ \frac{G}{1000} \left[1 + \delta \left(T_{cell} - 25 \right) \right] \right\} \tag{7.6}$$

Note that the power–temperature coefficient, δ, and the parameters related to standard test conditions, $I_{SC} (STC)$, ϵ_{STC}, and P_{STC}, are provided by the manufacturing company.

7.3.2 Wind Energy

Wind energy is the kinetic energy of air in motion. Wind power is the conversion of wind energy into a useful form of energy, such as using wind turbines to produce electrical power. Wind turbines harvest kinetic energy from the wind flow and convert it into usable power. The kinetic energy associated with a cube of air with mass m moving at a speed v (Figure 7.5) is [48]:

$$E_{kinetic} = \frac{1}{2} m v^2 \tag{7.7}$$

Thus, the power associated with such mass of air moving at a speed v through area A_c, which is depicted in Figure 7.6, is as follows:

$$P_{kinetic} = \frac{1}{2} \dot{m} v^2 \tag{7.8}$$

where:
   the mass flow rate $\dot{m}$ through area A_c can be formulated as the product of air density ρ, cross-sectional area A_c, and speed v

$$\dot{m} = \rho A_c v \tag{7.9}$$

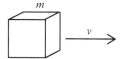

FIGURE 7.5
Cube of moving air.

FIGURE 7.6
Air flow through a surface.

Combining Equations 7.8 and 7.9 yields the following relation:

$$P_{\text{wind}} = \frac{1}{2}\rho A_c v^3 \tag{7.10}$$

where:

P_{wind} represents the wind power output

The wind blows on the blades of the rotor, making it spin, and thereby converting some of the wind's kinetic energy into mechanical energy. A shaft connects the rotor to a generator, so when the rotor spins, so does the generator. The generator converts the mechanical energy into electrical energy.

The wind flow through a wind turbine is sketched in Figure 7.7, where the upstream velocity is v_u, the velocity of the wind through the plane of the rotor blades is v_b, and the downstream wind velocity is v_d. The power extracted by the blades, P_b, is equal to the difference in kinetic energy between the upwind and downwind air flows:

$$P_b = \frac{1}{2}\dot{m}\left(v_u^2 - v_d^2\right) \tag{7.11}$$

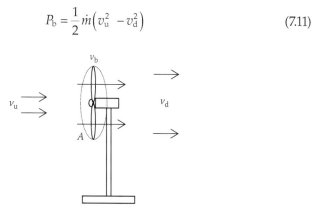

FIGURE 7.7
Air flow through a wind turbine.

The mass flow rate of air within the stream tube is defined as follows:

$$\dot{m} = \rho A_c v_b \tag{7.12}$$

Assuming that the velocity of the wind through the plane of the rotor is just the average of the upwind and downwind speeds, Equation 7.11 becomes:

$$P_b = \frac{1}{2}\rho A_c \left(\frac{v_u + v_d}{2}\right)\left(v_u^2 - v_d^2\right) \tag{7.13}$$

The ratio of downstream to upstream wind speed is defined as γ:

$$\gamma = \frac{v_d}{v_u} \tag{7.14}$$

Substituting Equation 7.14 into Equation 7.13 gives:

$$P_b = \frac{1}{2}\rho A_c \left(\frac{v_u + \gamma v_u}{2}\right)\left(v_u^2 - \gamma^2 v_u^2\right) = \frac{1}{2}\rho A_c v_u^3\left[\frac{1}{2}(1+\gamma)\left(1-\gamma^2\right)\right] \tag{7.15}$$

Equation 7.15 shows that the power extracted from wind is equal to the upstream power multiplied by the term in brackets, which is defined as the rotor efficiency, C_p:

$$C_p = \frac{1}{2}(1+\gamma)\left(1-\gamma^2\right) \tag{7.16}$$

The efficiency of the rotor can be represented versus the wind speed ratio, as shown in Figure 7.8. The maximum value of the rotor efficiency can be determined by deriving Equation 7.16 with respect to γ and equating to zero:

$$\frac{dC_p}{d\gamma} = \frac{1}{2}\left[(1+\gamma)(1-2\gamma)\left(1-\gamma^2\right)\right] = 0 \tag{7.17}$$

Solving Equation 7.17 yields:

$$\gamma = \frac{v_d}{v_u} = \frac{1}{3} \tag{7.18}$$

Substituting Equation 7.18 into Equation 7.16, the maximum rotor efficiency, $\bar{C}_p$, is attained:

$$\bar{C}_p = \frac{1}{2}\left(1+\frac{1}{3}\right)\left(1-\frac{1}{3^2}\right) = 59.3\% \tag{7.19}$$

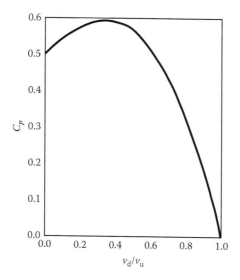

FIGURE 7.8
Rotor efficiency.

The result provided in Equation 7.19 is known as *Betz efficiency*.

7.3.2.1 Impact of Height

Wind power is proportional to the cube of wind speed. Therefore, a small variation in wind speed significantly modifies the power output. The following expression is used to characterize the wind speed profile and the impact of the roughness of the earth's surface on wind speed:

$$\left(\frac{v}{v_0}\right) = \left(\frac{H}{H_0}\right)^{\alpha} \tag{7.20}$$

Table 7.1 shows the value of the coefficient of friction, α, for different soil characteristics.

TABLE 7.1

Coefficient of Friction

Soil Characteristics	α
Smooth hard ground, calm water	0.10
Tall grass on level ground	0.15
High crops, hedges, and shrubs	0.20
Wooded countryside, many trees	0.25
Small town with trees and shrubs	0.30
Large city with tall buildings	0.40

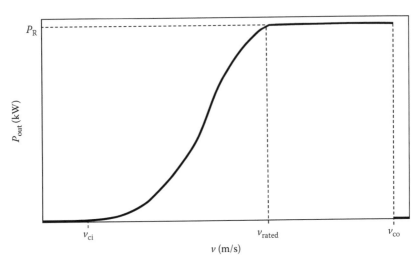

FIGURE 7.9
Power curve for a wind turbine.

7.3.2.2 Wind Generation Model

The wind generation model aims to convert wind speed values into levels of wind power. Turbine manufacturers provide power wind speed data in the form of a graph (Figure 7.9) or a table.

7.4 Uncertainty Modeling

The increasing penetration of stochastic resources in power systems requires accurately modeling the associated uncertainty for investment planning. Variability and uncertainty are not exclusive of stochastic generation resources. Similar challenges are posed by aggregated electricity demand and, to a certain extent, by conventional generation. Here, we propose characterizing the variability of demand, wind speed, and solar irradiation through a stochastic programming framework based on a set of scenarios. For the sake of simplicity, we consider that ambient temperature lies within a narrow range. Each scenario represents an operating condition for each time block b, in which the curves of historical data of demand, wind speed, and solar irradiation are discretized. Such operating condition is characterized by an average demand factor and the levels of wind power generation and PV power generation. Mathematically, the set of scenarios Ω_b is formulated as follows:

$$\Omega_b = \left\{ \mu_b^D(\omega), \left\{ \hat{G}_{iktb}^W(\omega) \right\}_{\forall i \in \Psi^W, \forall k \in K^W, \forall t \in T}, \left\{ \hat{G}_{iktb}^\Theta(\omega) \right\}_{\forall i \in \Psi^\Theta, \forall k \in K^\Theta, \forall t \in T} \right\}_{\forall \omega = 1 \dots n_\Omega} \quad (7.21)$$

$$\forall b \in B$$

Note that nodal demands at each scenario result from multiplying the forecasted values by the corresponding factor $\mu_b^D(\omega)$.

The set of scenarios is built on historical data of demand, wind speed, and solar irradiation through the methodology described in [50–52]. Demand, wind speed, and solar irradiation are not statistically independent magnitudes. Thus, the statistical interrelations among these three magnitudes are specifically accounted for while maintaining the correlation among them. The methodology proposed comprises seven steps that are described as follows.

- *Step 1*: Historical hourly data of demand, wind speed, and solar irradiation along a year are expressed as factors by dividing each by the corresponding maximum level (peak demand, maximum value of wind speed, and maximum value of solar irradiation). Hence, each set of factors represents the per-unit demand, wind speed, and solar irradiation profile.

- *Step 2*: Triplets of hourly factors for demand, wind speed, and solar irradiation are sorted by demand factor in descending order. Figure 7.10 shows an ordered demand factor curve and the corresponding profiles of wind speed and solar irradiation factors.

- *Step 3*: The factor curves resulting from step 2 are discretized into n_B time blocks. In order to accurately model the peak demand, which usually has a high influence on investment decisions, a relatively small time block related to such peak demand is defined. Figure 7.11 shows an example of such discretization with four blocks.

- *Step 4*: For each time block, the corresponding wind speed and solar irradiation factors are sorted in descending order as depicted in Figure 7.12.

- *Step 5*: For each time block, the cumulative distribution functions (cdf) of the ordered demand, wind speed, and solar irradiation factors are built. Figure 7.13 shows the cdfs corresponding to the curves shown in Figure 7.12.

- *Step 6*: The cdfs are divided into segments with their corresponding probabilities. The prespecified numbers of segments are denoted as n_s^D, n_s^W, and n_s^Θ for demand, wind speed, and solar irradiation factor curves, respectively. In addition to its prespecified probability, each segment is characterized by an average factor equal to the average value of the factors within such segment. As a consequence, pairs probability-average factor are generated for demand, namely $\pi_{sb}^D - \mu_{sb}^D$, for wind speed, namely $\pi_{sb}^W - \mu_{sb}^W$, and for solar irradiation, namely $\pi_{sb}^\Theta - \mu_{sb}^\Theta$, where s denotes the index for segments. As an example, Figure 7.14 shows the cdf of the first time block for the demand factor

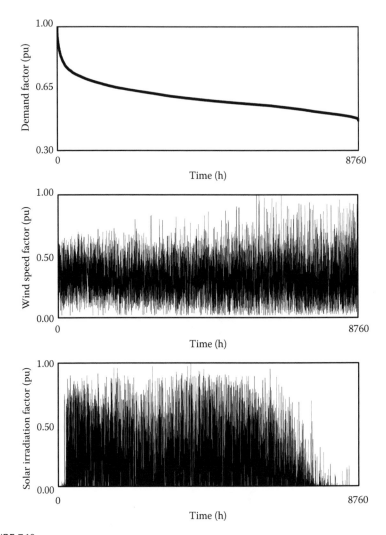

FIGURE 7.10
Ordered demand factor curve and the corresponding profiles of wind speed and solar irradiation factors.

curve in Figure 7.12, which is divided into three segments with probabilities, π_{11}^D, π_{21}^D, and π_{31}^D, equal to 0.4, 0.5, and 0.1, respectively, and average demand factors, μ_{11}^D, μ_{21}^D, and μ_{31}^D, equal to 0.67, 0.76, and 0.90, respectively.

- *Step 7*: Scenarios for each time block result from combining all pairs $\pi_{sb}^D - \mu_{sb}^D$, $\pi_{sb}^W - \mu_{sb}^W$, and $\pi_{sb}^\Theta - \mu_{sb}^\Theta$. Note that μ_{sb}^W and μ_{sb}^Θ are considered constant for all candidate nodes in Ψ^W and Ψ^Θ, respectively, along

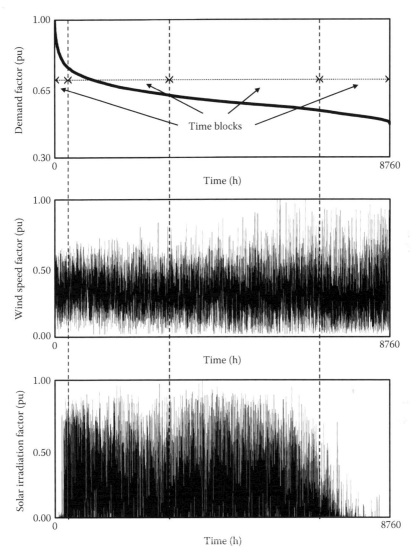

FIGURE 7.11
Time block discretization.

the planning horizon. For each scenario, average factors μ_{sb}^{W} and μ_{sb}^{Θ} are converted to wind speed and irradiation levels. Subsequently, those levels are fed to the corresponding production models described in Section 7.3 in order to yield the maximum levels of wind and PV power generation, $\hat{G}_{iktb}^{W}(\omega)$ and $\hat{G}_{iktb}^{\Theta}(\omega)$. Thus, as formulated

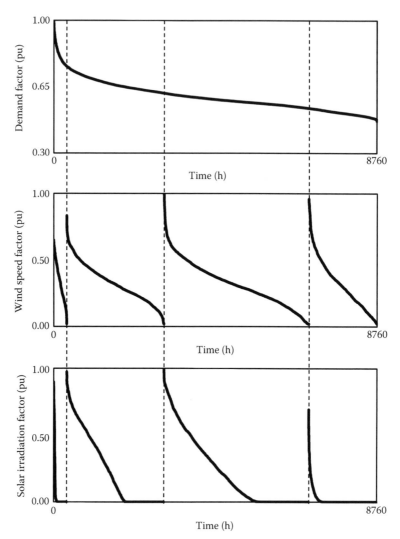

FIGURE 7.12
Ordered demand, wind speed, and solar irradiation factor curves.

in Equation 7.21 for each time block b, a scenario ω comprises an average demand factor, $\mu_b^D(\omega)$, a vector of maximum levels of wind power generation, $\hat{G}_{iktb}^W(\omega)$, and a vector of maximum levels of PV power generation, $\hat{G}_{iktb}^{\Theta}(\omega)$. In addition, each scenario is associated with a probability, $\pi_b(\omega)$, that is equal to the product of the probabilities of the pairs involved. The number of scenarios in each time block, n_{Ω}, is equal to $n_s^D \times n_s^W \times n_s^{\Theta}$. Considering n_B time blocks, the number of operating conditions at each stage is thus equal to $n_B \times n_{\Omega}$.

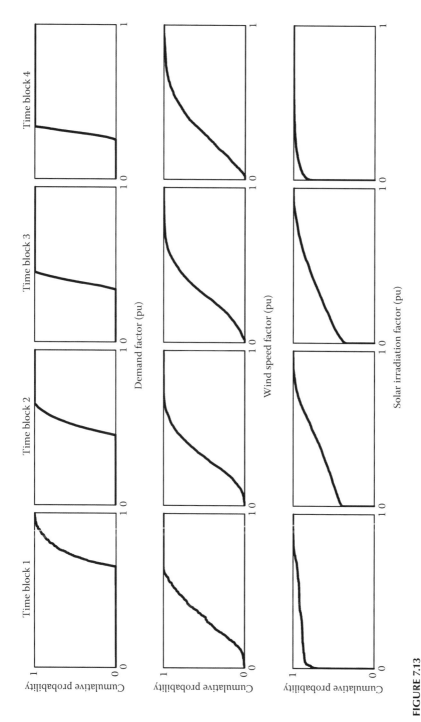

FIGURE 7.13
Cumulative distribution functions of demand, wind speed, and solar irradiation factors at each time block.

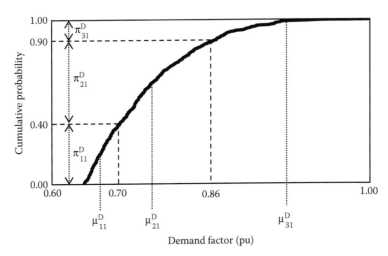

FIGURE 7.14
Cumulative distribution function for the first block of the demand factor curve.

7.5 Formulation of the Optimization Problem

In this section, the mathematical formulation of the expansion planning problem is presented. The model proposed is built on the distribution system planning models described in [3,8–10,31] wherein: (1) a multistage planning framework is adopted, (2) the annual load curve is discretized into several time blocks, (3) radial operation of the distribution network is explicitly imposed, (4) an approximate network model is used, (5) the costs of losses are included in the objective function, and (6) several investment alternatives exist for each asset. Besides, the uncertainty of demand, wind power, and PV power is characterized by a set of scenarios. This problem is solved through a stochastic optimization model [53], where investment variables are referred to as *here-and-now* decisions and operational variables are denoted as *wait-and-see* decisions.

The resulting model is formulated as a mixed-integer nonlinear programming problem, where nonlinearities are related to (1) quadratic energy losses in the objective function and (2) bilinear terms involving the products of continuous and binary decision variables in the equations associated with Kirchhoff's voltage law. Both nonlinearities are recast as linear expressions using a piecewise linear approximation for energy losses and integer algebra results for the bilinear terms. The solution approach yields a mixed-integer linear program for which effective off-the-shelf branch-and-cut software is available [54]. Note that mixed-integer linear programming guarantees finite convergence to optimality, while providing a measure of the distance to optimality along the solution process [55].

7.5.1 Objective Function

According to [3,10], the objective function to be minimized represents the present value of the expected total cost:

$$c^{TPV} = \sum_{t \in T} \frac{(1+I)^{-t}}{I} c_t^I + \sum_{t \in T} \left[(1+I)^{-t} \left(c_t^M + c_t^E + c_t^R + c_t^U \right) \right]$$
$$+ \frac{(1+I)^{-n_T}}{I} \left(c_{n_T}^M + c_{n_T}^E + c_{n_T}^R + c_{n_T}^U \right)$$

(7.22)

As done in [3,10], Equation 7.22 includes three terms. The first term corresponds to the present worth value of the investment cost under the assumption of a perpetual or infinite planning horizon [56]. In other words, the investment cost is amortized in annual installments throughout the lifetime of the installed equipment, considering that after the equipment lifetime has expired, there is a reinvestment in an identical piece of equipment. The second term is the present value of the operating costs throughout the time stages. Finally, the third term represents the present value of the operating costs incurred after the last time stage. As can be noted, such term relies on the operating costs at the last time stage and also assumes a perpetual planning horizon.

The total cost in Equation 7.22 comprises amortized investment, maintenance, production, energy losses, and unserved energy costs, which are formulated as follows:

$$c_t^I = \sum_{l \in \{NRF,NAF\}} RR^l \sum_{k \in K^l} \sum_{(i,j) \in Y^l} C_k^{I,l} \ell_{ij} x_{ijkt}^l + RR^{SS} \sum_{i \in \Psi^{SS}} C_i^{I,SS} x_{it}^{SS}$$
$$+ RR^{NT} \sum_{k \in K^{NT}} \sum_{i \in \Psi^{SS}} C_k^{I,NT} x_{ikt}^{NT} + \sum_{p \in P} RR^p \sum_{k \in K^p} \sum_{i \in \Psi^p} C_k^{I,p} pf \bar{G}_k^p x_{ikt}^p ; \quad \forall t \in T$$

(7.23)

where:

$$RR^l = \frac{I(1+I)^{n^l}}{(1+I)^{n^l} - 1}, \quad \forall l \in \{NRF,NAF\}$$

$$RR^{NT} = \frac{I(1+I)^{n^{NT}}}{(1+I)^{n^{NT}} - 1}$$

$$RR^p = \frac{I(1+I)^{n^p}}{(1+I)^{n^p} - 1}, \quad \forall p \in P$$

and

$$RR^{SS} = \frac{I(1+I)^{n^{SS}}}{(1+I)^{n^{SS}} - 1}$$

$$c_t^M = \sum_{l \in L} \sum_{k \in K^l} \sum_{(i,j) \in Y^l} C_k^{M,l} \left(y_{ijkt}^l + y_{jikt}^l \right) + \sum_{tr \in TR} \sum_{k \in K^{tr}} \sum_{i \in \Psi^{SS}} C_k^{M,tr} y_{ikt}^{tr}$$

$$+ \sum_{p \in P} \sum_{k \in K^p} \sum_{i \in \Psi^p} C_k^{M,p} y_{ikt}^p; \quad \forall t \in T \tag{7.24}$$

$$c_t^E = \sum_{b \in B} \sum_{\omega=1}^{n_\Omega} \pi_b(\omega) \Delta_b pf \left[\sum_{tr \in TR} \sum_{k \in K^{tr}} \sum_{i \in \Psi^{SS}} C_b^{SS} g_{iktb}^{tr}(\omega) + \sum_{p \in P} \sum_{k \in K^p} \sum_{i \in \Psi^p} C_k^{E,p} g_{iktb}^p(\omega) \right]; \tag{7.25}$$

$$\forall t \in T$$

$$c_t^R = \sum_{b \in B} \sum_{\omega=1}^{n_\Omega} \pi_b(\omega) \Delta_b C_b^{SS} pf \left\{ \begin{array}{l} \sum_{l \in L} \sum_{k \in K^l} \sum_{(i,j) \in Y^l} Z_k^l \ell_{ij} \left[f_{ijktb}^l(\omega) + f_{jiktb}^l(\omega) \right]^2 \\ + \sum_{tr \in TR} \sum_{k \in K^{tr}} \sum_{i \in \Psi^{SS}} Z_k^{tr} \left[g_{iktb}^{tr}(\omega) \right]^2 \end{array} \right\}; \quad \forall t \in T \tag{7.26}$$

$$c_t^U = \sum_{b \in B} \sum_{\omega=1}^{n_\Omega} \sum_{i \in \Psi_t^{LN}} \pi_b(\omega) \Delta_b C^U pf d_{itb}^U(\omega); \quad \forall t \in T \tag{7.27}$$

In Equation 7.23, the amortized investment cost at each stage is formulated as the sum of terms related to (1) replacement and addition of feeders, (2) reinforcement of existing substations and construction of new ones, (3) installation of new transformers, and (4) installation of renewable generators. It should be emphasized that the investment cost for substations corresponds to the cost associated with the upgrading or construction of this infrastructure excluding the cost of transformers, which is explicitly considered in the third term of Equation 7.23. Equation 7.24 models the maintenance costs of feeders, transformers, and generators at each stage. The expected production costs associated with substations and generators are characterized in Equation 7.25. Analogously, the expected costs of energy losses in feeders and transformers are modeled in Equation 7.26. Similar to [3,10], energy losses are formulated as quadratic terms. Such nonlinearities can be accurately approximated by a set of tangent lines. This approximation yields piecewise linear functions, which, for practical purposes, are indistinguishable from the nonlinear models if enough segments are used. Finally, Equation 7.27 corresponds to the penalty cost of the expected unserved energy.

It is worth emphasizing that, for each time stage, a single binary variable per conductor in the feeder connecting nodes i and j is used to model the associated investment decision, namely, x_{ijkt}^l. In contrast, two binary variables, y_{ijkt}^l and y_{jikt}^l, as well as two continuous variables, $f_{ijktb}^l(\omega)$ and $f_{jiktb}^l(\omega)$, are associated with each feeder in order to model its utilization and current flow, respectively. Note that $f_{ijktb}^l(\omega)$ is greater than 0 and equal to the current flow through the feeder between nodes i and j measured at node i only when the current flows from i to j, being 0 otherwise.

7.5.2 Constraints

The joint expansion planning problem includes the constraints described next.

7.5.2.1 Integrality Constraints

Investments in new assets are modeled by binary variables as follows:

$$x_{ijkt}^l \in \{0,1\}; \quad \forall l \in \{\text{NRF,NAF}\}, \forall (i,j) \in Y^l, \forall k \in K^l, \forall t \in T \qquad (7.28)$$

$$x_{it}^{SS} \in \{0,1\}; \quad \forall i \in \Psi^{SS}, \forall t \in T \qquad (7.29)$$

$$x_{ikt}^{NT} \in \{0,1\}; \quad \forall i \in \Psi^{SS}, \forall k \in K^{NT}, \forall t \in T \qquad (7.30)$$

$$x_{ikt}^p \in \{0,1\}; \quad \forall p \in P, \forall i \in \Psi^p, \forall k \in K^p, \forall t \in T \qquad (7.31)$$

where such variables are equal to 1 when the corresponding investment is made, being 0 otherwise.

Utilization decisions are also modeled by binary variables:

$$y_{ijkt}^l \in \{0,1\}; \quad \forall l \in L, \forall i \in \Psi_j^l, \forall j \in \Psi^N, \forall k \in K^l, \forall t \in T \qquad (7.32)$$

$$y_{ikt}^{tr} \in \{0,1\}; \quad \forall tr \in TR, \forall i \in \Psi^{SS}, \forall k \in K^{tr}, \forall t \in T \qquad (7.33)$$

$$y_{ikt}^p \in \{0,1\}; \quad \forall p \in P, \forall i \in \Psi^p, \forall k \in K^p, \forall t \in T \qquad (7.34)$$

where the utilization of a distribution asset is indicated by 1, whereas 0 denotes nonutilization.

7.5.2.2 Kirchhoff's Laws

Based on the linearized network model that was first proposed by Haffner et al. [8] and successfully applied in [3,10,20], the effect of the distribution network is represented by the following constraints:

$$\sum_{l \in L} \sum_{k \in K^l} \sum_{j \in \Psi_i^l} \left[f_{ijktb}^l(\omega) - f_{jiktb}^l(\omega) \right] = \sum_{tr \in TR} \sum_{k \in K^{tr}} g_{iktb}^{tr}(\omega) + \sum_{p \in P} \sum_{k \in K^p} g_{iktb}^p(\omega)$$

$$- \mu_b^D(\omega) D_{it} + d_{itb}^U(\omega); \qquad (7.35)$$

$$\forall i \in \Psi^N, \forall t \in T, \forall b \in B, \forall \omega = 1 \ldots n_\Omega$$

$$y^l_{ijkt} \left\{ Z^l_k \ell_{ij} \, f^l_{ijktb}(\omega) - \left[v_{itb}(\omega) - v_{jtb}(\omega) \right] \right\} = 0;$$

$$\forall l \in L, \forall i \in \Psi^l_j, \forall j \in \Psi^N, \forall k \in K^l, \forall t \in T, \forall b \in B, \forall \omega = 1...n_\Omega$$

(7.36)

Equation 7.35 represents the nodal current balance equations, that is, Kirchhoff's current law. These constraints model that the algebraic sum of all outgoing and incoming currents at node i must be equal to 0 for each stage t, time block b, and scenario ω. Equation 7.36 models Kirchhoff's voltage law for all feeders in use.

As described in [8], the linearized network model is an adapted version of the dc model used for the transmission network that is based on two assumptions: (1) all current injections and flows have the same power factor and (2) the per-unit voltage drop across a branch is equal to the difference between the per-unit magnitudes of the nodal voltages at both ends of the branch. Assumption 1 allows expressing Kirchhoff's current law as a set of linear scalar equalities in terms of current magnitudes, giving rise to Equation 7.35. In addition, assumption 2 allows formulating Kirchhoff's voltage law for each feeder in use as a linear expression relating the magnitudes of currents, nodal voltages, and branch impedances. Equation 7.36 extends this result to account for the utilization state of all feeders. Such extension yields nonlinearities involving the products of binary variables y^l_{ijkt} and continuous variables $f^l_{ijktb}(\omega)$ and $v_{itb}(\omega)$.

7.5.2.3 Voltage Limits

The magnitudes of nodal voltages are constrained by upper and lower limits. Mathematically, these bounds are formulated as follows:

$$\underline{V} \le v_{itb}(\omega) \le \overline{V}; \quad \forall i \in \Psi^N, \forall t \in T, \forall b \in B, \forall \omega = 1...n_\Omega \qquad (7.37)$$

7.5.2.4 Operational Bounds for Feeders

Current flows through feeders are nonnegative variables with upper bounds:

$$0 \le f^l_{ijktb}(\omega) \le y^l_{ijkt} \overline{F}^l_k; \quad \forall l \in L, \forall i \in \Psi^l_j, \forall j \in \Psi^N,$$

$$\forall k \in K^l, \forall t \in T, \forall b \in B, \forall \omega = 1...n_\Omega \qquad (7.38)$$

Equation 7.38 sets the bounds on current flows through the feeders in use. If a feeder is not used, that is, $y^l_{ijkt} = 0$, the corresponding current flow is 0.

7.5.2.5 Operational Bounds for Transformers

Similar to feeders, current injections by transformers at substations are also nonnegative variables limited by upper bounds:

$$0 \leq g_{iktb}^{tr}(\omega) \leq y_{ikt}^{tr}\bar{G}_k^{tr}; \quad \forall tr \in TR, \forall i \in \Psi^{SS},$$

$$\forall k \in K^{tr}, \forall t \in T, \forall b \in B, \forall \omega = 1 \ldots n_\Omega \tag{7.39}$$

Equation 7.39 sets the bounds for the current levels injected by the transformers in use. If a transformer is not used, that is, $y_{ikt}^{tr} = 0$, the corresponding current injection is 0.

7.5.2.6 Operational Bounds for Generators

The upper and lower limits for renewable generation are set as follows:

$$0 \leq g_{iktb}^{p}(\omega) \leq y_{ikt}^{p}\hat{G}_{iktb}^{p}(\omega); \quad \forall p \in P, \forall i \in \Psi^{p},$$

$$\forall k \in K^{p}, \forall t \in T, \forall b \in B, \forall \omega = 1 \ldots n_\Omega \tag{7.40}$$

The upper bound for renewable power generation is the available power associated with the corresponding generation technology. Thus, the available power for wind generators depends on the level of wind speed, whereas for PV units, such availability is determined by the level of solar irradiation. Note that if a unit is not in use, that is, $y_{ikt}^{p} = 0$, the current injection is set equal to 0.

7.5.2.7 Unserved Energy

Unserved energy is modeled by a continuous and nonnegative variable denoted by $d_{itb}^{U}(\omega)$ that is bounded as follows:

$$0 \leq d_{itb}^{U}(\omega) \leq \mu_b^{D}(\omega)D_{it}; \quad \forall i \in \Psi_t^{LN}, \forall t \in T, \forall b \in B, \forall \omega = 1 \ldots n_\Omega \tag{7.41}$$

As can be seen, the maximum level of unserved energy is equal to the corresponding demand level.

7.5.2.8 DG Penetration Limit

The level of penetration of DG is limited to a fraction ξ of the demand. This is mathematically formulated as follows:

$$\sum_{p\in P}\sum_{k\in K^p}\sum_{i\in\Psi^p} g_{iktb}^{p}(\omega) \leq \xi \sum_{i\in\Psi_t^{LN}} \mu_b^{D}(\omega)D_{it}; \quad \forall t \in T, \forall b \in B, \forall \omega = 1 \ldots n_\Omega \tag{7.42}$$

7.5.2.9 Investment Constraints

Investment decisions are constrained according to the following expressions:

$$\sum_{t\in T}\sum_{k\in K^l} x_{ijkt}^{l} \leq 1; \quad \forall l \in \{NRF, NAF\}, \forall (i,j) \in Y^l \tag{7.43}$$

$$\sum_{t \in T} x_{it}^{SS} \le 1; \quad \forall i \in \Psi^{SS} \tag{7.44}$$

$$\sum_{t \in T} \sum_{k \in K^{NT}} x_{ikt}^{NT} \le 1; \quad \forall i \in \Psi^{SS} \tag{7.45}$$

$$\sum_{t \in T} \sum_{k \in K^{P}} x_{ikt}^{p} \le 1; \quad \forall p \in P, \forall i \in \Psi^{p} \tag{7.46}$$

$$x_{ikt}^{NT} \le \sum_{\tau=1}^{t} x_{i\tau}^{SS}; \quad \forall i \in \Psi^{SS}, \forall k \in K^{NT}, \forall t \in T \tag{7.47}$$

As per Equations 7.43 through 7.46, a maximum of one reinforcement, replacement, or addition is allowed for each system component and location along the planning horizon. Equation 7.47 guarantees that new transformers can only be added in substations that have been previously expanded or built.

7.5.2.10 Utilization Constraints

Candidate assets for reinforcement, replacement, or installation can only be used once the investment has been made. Mathematically, this is formulated as follows:

$$y_{ijkt}^{EFF} + y_{jikt}^{EFF} = 1; \quad \forall(i,j) \in Y^{EFF}, \forall k \in K^{EFF}, \forall t \in T \tag{7.48}$$

$$y_{ijkt}^{l} + y_{jikt}^{l} = \sum_{\tau=1}^{t} x_{ijk\tau}^{l}; \quad \forall l \in \{NRF,NAF\}, \forall(i,j) \in Y^{l}, \forall k \in K^{l}, \forall t \in T \tag{7.49}$$

$$y_{ijkt}^{ERF} + y_{jikt}^{ERF} = 1 - \sum_{\tau=1}^{t} \sum_{\kappa \in K^{NRF}} x_{ijk\tau}^{NRF}; \quad \forall(i,j) \in Y^{ERF}, \forall k \in K^{ERF}, \forall t \in T \tag{7.50}$$

$$y_{ikt}^{NT} \le \sum_{\tau=1}^{t} x_{ik\tau}^{NT}; \quad \forall i \in \Psi^{SS}, \forall k \in K^{NT}, \forall t \in T \tag{7.51}$$

$$y_{ikt}^{p} \le \sum_{\tau=1}^{t} x_{ik\tau}^{p}; \quad \forall p \in P, \forall i \in \Psi^{p}, \forall k \in K^{p}, \forall t \in T \tag{7.52}$$

Equations 7.48 through 7.50 model the utilization of feeders while explicitly characterizing the direction of current flows. Under the assumption that the existing network is radial, Equations 7.48 through 7.50 impose the utilization of both existing and newly installed feeders in order to keep radiality. The utilization of new transformers is formulated in Equation 7.51, whereas the utilization of newly installed generators is modeled in Equation 7.52.

7.5.2.11 Investment Limit

The total investment cost at each stage t is limited by an upper bound:

$$\sum_{l\in\{NRF,NAF\}}\sum_{k\in K^l}\sum_{(i,j)\in Y^l}C_k^{l,l}\ell_{ij}x_{ijkt}^l + \sum_{i\in\Psi^{SS}}C_i^{l,SS}x_{it}^{SS} + \sum_{k\in K^{NT}}\sum_{i\in\Psi^{SS}}C_k^{l,NT}x_{ikt}^{NT}$$

$$+\sum_{p\in P}\sum_{k\in K^p}\sum_{i\in\Psi^p}C_k^{l,p}pf\overline{G}_k^p x_{ikt}^p \le IB_t; \quad \forall t \in T \tag{7.53}$$

7.5.2.12 Radiality Constraints

The proposed model includes the following traditional radiality constraints [3,17]:

$$\sum_{l\in L}\sum_{i\in\Psi_j^l}\sum_{k\in K^l}y_{ijkt}^l = 1; \quad \forall j \in \Psi_t^{LN}, \forall t \in T \tag{7.54}$$

$$\sum_{l\in L}\sum_{i\in\Psi_j^l}\sum_{k\in K^l}y_{ijkt}^l \le 1; \quad \forall j \notin \Psi_t^{LN}, \forall t \in T \tag{7.55}$$

Equation 7.54 imposes load nodes to have a single input flow, whereas Equation 7.55 sets a maximum of one input flow for the remaining nodes. It is worth mentioning that, in the absence of DG, this formulation does not feature the shortcomings described in [8,22] and is suitable to impose radial operation even when transfer nodes are part of the distribution network. However, when DG is considered, Equations 7.54 and 7.55 are insufficient to prevent the existence of areas exclusively supplied by DG and thereby topologically disconnected from all substations. This difficulty is overcome by adding the following set of radiality constraints [22]:

$$\sum_{l\in L}\sum_{k\in K^l}\sum_{j\in\Psi_i^l}\left(\tilde{f}_{ijkt}^l - \tilde{f}_{jikt}^l\right) = \tilde{g}_{it}^{SS} - \tilde{D}_{it}; \quad \forall i \in \Psi^N, \forall t \in T \tag{7.56}$$

$$0 \le \tilde{f}_{ijkt}^{EFF} \le n_{DG}; \quad \forall i \in \Psi_j^{EFF}, \forall j \in \Psi^N, \forall k \in K^{EFF}, \forall t \in T \tag{7.57}$$

$$0 \le \tilde{f}_{ijkt}^{ERF} \le \left(1 - \sum_{\tau=1}^{t}\sum_{\kappa\in K^{NRF}}x_{ij\kappa\tau}^{NRF}\right)n_{DG}; \quad \forall(i,j)\in Y^{ERF}, \forall k \in K^{ERF}, \forall t \in T \tag{7.58}$$

$$0 \le \tilde{f}_{jikt}^{ERF} \le \left(1 - \sum_{\tau=1}^{t}\sum_{\kappa\in K^{NRF}}x_{ij\kappa\tau}^{NRF}\right)n_{DG}; \quad \forall(i,j)\in Y^{ERF}, \forall k \in K^{ERF}, \forall t \in T \tag{7.59}$$

$$0 \le \tilde{f}_{ijkt}^{l} \le \left(\sum_{\tau=1}^{t} x_{ijk\tau}^{l} \right) n_{DG}; \quad \forall l \in \{\text{NRF,NAF}\}, \forall (i,j) \in Y^{l}, \forall k \in K^{l}, \forall t \in T \quad (7.60)$$

$$0 \le \tilde{f}_{jikt}^{l} \le \left(\sum_{\tau=1}^{t} x_{ijk\tau}^{l} \right) n_{DG}; \quad \forall l \in \{\text{NRF,NAF}\}, \forall (i,j) \in Y^{l}, \forall k \in K^{l}, \forall t \in T \quad (7.61)$$

$$0 \le \tilde{g}_{it}^{SS} \le n_{DG}; \quad \forall i \in \Psi^{SS}, \forall t \in T \quad (7.62)$$

As described in [22], the existence of isolated generators is avoided by Equations 7.56 through 7.62, which model a fictitious system with fictitious demands. According to [22], the fictitious demand at load nodes that are candidate locations for DG installation is equal to 1 pu, whereas the fictitious demand at the remaining nodes is set to 0. Mathematically:

$$\tilde{D}_{it} = \begin{cases} 1; & \forall i \in \left[\left(\Psi^{W} \cup \Psi^{\Theta} \right) \cap \Psi_{t}^{LN} \right], \forall t \in T \\ 0; & \forall i \notin \left[\left(\Psi^{W} \cup \Psi^{\Theta} \right) \cap \Psi_{t}^{LN} \right], \forall t \in T \end{cases} \quad (7.63)$$

Such fictitious nodal demands can only be supplied by fictitious substations located at the original substation nodes, which inject fictitious energy through the actual feeders.

Equation 7.56 represents the nodal balance equations for fictitious currents. Equations 7.57 through 7.61 bound the fictitious flows through feeders. Equation 7.62 sets the limits for the fictitious currents injected by the fictitious substations.

7.5.3 Linearizations

The optimization model for the joint expansion planning of DG and distribution network presented in the previous section includes nonlinearities that drastically complicate the attainment of the optimal solution. Instead of directly addressing the above instance of mixed-integer nonlinear programming, we propose transforming the original problem into a mixed-integer linear program. Note that mixed-integer linear programming guarantees finite convergence to the optimal solution, while providing a measure of the distance to optimality along the solution process [55]. In addition, effective off-the-shelf branch-and-cut software is available [54].

The nonlinearities are related to (1) quadratic energy losses in the objective function and (2) bilinear terms involving the products of continuous and binary decision variables in the equations associated with Kirchhoff's voltage law. Both nonlinearities are recast as linear expressions using a piecewise linear approximation for energy losses and integer algebra results for the bilinear terms.

7.5.3.1 Linearized Energy Losses

Energy losses are modeled in Equation 7.26 by quadratic expressions, which can be accurately approximated by a set of tangent lines. Such approximation yields piecewise linear functions, which, for practical purposes, are indistinguishable from the original nonlinear models if enough blocks are used. As an example, Figure 7.15 depicts a three-piece linear approximation of a general quadratic function relating energy losses and current. The general formulation of such piecewise linear approximation is as follows [57]:

$$\text{Current} = \sum_{v=1}^{n_v} \delta_v \tag{7.64}$$

$$\text{Losses} = \sum_{v=1}^{n_v} M_v \delta_v \tag{7.65}$$

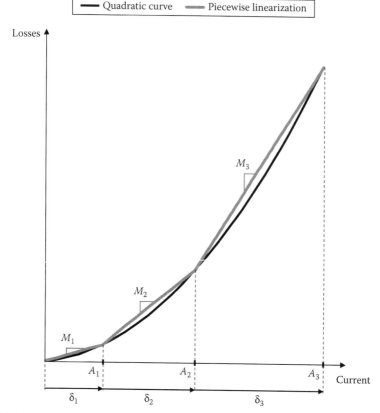

FIGURE 7.15
Quadratic energy losses and piecewise linear approximation.

$$0 \le \delta_v \le A_v; \quad \forall v = 1...n_v \tag{7.66}$$

This piecewise linearization technique has been used to approximate the energy losses in existing fixed feeders, existing replaceable feeders, new replacement feeders, new added feeders, existing transformers, and new transformers. Therefore, Equation 7.26 is replaced by:

$$c_t^R = \sum_{b \in B} \sum_{\omega=1}^{n_\Omega} \pi_b(\omega) \Delta_b C_b^{SS} pf \left\{ \sum_{l \in L} \sum_{k \in K^l} \sum_{(i,j) \in Y^l} \sum_{v=1}^{n_v} M_{kv}^l \ell_{ij} \left[\delta_{ijktbv}^l(\omega) + \delta_{jiktbv}^l(\omega) \right] \right.$$

$$\left. + \sum_{tr \in TR} \sum_{k \in K^{tr}} \sum_{i \in \Psi^{SS}} \sum_{v=1}^{n_v} M_{kv}^{tr} \delta_{iktbv}^{tr}(\omega) \right\}; \quad \forall t \in T \tag{7.67}$$

$$f_{ijktb}^l(\omega) = \sum_{v=1}^{n_v} \delta_{ijktbv}^l(\omega); \quad \forall l \in L, \forall i \in \Psi_j^l, \forall j \in \Psi^N,$$

$$\forall k \in K^l, \forall t \in T, \forall b \in B, \forall \omega = 1...n_\Omega \tag{7.68}$$

$$0 \le \delta_{ijktbv}^l(\omega) \le A_{kv}^l; \quad \forall l \in L, \forall i \in \Psi_j^l, \forall j \in \Psi^N, \forall k \in K^l, \forall t \in T,$$

$$\forall b \in B, \forall \omega = 1...n_\Omega, \forall v = 1...n_v \tag{7.69}$$

$$g_{iktb}^{tr}(\omega) = \sum_{v=1}^{n_v} \delta_{iktbv}^{tr}(\omega); \quad \forall tr \in TR, \forall i \in \Psi^{SS}, \forall k \in K^{tr}, \forall t \in T,$$

$$\forall b \in B, \forall \omega = 1...n_\Omega \tag{7.70}$$

$$0 \le \delta_{iktbv}^{tr}(\omega) \le A_{kv}^{tr}; \quad \forall tr \in TR, \forall i \in \Psi^{SS}, \forall k \in K^{tr}, \forall t \in T,$$

$$\forall b \in B, \forall \omega = 1...n_\Omega, \forall v = 1...n_v \tag{7.71}$$

where Equation 7.67 models the linearized expected costs of energy losses, while Equations 7.68 and 7.69 and Equations 7.70 and 7.71 are related to the linearization of energy losses in feeders and transformers, respectively.

7.5.3.2 Linearized Kirchhoff's Voltage Law

Equation 7.36 models Kirchhoff's voltage law for all feeders in use considering existing fixed feeders, existing replaceable feeders, new replacement feeders, and newly added feeders. Accounting for the utilization state of feeders leads to nonlinear expressions involving the products of binary variables and continuous variables. Based on [58], nonlinear Equation 7.36 has the following linear equivalent:

$$-J\left(1-y^l_{ijkt}\right) \le Z^l_k \ell_{ij} f^l_{ijktb}(\omega) - \left[v_{itb}(\omega) - v_{jtb}(\omega)\right] \le J\left(1-y^l_{ijkt}\right);$$

$$\forall l \in L, \forall i \in \Psi^l_j, \forall j \in \Psi^N, \forall k \in K^l, \forall t \in T, \forall b \in B, \forall \omega = 1 \ldots n_\Omega$$

(7.72)

where:
  J is a sufficiently large positive constant

If y^l_{ijkt} is equal to 1, Equation 7.72 is equivalent to Equation 7.36. Conversely, if y^l_{ijkt} is equal to 0, no limitation is imposed on the expression in brackets in Equation 7.36. Thus, the equivalence between Equation 7.72 and Equation 7.36 when $y^l_{ijkt} = 0$ is also attained by properly selecting J in order to prevent Equation 7.72 from being binding.

7.6 Reliability Calculation

Among the multiple definitions for reliability, this term is here associated with the continuity of service along time. Valuing reliability is a difficult task that is typically performed through surveys [3,7,59]. Distribution companies are monitored by the regulator, whose goal is to ensure a reliable service to customers by setting minimum reliability levels. Distribution companies not meeting those levels are economically penalized. The levels that are considered acceptable for the continuity of service from the regulator's standpoint should be based on the explicit knowledge of the perception of the customers' tolerance levels and the economic losses associated with interruptions [3].

In this work, a customer-based approach is incorporated into the proposed model to provide the decision maker with the available information to calculate the cost to achieve the reliability targets fixed by the regulator. Based on [3,7,59], several reliability indices and their corresponding costs are calculated in order to quantitatively measure system reliability and its economic impact.

As proposed in [3], the calculation of reliability indices and their associated costs considers (1) failure rates of system components and (2) duration of the interruptions, which depends on the repair time and the reconfiguration time. In addition, costs related to reliability are calculated for a particular loading condition and a given network topology.

7.6.1 Reliability Indices

Reliability indices are typically averages that weigh each customer equally, that is, a small residential customer has just as much importance as a large industrial customer. Albeit featuring limitations, such indices are generally considered good measures of reliability and are often used as reliability benchmarks and improvement targets [7].

According to [3], reliability indices are calculated on the basis of two assumptions, namely, (1) each feeder connected to a substation has a circuit breaker without a recloser at the output of the substation and (2) each section between two nodes has a switch that enables the reconfiguration of the system after a fault in order to meet the demand in the most efficient way [13,28]. Thus, once a fault has occurred, the circuit breaker in the corresponding circuit trips. Subsequently, the system topology is reconfigured by operating switches and circuit breakers to reduce the nonsupplied energy. Finally, complete service is restored once the fault is cleared. Only sustained interruptions are considered in the index definition.

In this work, only single outages of feeders are considered in order to calculate the reliability indices and their associated costs. Such outages may be either due to a fault in the feeder itself or due to a contingency in the load node downstream. DG is characterized by the installed capacity at the corresponding node. Thus, for the purpose of reliability calculation, a new parameter referred to as residual demand is defined as follows:

$$D_{itb}^{R}(\omega) = \max\left\{0, \mu_{b}^{D}(\omega)D_{it} - \sum_{k \in K^{P}}\sum_{\tau=1}^{t}\hat{G}_{iktb}^{p}(\omega)x_{ik\tau}^{p}\right\}; \quad \forall i \in \Psi_{t}^{LN}, \forall t \in T,$$

$$\forall b \in B, \forall \omega = 1\ldots n_{\Omega}$$

(7.73)

The following indices are the most commonly used metrics for distribution system reliability [3,47,60]:

- Customer interruption frequency (CIF) at each load node and for each stage:

$$\text{CIF}_{it} = \lambda\sum_{z \in Z}\text{NI}_{itz}$$

(7.74)

- Customer interruption duration (CID) at each load node and for each stage:

$$\text{CID}_{it} = \lambda\sum_{z \in Z}\text{NI}_{itz}\,\text{DI}_{z}$$

(7.75)

- System average interruption frequency index (SAIFI) for each stage, which is a measure of how many sustained interruptions an average customer will experience at each stage:

$$\text{SAIFI}_{t} = \frac{\lambda\sum_{i \in \Psi_{t}^{LN}}\text{NC}_{it}\sum_{z \in Z}\text{NI}_{itz}}{\sum_{i \in \Psi_{t}^{LN}}\text{NC}_{it}}$$

(7.76)

For a fixed number of customers, the only way to improve SAIFI is to reduce the number of sustained interruptions experienced by customers.

- System average interruption duration index (SAIDI) for each stage, which is a measure of how many interruption hours an average customer will experience at each stage:

$$\mathrm{SAIDI}_t = \frac{\lambda \sum_{i \in \Psi_t^{\mathrm{LN}}} \mathrm{NC}_{it} \sum_{z \in Z} \mathrm{NI}_{itz} \, \mathrm{DI}_z}{\sum_{i \in \Psi_t^{\mathrm{LN}}} \mathrm{NC}_{it}} \qquad (7.77)$$

For a fixed number of customers, SAIDI and hence reliability can be improved by reducing either the number or the duration of interruptions.

- Average system availability index (ASAI) for each stage, which is defined as the number of hours that the service was available during a year for an average customer divided by 8760, that is, the number of hours in a year. The relation between the duration of the available service and the duration of interruptions allows calculating ASAI in terms of SAIDI as follows:

$$\mathrm{ASAI}_t = \left(1 - \frac{\mathrm{SAIDI}_t}{8760} \right) \qquad (7.78)$$

Thus, higher ASAI values reflect higher levels of system reliability.

- Expected energy not supplied (EENS) for each stage:

$$\mathrm{EENS}_t = \sum_{b \in B} \frac{\Delta_b}{8760} \sum_{i \in \Psi_t^{\mathrm{LN}}} \mathrm{CID}_{it} \sum_{\omega=1}^{n_\Omega} \pi_b(\omega) D_{itb}^{\mathrm{R}}(\omega) \qquad (7.79)$$

EENS relies on CID and the residual demand, $D_{itb}^{\mathrm{R}}(\omega)$.

The expressions for the reliability indices (Equations 7.74 through 7.79) depend on the number and duration of interruptions, that is, feeder outages, that may affect each node. Thus, to determine the reliability of the network in terms of continuity of supply, it is necessary to analyze the impact of feeder outages on the unserved energy at each load node. Such analysis is illustrated with the system depicted in Figure 7.16 comprising five load nodes, one substation node, and two circuits supplied by the substation with three and two feeders, respectively. As an example, let us consider a single fault at load node 3. Consequently, circuit breaker 1 trips and loads connected at nodes 1–3 will, therefore, not be met. Next, switch 3 is opened to repair the fault and circuit breaker 1 is manually closed. Therefore, during the repair

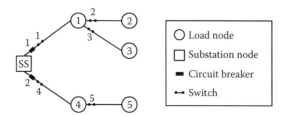

FIGURE 7.16
Illustrative example.

time, all loads are supplied except load 3. When the fault is cleared, switch 3 is closed and normal operation is restored. As a result, loads 1 and 2 have not been met during a time equal to the reconfiguration time, load 3 has not been met during a time equal to the repair time, and loads 4 and 5 are not affected by this fault.

If the fault occurs at load node 1, loads 1–3 will not be met during a time equal to the repair time since loads 2 and 3 are downstream of the fault. Loads 4 and 5 are also unaffected by this fault. The same process is repeated for all single outages under consideration. For the sake of simplicity, both repair and reconfiguration times have been considered to be, respectively, equal for all outages. Such simplification allows quantifying the impact of outages on load nodes in a compact way in terms of both the number of interruptions associated with repairs, $NI_{itRepair}$, and the number of interruptions related to reconfigurations, $NI_{itReconfiguration}$.

Table 7.2 reports such impact quantification for the illustrative example after analyzing all single outages.

7.6.2 Reliability Costs

According to the aforementioned indices, several costs of reliability can be defined such as the cost of customer interruption frequency (CIFC), the cost of customer interruption duration (CIDC), the cost of system average interruption (SAIC), and the cost of expected energy not supplied (EENSC). The definition of these cost terms is dependent on regulation. It is assumed that distribution companies compensate customers for the violation of reliability indices. Such

TABLE 7.2

Number of Potential Interruptions Per Node

Node i	$NI_{itRepair}$	$NI_{itReconfiguration}$
1	1	2
2	2	1
3	2	1
4	1	1
5	2	0

compensation is proportional to the violation level according to the cost of energy supplied by substations, C_b^{SS}, and to a penalty factor set by the regulator.

The costs related to CIF and CID indices at each stage are as follows:

$$\text{CIFC}_t = \chi \sum_{i \in \Psi_t^{LN}} \left[\left(\text{CIF}_{it} - \overline{\text{CIF}} \right) \sum_{b \in B} \sum_{\omega=1}^{n_\Omega} \pi_b(\omega) \frac{\Delta_b}{8760} C_b^{SS} D_{itb}^R(\omega) \right] \text{if } \text{CIF}_{it} > \overline{\text{CIF}} \quad (7.80)$$

$$\text{CIDC}_t = \chi \sum_{i \in \Psi_t^{LN}} \left[\left(\text{CID}_{it} - \overline{\text{CID}} \right) \sum_{b \in B} \sum_{\omega=1}^{n_\Omega} \pi_b(\omega) \frac{\Delta_b}{8760} C_b^{SS} D_{itb}^R(\omega) \right] \text{if } \text{CID}_{it} > \overline{\text{CID}} \quad (7.81)$$

where:

χ is the penalty factor

The largest of both values is passed on to the affected customers in proportion to their individual bills.

The economic valuation of not complying with the average frequency values or the duration of faults at a specific stage is also dependent on regulation. In this work, we consider the penalty applied by the regulator as a percentage ς of the cost of energy supplied by substations, C_b^{SS}. Thus, the cost associated with either SAIFI or SAIDI at each stage is as follows:

$$\text{SAIC}_t = \varsigma \sum_{b \in B} \sum_{\omega=1}^{n_\Omega} \sum_{i \in \Psi_t^{LN}} \pi_b(\omega) \Delta_b C_b^{SS} D_{itb}^R(\omega)$$

$$\text{if } \left(\text{SAIFI}_t > \overline{\text{SAIFI}} \right) \text{ or } \left(\text{SAIDI}_t > \overline{\text{SAIDI}} \right) \quad (7.82)$$

The cost of EENS for each stage is calculated as follows:

$$\text{EENSC}_t = \sum_{b \in B} \frac{\Delta_b}{8760} C_b^{SS} \sum_{i \in \Psi_t^{LN}} \text{CID}_{it} \sum_{\omega=1}^{n_\Omega} \pi_b(\omega) D_{itb}^R(\omega) \quad (7.83)$$

In order to properly consider the above cost terms in the framework of multistage expansion planning, their present values are calculated under the same assumptions adopted for the operating cost terms (Equation 7.22):

$$\text{CIFC}^{PV} = \sum_{t \in T} \left[(1+I)^{-t} \text{CIFC}_t \right] + \frac{(1+I)^{-n_T}}{I} \text{CIFC}_{n_T} \quad (7.84)$$

$$\text{CIDC}^{PV} = \sum_{t \in T} \left[(1+I)^{-t} \text{CIDC}_t \right] + \frac{(1+I)^{-n_T}}{I} \text{CIDC}_{n_T} \quad (7.85)$$

$$\text{SAIC}^{\text{PV}} = \sum_{t \in T} \left[(1+I)^{-t} \text{SAIC}_t \right] + \frac{(1+I)^{-n_T}}{I} \text{SAIC}_{n_T} \qquad (7.86)$$

$$\text{EENSC}^{\text{PV}} = \sum_{t \in T} \left[(1+I)^{-t} \text{EENSC}_t \right] + \frac{(1+I)^{-n_T}}{I} \text{EENSC}_{n_T} \qquad (7.87)$$

7.7 Proposed Algorithm

The difficulty of incorporating reliability into distribution expansion models stems from the need to know (1) the network topology in order to calculate the reliability indices that characterize the system and (2) the operation under contingency for each topology. However, the optimal network topology is an outcome of the optimization process characterizing the expansion planning problem wherein reliability is part of the model. As a consequence of this catch-22 conundrum, methods based on standard mathematical programming are currently unavailable in the technical literature and hence new approaches are yet to be explored. Here, we propose a novel algorithm that extends the methodology described in [3], wherein DG was neglected. The proposed algorithm relies on the generation of a pool of candidate expansion plans, which are determined disregarding reliability. For each expansion plan, reliability indices and their associated costs are computed. The information on investment, operation, and reliability costs allows the planner to make informed decisions on the most suitable expansion plans.

As sketched in Figure 7.17, the proposed planning algorithm consists of the following steps:

1. Obtain a pool of solutions with different topologies by iteratively solving the expansion planning problem described in Section 7.5. At each iteration m, the following constraint [61] is added to the optimization problem:

$$\sum_{(i,j) \in \Lambda_\varrho^0} \left(\sum_{k \in K^{\text{NAF}}} \sum_{t \in T} x_{ijkt}^{\text{NAF}} \right)$$

$$+ \sum_{(i,j) \in \Lambda_\varrho^1} \left(1 - \sum_{k \in K^{\text{NAF}}} \sum_{t \in T} x_{ijkt}^{\text{NAF}} \right) \geq n_d; \quad \forall \varrho = 1 \ldots m-1 \qquad (7.88)$$

where:

$$\Lambda_\varrho^0 = \left\{ (i,j) \in Y^{\text{NAF}} \mid \left[\sum_{k \in K^{\text{NAF}}} \sum_{t \in T} x_{ijkt}^{\text{NAF}(\varrho)} = 0 \right] \right\}$$

and

$$\Lambda_\varrho^1 = \left\{ (i,j) \in Y^{NAF} \mid \left[\sum_{k \in K^{NAF}} \sum_{t \in T} x_{ijkt}^{NAF(\varrho)} = 1 \right] \right\}$$

Equation 7.88 allows obtaining at each iteration a new expansion plan with a different topology at the last stage. At each iteration ϱ, Λ_ϱ^0 includes the indices of branches with candidate feeders for addition where no expansion has been planned. Analogously, Λ_ϱ^1 stores the indices of branches with candidate feeders for addition where investments have been made. In addition, n_d denotes the minimum number of investments in newly added feeders by which the expansion plan identified at each iteration must differ from solutions of previous iterations.

2. Evaluate the reliability indices of the topology associated with each expansion plan resulting from step 1.

3. Calculate the reliability costs associated with each expansion plan resulting from step 1.

4. Select the investment decision on the basis of the comparison of the topologies of the different solutions of the pool and the consideration of the associated costs.

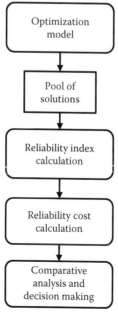

FIGURE 7.17
Flowchart of the proposed planning algorithm.

7.8 Case Study

In this section, a case study based on the distribution network of the pilot site of La Graciosa is presented. La Graciosa is a small island belonging to the Canary Islands, Spain, and located near Lanzarote.

7.8.1 Data

The test system consists of 26 nodes and 37 branches. Figure 7.18 shows the one-line diagram of the distribution network of La Graciosa over a map of the island. For the sake of clarity, the topology of such network is also represented in Figure 7.19. Base power and base voltage are 1 MVA and 20 kV, respectively. The system power factor *pf* is set equal to 0.9 and a three-block piecewise linearization is used to approximate energy losses.

Investment decisions are made over a three-year planning horizon divided into yearly stages considering a 10% interest rate and an investment budget equal to €120,000 per stage.

Nodal peak demands are presented in Table 7.3. Upper and lower bounds for voltages at load nodes are equal to 1.05 and 0.95 pu, respectively. For the purposes of calculation of reliability indices, Table 7.4 provides the number of customers per node at each stage.

Branch lengths are listed in Table 7.5. The capacity and unitary imped-ance of existing feeders are 0.1 MVA and 0.4522 Ω/km, respectively. Tables 7.6 and 7.7 include data for candidate feeders in branches subject to replacement and in nonexisting branches, respectively. Note that two con-ductor alternatives are available per branch. For simplicity, maintenance costs and lifetimes for all feeders are equal to €450/year and 25 years, respectively. Moreover, all feeders are characterized by 0.8 failures per year with interruption durations equal to 2 hours for repair and 0.25 hours for reconfiguration.

Existing substations, which are located at nodes 1 and 2, include a 0.4-MVA transformer characterized by an impedance equal to 0.277 Ω and a mainte-nance cost equal to €1000. Node 3 is the location of a candidate substation. Investment decisions consist of (1) expanding existing substations by adding a new transformer and (2) building a new substation from scratch. The costs of energy supplied by all substations, C_b^{SS}, are identical and equal to €225.33/MWh, €182.72/MWh, €154.43/MWh, and €81.62/MWh, for time blocks 1–4, respectively. The cost of unserved energy, C^U, is €2000/MWh. Expansion costs of substations at nodes 1–3 are €4000, €4000, and €6000, respectively. Data for candidate transformers are listed in Table 7.8, where two alternatives are available for each substation. The lifetime of all candidate transformers is 15 years. It is assumed that η^{SS} is considerably larger than the lifetimes of

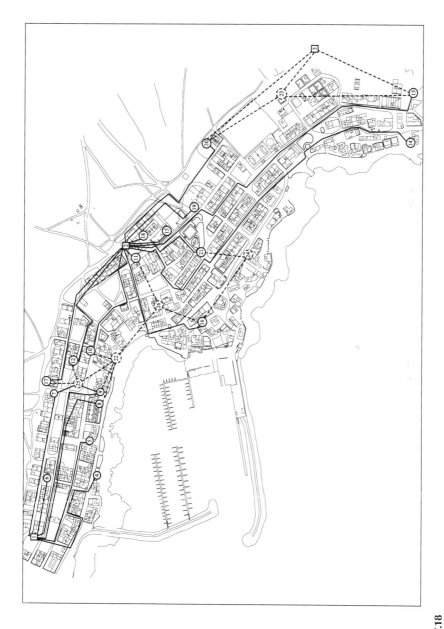

FIGURE 7.18
Map of La Graciosa and one-line diagram of the distribution network.

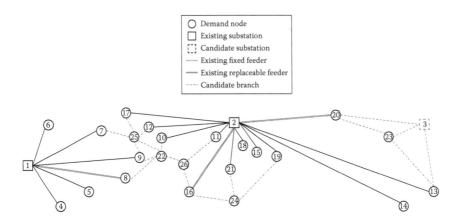

FIGURE 7.19
One-line diagram of the distribution network.

TABLE 7.3

Nodal Peak Demands (kVA)

	Stage				Stage		
Node	1	2	3	Node	1	2	3
4	36	39	40	16	116	125	130
5	31	33	38	17	26	27	29
6	16	20	25	18	6	10	13
7	44	47	49	19	3	4	7
8	56	58	59	20	1	2	6
9	2	5	6	21	1	2	3
10	75	79	82	22	106	107	109
11	20	23	27	23	56	58	63
12	26	29	31	24	0	8	10
13	71	74	78	25	0	26	32
14	54	55	60	26	0	0	41
15	33	35	37				

the other distribution assets. Thus, $RR^{SS} = I$. Voltages at substation nodes are set to 1.05 pu.

Investment in DG is allowed with a penetration limit, ξ, set to 25%. Candidate nodes for installation of wind generators are 8, 9, 10, 11, 13, and 16. Candidate nodes for installation of PV generators are 7, 12, 13, 14, 16, and 21. The economic and technical features of candidate DG units are presented in Table 7.9, where two alternatives are considered for each technology. Maintenance costs for DG units are set as $C_{ik}^{M,p} = 0.05 C_{ik}^{I,p} pf \overline{G}_k^p$. The power output curves of alternatives 1 and 2 for wind generators are presented in Figures 7.20 and 7.21, respectively. Analogously, the power

TABLE 7.4

Number of Customers Per Node

Node	Stage			Node	Stage		
	1	2	3		1	2	3
4	35	37	38	16	104	106	107
5	30	32	37	17	26	27	28
6	17	21	25	18	9	12	14
7	42	44	46	19	6	7	9
8	52	54	55	20	1	2	4
9	3	8	9	21	4	5	5
10	69	72	74	22	96	97	97
11	21	23	26	23	52	54	57
12	26	29	30	24	0	10	10
13	65	68	70	25	0	26	30
14	50	51	55	26	0	0	38
15	32	34	35				

TABLE 7.5

Branch Lengths (m)

Branch			Branch			Branch		
i	j	ℓ_{ij}	i	j	ℓ_{ij}	i	j	ℓ_{ij}
1	4	200	2	17	230	10	22	50
1	5	185	2	18	30	11	26	112
1	6	90	2	19	160	12	25	43
1	7	190	2	20	240	13	23	270
1	8	260	2	21	150	16	24	155
1	9	250	3	13	220	16	26	90
2	10	200	3	20	290	17	25	66
2	11	20	3	23	120	19	24	220
2	12	220	7	25	53	20	23	170
2	13	500	8	22	82	21	24	93
2	14	550	9	22	64	22	25	85
2	15	50	9	25	46	22	26	104
2	16	220						

output curves of alternatives 1 and 2 for PV generators are presented in Figures 7.22 and 7.23, respectively. A 20-year lifetime is considered for all units.

The characterization of uncertainty relies on historical data corresponding to 2012 with maximum levels of wind speed and solar irradiation equal

TABLE 7.6

Data for Candidate Replacement Conductors

	Alternative 1			Alternative 2	
$\bar{F}_{ij1}^{NRF}$ (MVA)	Z_{ij1}^{NRF} (Ω/km)	$C_{ij1}^{I,NRF}$ (€/km)	$\bar{F}_{ij2}^{NRF}$ (MVA)	Z_{ij2}^{NRF} (Ω/km)	$C_{ij2}^{I,NRF}$ (€/km)
0.3	0.2333	30,200	0.5	0.1363	35,300

TABLE 7.7

Data for Candidate Conductors in Nonexisting Branches

	Alternative 1			Alternative 2	
$\bar{F}_{ij1}^{NAF}$ (MVA)	Z_{ij1}^{NAF} (Ω/km)	$C_{ij1}^{I,NRF}$ (€/km)	$\bar{F}_{ij2}^{NAF}$ (MVA)	Z_{ij2}^{NAF} (Ω/km)	$C_{ij2}^{I,NRF}$ (€/km)
0.1	0.4522	32,300	0.3	0.2333	38,700

TABLE 7.8

Data for Candidate Transformers

	Alternative 1				Alternative 2			
Node i	$\bar{G}_{i1}^{NT}$ (MVA)	Z_{i1}^{NT} (Ω)	$C_{i1}^{M,NT}$ (€)	$C_{i1}^{I,NT}$ (€)	$\bar{G}_{i2}^{NT}$ (MVA)	Z_{i2}^{NT} (Ω)	$C_{i2}^{M,NT}$ (€)	$C_{i2}^{I,NT}$ (€)
1	0.6	0.177	200	25,000	0.8	0.144	300	32,000
2	0.6	0.177	200	25,000	0.8	0.144	300	32,000
3	0.6	0.177	200	25,000	0.8	0.144	300	32,000

TABLE 7.9

Data for Candidate DG Units

Alternative k	$\bar{G}_k^W$ (MVA)	$C_k^{I,W}$ (€/MW)	$C_k^{E,W}$ (€/MWh)	$\bar{G}_k^\Theta$ (MVA)	$C_k^{I,\Theta}$ (€/MW)	$C_k^{E,\Theta}$ (€/MWh)
1	0.0175	1,010	5	0.024	500	4
2	0.0375	1,000	7	0.048	500	4

to 17.08 m/s and 1114.21 W/m², respectively. Moreover, four time blocks are considered with durations equal to 350, 2650, 3900, and 1860 hours/year, respectively. In addition, the cdfs for demand, wind speed, and solar irradiation are divided into three equiprobable segments. Thus, as reported in Table 7.10, for each time block, three different conditions of demand, wind speed, and solar irradiation are considered, according to the procedure described in Section 7.4. As a result, 27 equiprobable scenarios are generated for each time block, thereby totaling 108 scenarios for each stage.

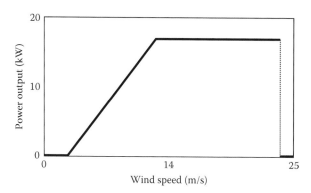

FIGURE 7.20
Power output curve of the alternative 1 for wind generation.

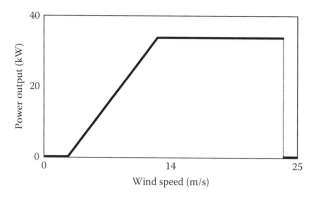

FIGURE 7.21
Power output curve of the alternative 2 for wind generation.

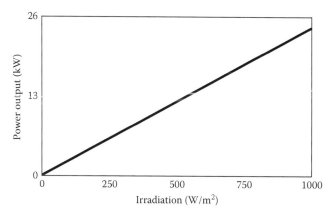

FIGURE 7.22
Power output curve of the alternative 1 for photovoltaic generation.

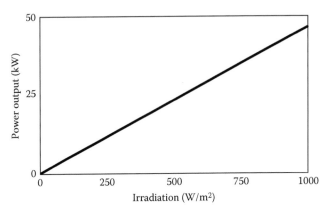

FIGURE 7.23
Power output curve of the alternative 2 for photovoltaic generation.

TABLE 7.10

Operational Conditions Used for Scenario Generation (pu)

Time Block	Average Demand Factor	Average Wind Speed Factor	Average Solar Irradiation Factor
1	0.8321	0.5036	0.2166
	0.7211	0.3208	0.0000
	0.6703	0.1561	0.0000
2	0.5886	0.4802	0.5802
	0.5142	0.2765	0.1424
	0.4700	0.0481	0.0000
3	0.4262	0.4971	0.5402
	0.3893	0.2782	0.1149
	0.3578	0.0590	0.0000
4	0.3258	0.5249	0.0610
	0.3013	0.2608	0.0000
	0.2753	0.0585	0.0000

Based on [3], the penalty factors, χ and ς, are set at 30 and 0.02, respectively. As for acceptable reliability levels, the following values have been adopted: $\overline{CIF} = 3$ interruptions/year, $\overline{CID} = 3$ hours/year, $\overline{SAIFI} = 2$ interruptions/year, and $\overline{SAIDI} = 2$ hours/year.

7.8.2 Results

The proposed approach has been implemented on a Dell PowerEdge R910X64 with four Intel Xeon E7520 processors at 1.866 GHz and 32 GB of RAM using CPLEX 12.6 [54] and GAMS 24.2 [62]. The algorithm has been run for three iterations with parameter n_d equal to 8, so that three sufficiently different solutions are obtained, namely, solution 1, solution 2, and solution 3. The stopping criterion for the branch-and-cut algorithm of CPLEX is based on an optimality gap equal to 0.1%. Under this stopping criterion, the pool of solutions was obtained in 4099.84 seconds. Figure 7.24 describes the symbols used to represent the solutions in Figures 7.25 through 7.27.

As can be observed, the topologies of the three expansion plans feature differences related to investment decisions. The construction of a substation at node 3 is only planned in solution 3. Besides, the three solutions differ in the branches used to install conductors to connect new load nodes to the distribution system. As an example, the demand at node 22 is connected by branch 9-22 for solutions 1 and 3, whereas for solution 2, it is connected by branch 8-22. Analogously, for solutions 1 and 2, the demand at node 23 is supplied by the substation located at node 2 through branch 20-23, whereas, for solution 3, such demand is connected to the substation located at node 3 by branch 3-23. Further differences arise among branches connecting nodes 24–26, thereby revealing the effect of Equation 7.88.

Regarding DG installation, it can be seen that, for the three expansion plans, the generation technologies used at each node are identical. Notwithstanding, different technology alternatives and installation times can be observed.

Finally, Tables 7.11 and 7.12 provide cost information for the three solutions. Table 7.11 shows the present value of investment, maintenance, production, losses, and unserved energy costs, whereas Table 7.12 lists the present value of the costs associated with the reliability indices described in Section 7.6.

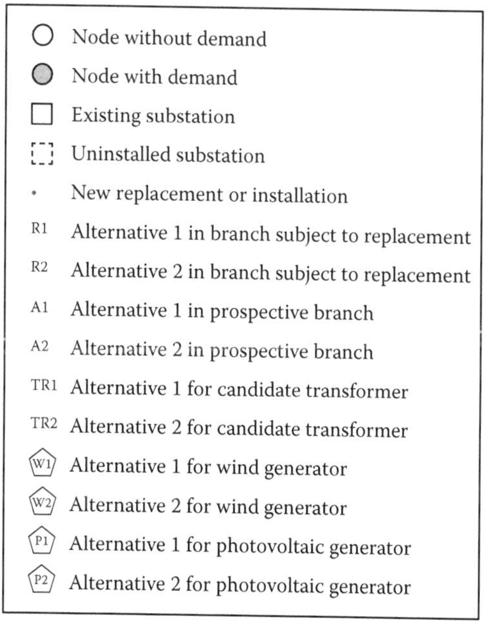

◯	Node without demand
◉	Node with demand
☐	Existing substation
⌐⌐	Uninstalled substation
·	New replacement or installation
R1	Alternative 1 in branch subject to replacement
R2	Alternative 2 in branch subject to replacement
A1	Alternative 1 in prospective branch
A2	Alternative 2 in prospective branch
TR1	Alternative 1 for candidate transformer
TR2	Alternative 2 for candidate transformer
W1	Alternative 1 for wind generator
W2	Alternative 2 for wind generator
P1	Alternative 1 for photovoltaic generator
P2	Alternative 2 for photovoltaic generator

FIGURE 7.24
List of symbols.

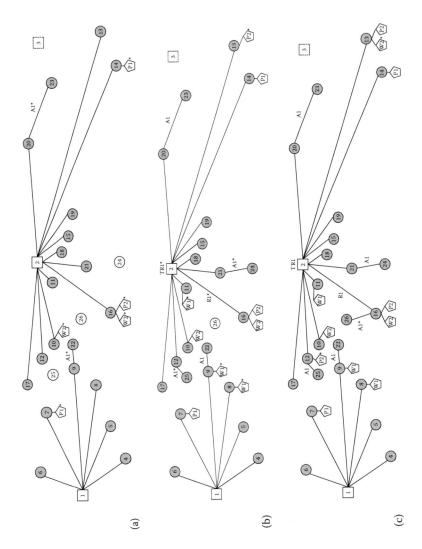

FIGURE 7.25
Solution 1: (a) stage 1, (b) stage 2, and (c) stage 3.

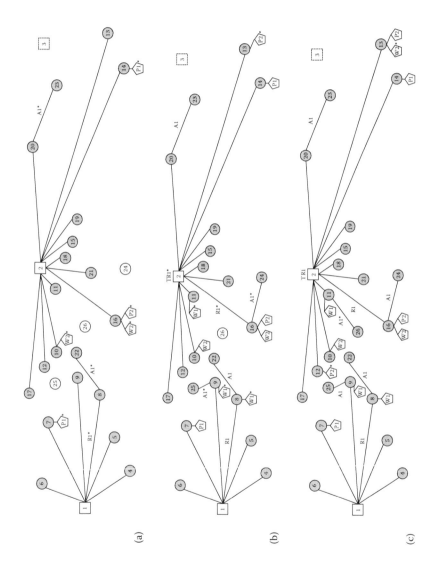

FIGURE 7.26
Solution 2: (a) stage 1, (b) stage 2, and (c) stage 3.

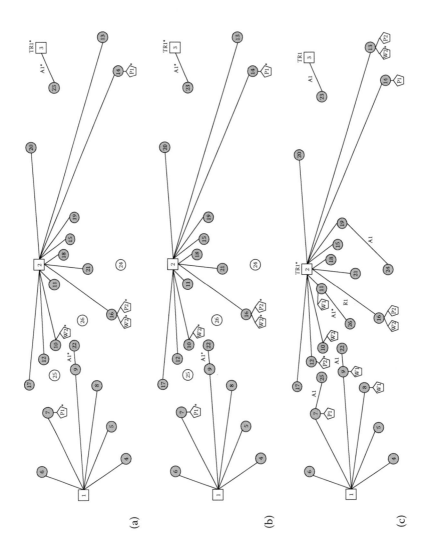

FIGURE 7.27
Solution 3: (a) stage 1, (b) stage 2, and (c) stage 3.

TABLE 7.11

Present Value of Investment and Operational Costs (€)

Costs	Solution 1	Solution 2	Solution 3
Investment	284,842	295,104	307,964
Maintenance	214,542	214,222	210,633
Production	4,290,901	4,292,231	4,312,611
Losses	555	552	516
Unserved energy	1159	2571	0
Total	4,791,999	4,804,680	4,831,724

TABLE 7.12

Present Value of Reliability Costs (€)

	Solution 1	Solution 2	Solution 3
$CIFC^{PV}$	0	0	0
$CIDC^{PV}$	984	984	731
$SAIC^{PV}$	77,968	77,414	0
$EENSC^{PV}$	1119	1119	1067

The results presented in Table 7.11 show that the costs related to maintenance and energy losses of solution 2 are slightly lower than those of solution 1. However, the investment and unserved energy costs of solution 2 are significantly higher than those of solution 1. Thus, solution 1 is cheaper than solution 2 when reliability is disregarded. From Table 7.12 it can be observed that both solutions are characterized by identical reliability costs except for $SAIC^{PV}$, which is lower for solution 2. However, the reduction in $SAIC^{PV}$ does not outweigh the larger investment and operational costs associated with solution 2. Thus, solution 1 outperforms solution 2, which is discarded by the planner.

Table 7.11 reveals that solution 3 incurs significantly higher investment and production costs as compared with solution 1. In contrast, as can be seen in Table 7.12, $CIDC^{PV}$, $EENSC^{PV}$, and particularly $SAIC^{PV}$, are lower for solution 3. It is worth noting that the difference between the total operational and investment costs of both solutions is smaller than the difference between their levels of $SAIC^{PV}$. Therefore, for this example, solution 3 is the most suitable expansion plan taking into account both economic and reliability aspects.

7.9 Conclusions

This chapter has addressed the incorporation of uncertainty and reliability in the joint expansion planning of distributed network assets and renewable DG. In the absence of optimization models considering both economic and reliability aspects in expansion planning, this chapter presents a novel algorithm combining stochastic programming and simulation-based reliability assessment. The proposed algorithm provides a pool of cost-effective candidate expansion plans characterized by their corresponding reliability indices and costs.

The proposed tool is useful for the distribution planner to balance investment and operational costs versus quality of supply. Numerical experience evidences the impact of considering reliability on the decision-making problem faced by the distribution planner. Such impact may lead to selecting an expansion plan different from that with minimum investment and operational costs. Results from a case study based on a real-life insular system illustrate the effective performance of the proposed methodology.

Summary

This chapter describes the incorporation of uncertainty and reliability in the joint expansion planning of distribution network assets and renewable DG. This tool is of particular interest for the planner of an insular distribution system wherein renewable-based DG may play a crucial role to face the forecasted load growth. First, models for PV and wind generation are presented. Next, the uncertainty related to the demand and RESs is modeled. The co-optimized planning of DG and distribution network expansion is analyzed using an iterative algorithm that yields a pool of candidate solutions. Such prospective expansion plans are attained by a multistage stochastic programming model, where the total investment and operational costs are minimized. The associated scenario-based deterministic equivalent is formulated as a mixed-integer linear program. In addition, reliability is explicitly characterized through appropriate metrics. A case study from La Graciosa in the Canary Islands illustrates the application of the proposed approach to an actual insular system.

Nomenclature

Sets

B	set of time blocks
K^l	set of available alternatives for feeders of type l
K^p	set of available alternatives for generators of type p
K^{tr}	set of available alternatives for transformers of type tr

L	set of feeder types where $L = \{\text{EFF, ERF, NRF, NAF}\}$
P	set of generator types where $P = \{W, \Theta\}$
T	set of time stages
TR	set of transformer types where $TR = \{\text{ET, NT}\}$
Y^l	set of branches with feeders of type l
Z	set of interruption types where $Z = \{\text{Repair, Reconfiguration}\}$
Λ_ϱ^0	set of indices of branches with candidate feeders for addition where no investment has been made at iteration ϱ
Λ_ϱ^1	set of indices of branches with candidate feeders for addition where investments have been made at iteration ϱ
Ψ^N	set of system nodes
Ψ^{SS}	set of substation nodes
Ψ_i^l	set of nodes connected to node i by a feeder of type l
Ψ^p	set of candidate nodes for distributed generators of type p
Ψ_t^{LN}	set of load nodes at stage t
Ω_b	set of scenarios for time block b

Miscellaneous

EFF	existing fixed feeder
ERF	existing replaceable feeder
ET	existing transformer
NAF	newly added feeder
NRF	new replacement feeder
NT	new transformer
W	wind generator
Θ	PV generator

Indices

b	index for time blocks
i, j	indices for nodes
k, κ	indices for available investment alternatives
l	index for feeder types
p	index for generator types
s	index for segments of the cumulative distribution functions
t, τ	indices for time stages
tr	index for transformer types
v	index for the blocks used in the piecewise linearization for energy losses
z	index for interruption types
ω	index for scenarios
ϱ	index for iterations

Parameters

A_c	cross-sectional area
A_p	area of the PV panel
A_v	width of block v of the piecewise linear energy losses
A_{kv}^l	width of block v of the piecewise linear energy losses for alternative k of feeder type l
A_{kv}^{tr}	width of block v of the piecewise linear energy losses for alternative k of transformer type tr
ASAI_t	average system availability index for stage t
C^U	unserved energy cost coefficient
C_b^{SS}	cost coefficient of energy supplied by substations at time block b
$C_i^{I,SS}$	investment cost coefficient of the substation located at node i
$C_k^{E,p}$	cost coefficient of energy supplied by alternative k of generator type p
$C_k^{I,l}$	unitary investment cost coefficient of alternative k of feeder type l
$C_k^{I,NT}$	investment cost coefficient of alternative k of new transformers
$C_k^{I,p}$	investment cost coefficient of alternative k of generator type p
$C_k^{M,l}$	maintenance cost coefficient of alternative k of feeder type l
$C_k^{M,p}$	maintenance cost coefficient of alternative k of generator type p
$C_k^{M,tr}$	maintenance cost coefficient of alternative k of transformer type tr
C_p	rotor efficiency
$\bar{C_p}$	maximum rotor efficiency, also known as Betz efficiency
$\overline{\text{CID}}$	target for the index of customer interruption duration
CID_{it}	index of customer interruption duration at node i and stage t
CIDC_t	cost of customer interruption duration at stage t
CIDC^{PV}	present value of the cost of customer interruption duration
$\overline{\text{CIF}}$	target for the index of customer interruption frequency
CIF_{it}	index of customer interruption frequency at node i and stage t
CIFC_t	cost of customer interruption frequency at stage t
CIFC^{PV}	present value of the cost of customer interruption frequency
D_{it}	actual nodal peak demand at node i and stage t
$\tilde{D}_{it}$	fictitious nodal demand in the substation at node i and stage t
$D_{itb}^R(\omega)$	residual demand at node i and stage t for time block b and scenario ω
DI_z	duration of interruption type z
E_{kinetic}	kinetic energy
EENS_t	expected energy not supplied at stage t associated with feeder outages
EENSC_t	cost of the expected energy not supplied at stage t associated with feeder outages
EENSC^{PV}	present value of the cost of the expected energy not supplied associated with feeder outages

$\bar{F}_k^l$	upper limit for the actual current flow through alternative k of feeder type l
G	solar irradiance
$\hat{G}_{iktb}^p(\omega)$	maximum power availability for alternative k of generator type p at node i and stage t for time block b and scenario ω
$\bar{G}_k^p$	rated capacity for alternative k of generator type p
$\bar{G}_k^{tr}$	upper limit for the current injected by alternative k of transformer type tr
H	height
H_0	reference height
I	annual interest rate
I_0	reverse saturation current
I_{MPP}	current at maximum power output
I_{out}	current output
I_{SC}	short-circuit current
$I_{SC}(STC)$	short-circuit current at standard test conditions
IB_t	investment budget for stage t
J	sufficiently large positive constant
k_b	Boltzmann's constant (1.38E–23 J K^{-1})
ℓ_{ij}	length of the branch connecting nodes i and j
m	air mass
$\dot{m}$	air mass flow rate
M_{kv}^l	slope of block v of the piecewise linear energy losses for alternative k of feeder type l
M_{kv}^{tr}	slope of block v of the piecewise linear energy losses for alternative k of transformer type tr
M_v	slope of block v of the piecewise linear energy losses
n_B	number of time blocks
n_d	minimum number of investments in newly added feeders by which an expansion plan must differ from solutions of previous iterations
n_{DG}	number of candidate nodes for installation of DG
n_s^D	number of segments for demand factors at each time block
n_s^p	number of segments for factors for generation of type p at each time block
n_T	number of time stages
n_v	number of blocks of the piecewise linear energy losses
n_Ω	number of scenarios at each time block
NC_{it}	total number of customers at node i and stage t
NI_{itz}	number of interruptions of type z that can affect node i at stage t
P_b	power extracted by the blades
$P_{kinetic}$	power corresponding to the kinetic energy
P_{MAX}	maximum power
P_{out}	power output

P_R	rated electrical power
P_{STC}	power output at standard test conditions
P_{wind}	wind power output
pf	system power factor
q	electron charge (1.6E-19 C)
R_p	parallel resistance
RR^l	capital recovery rate for investment in feeders of type l
RR^{NT}	capital recovery rate for investment in new transformers
RR^p	capital recovery rate for investment in generators of type p
RR^{SS}	capital recovery rate for investment in substations
$SAIC_t$	cost of system average interruption at stage t
$SAIC^{PV}$	present value of the cost of system average interruption
$\overline{SAIDI}$	target for the index of system average interruption duration
$SAIDI_t$	index of system average interruption duration at stage t
$\overline{SAIFI}$	target for the index of system average interruption frequency
$SAIFI_t$	index of system average interruption frequency at stage t
T_{amb}	ambient temperature (°C)
T_{cell}	cell temperature (°C)
T_{NOC}	temperature at nominal operating cell conditions (°C)
T_p	temperature (K)
v	wind speed
v_0	wind speed at height H_0
v_b	wind speed through the plane of the rotor blades
v_{ci}	cut-in speed
v_{co}	cut-out speed
v_d	downstream wind speed
v_{rated}	rated speed
v_u	upstream wind speed
$\underline{V}$	lower bound for nodal voltages
$\overline{V}$	upper bound for nodal voltages
V_d	voltage across the diode terminals
V_{MPP}	voltage at maximum power output
V_{OC}	open-circuit voltage
$x_{ijkt}^{NAF(\varrho)}$	value of variable x_{ijkt}^{NAF} at iteration ϱ
Z_k^l	unitary impedance magnitude of alternative k of feeder type l
Z_k^{tr}	impedance magnitude of alternative k of transformer type tr
α	coefficient of friction
γ	ratio of downstream to upstream wind speed
δ	power–temperature coefficient
Δ_b	duration of time block b
ϵ_{out}	efficiency under a particular operating condition
ϵ_{STC}	efficiency at standard test conditions
η^l	lifetime of feeders of type l

η^{NT}	lifetime of new transformers
η^{p}	lifetime of generators of type p
η^{SS}	lifetime of substation assets other than transformers
μ_{sb}^{D}	average demand factor at segment s and time block b
μ_{sb}^{p}	average factor of generation type p at segment s and time block b
$\mu_{b}^{D}(\omega)$	average demand factor of time block b and scenario ω
λ	average failure rate of feeders
π_{sb}^{D}	probability of the average demand factor in segment s and time block b
π_{sb}^{p}	probability of the average factor of generation type p in segment s and time block b
$\pi_{b}(\omega)$	probability of scenario ω of time block b
ρ	air density
ξ	penetration limit for DG
ς	penalty factor associated with the violation of SAIFI or SAIDI
χ	penalty factor associated with the violation of CIF or CID

Variables

c_{t}^{E}	expected production cost at stage t
c_{t}^{I}	amortized investment cost at stage t
c_{t}^{M}	maintenance cost at stage t
c_{t}^{R}	expected cost of energy losses at stage t
c^{TPV}	present value of the total investment and operational cost
c_{t}^{U}	expected cost of the unserved energy at stage t
$d_{itb}^{U}(\omega)$	unserved energy at node i and stage t for time block b and scenario ω
$\tilde{f}_{ijkt}^{l}$	fictitious current flow through alternative k of feeder type l connecting nodes i and j at stage t, measured at node i
$f_{ijktb}^{l}(\omega)$	actual current flow through alternative k of feeder type l connecting nodes i and j at stage t for time block b and scenario ω, measured at node i
$g_{iktb}^{p}(\omega)$	current injection by alternative k of generator type p installed at node i and stage t for time block b and scenario ω
$g_{iktb}^{tr}(\omega)$	current injection by alternative k of transformer type tr installed at substation node i and stage t for time block b and scenario ω
$\tilde{g}_{it}^{SS}$	fictitious current injection at substation node i and stage t
$v_{itb}(\omega)$	nodal voltage magnitude at node i and stage t for time block b and scenario ω
x_{ijkt}^{l}	binary variable representing the investment in alternative k of feeder type l connecting nodes i and j at stage t
x_{ikt}^{NT}	binary variable representing the investment in alternative k of new transformers at substation node i and stage t
x_{ikt}^{p}	binary variable representing the investment in alternative k of generator type p at node i and stage t

x_{it}^{SS} binary variable representing the investment decision at substation node i and stage t

y_{ijkt}^{l} binary variable representing the utilization of alternative k of feeder type l connecting nodes i and j at stage t

y_{ikt}^{p} binary variable representing the utilization of alternative k of generator type p at node i and stage t

y_{ikt}^{tr} binary variable representing the utilization of alternative k of transformer type tr at substation node i and stage t

δ_{v} current flow corresponding to block v of the piecewise linear energy losses

$\delta_{ijktbv}^{l}(\omega)$ current flow corresponding to block v of the piecewise linear energy losses for alternative k of feeder type l connecting nodes i and j at stage t for time block b and scenario ω

$\delta_{iktbv}^{tr}(\omega)$ current injection corresponding to block v of the piecewise linear energy losses for alternative k of transformer type tr at substation node i and stage t for time block b and scenario ω

References

1. A. Gómez-Expósito, A. J. Conejo, and C. Cañizares, *Electric Energy Systems. Analysis and Operation.* Boca Raton, FL: CRC Press, 2009.
2. W. H. Kersting, *Distribution System Modeling and Analysis, 3rd ed.* Boca Raton, FL: CRC Press, 2012.
3. R. C. Lotero and J. Contreras, Distribution system planning with reliability. *IEEE Trans. Power Deliv.*, vol. 26, no. 4, pp. 2552–2562, 2011.
4. H. L. Willis, *Power Distribution Planning Reference Book, 2nd ed.* New York, NY: Marcel Dekker, Inc., 2004.
5. W. El-Khattam, K. Bhattacharya, Y. Hegazy, and M. M. A. Salama, Optimal investment planning for distributed generation in a competitive electricity market. *IEEE Trans. Power Syst.*, vol. 19, no. 3, pp. 1674–1684, 2004.
6. R. Viral and D. K. Khatod, Optimal planning of distributed generation systems in distribution system: A review. *Renew. Sust. Energ. Rev.*, vol. 16, no. 7, pp. 5146–5165, 2012.
7. R. E. Brown, *Electric Power Distribution Reliability, 2nd ed.* Boca Raton, FL: CRC Press, 2008.
8. S. Haffner, L. F. A. Pereira, L. A. Pereira, and L. S. Barreto, Multistage model for distribution expansion planning with distributed generation—Part I: Problem formulation. *IEEE Trans. Power Deliv.*, vol. 23, no. 2, pp. 915–923, 2008.
9. W. El-Khattam, Y. G. Hegazy, and M. M. A. Salama, An integrated distributed generation optimization model for distribution system planning. *IEEE Trans. Power Syst.*, vol. 20, no. 2, pp. 1158–1165, 2005.
10. G. Muñoz-Delgado, J. Contreras, and J. M. Arroyo, Joint expansion planning of distributed generation and distribution networks. *IEEE Trans. Power Syst.*, 2015. Available at: http://ieeexplore.ieee.org/stamp/stamp.jsp?arnumber=6966819.

11. T. Gönen and B. L. Foote, Distribution-system planning using mixed-integer programming. *IEE Proc. Gener. Transm. Distrib.*, vol. 128, no. 2, pp. 70–79, 1981.

12. M. Ponnavaikko, K. S. P. Rao, and S. S. Venkata, Distribution system planning through a quadratic mixed integer programming approach. *IEEE Trans. Power Deliv.*, vol. 2, no. 4, pp. 1157–1163, 1987.

13. Y. Tang, Power distribution system planning with reliability modeling and optimization. *IEEE Trans. Power Syst.*, vol. 11, no. 1, pp. 181–189, 1996.

14. S. K. Khator and L. C. Leung, Power distribution planning: A review of models and issues. *IEEE Trans. Power Syst.*, vol. 12, no. 3, pp. 1151–1159, 1997.

15. S. Bhowmik, S. K. Goswami, and P. K. Bhattacherjee, A new power distribution system planning through reliability evaluation technique. *Electr. Power Syst. Res.*, vol. 54, no. 3, pp. 169–179, 2000.

16. G. Celli, S. Mocci, F. Pilo, and R. Cicoria, Probabilistic optimization of MV distribution network in presence of distributed generation. In *Proceedings of the 14th Power Systems Computation Conference*, Seville, Spain, Session 1, Paper 1, pp. 1–7, June 24–28, 2002.

17. P. C. Paiva, H. M. Khodr, J. A. Domínguez-Navarro, J. M. Yusta, and A. J. Urdaneta, Integral planning of primary-secondary distribution systems using mixed integer linear programming. *IEEE Trans. Power Syst.*, vol. 20, no. 2, pp. 1134–1143, 2005.

18. E. G. Carrano, F. G. Guimarães, R. H. C. Takahashi, O. M. Neto, and F. Campelo, Electric distribution network expansion under load-evolution uncertainty using an immune system inspired algorithm. *IEEE Trans. Power Syst.*, vol. 22, no. 2, pp. 851–861, 2007.

19. S. Haffner, L. F. A. Pereira, L. A. Pereira, and L. S. Barreto, Multistage model for distribution expansion planning with distributed generation—Part II: Numerical results. *IEEE Trans. Power Deliv.*, vol. 23, no. 2, pp. 924–929, 2008.

20. M. Lavorato, M. J. Rider, A. V. Garcia, and R. Romero, A constructive heuristic algorithm for distribution system planning. *IEEE Trans. Power Syst.*, vol. 25, no. 3, pp. 1734–1742, 2010.

21. D. T.-C. Wang, L. F. Ochoa, and G. P. Harrison, Modified GA and data envelopment analysis for multistage distribution network expansion planning under uncertainty. *IEEE Trans. Power Syst.*, vol. 26, no. 2, pp. 897–904, 2011.

22. M. Lavorato, J. F. Franco, M. J. Rider, and R. Romero, Imposing radiality constraints in distribution system optimization problems. *IEEE Trans. Power Syst.*, vol. 27, no. 1, pp. 172–180, 2012.

23. A. M. Cossi, L. G. W. da Silva, R. A. R. Lázaro, and J. R. S. Mantovani, Primary power distribution systems planning taking into account reliability, operation and expansion costs. *IET Gener. Transm. Distrib.*, vol. 6, no. 3, pp. 274–284, 2012.

24. J. Salehi and M.-R. Haghifam, Long term distribution network planning considering urbanity uncertainties. *Int. J. Electr. Power Energ. Syst.*, vol. 42, no. 1, pp. 321–333, 2012.

25. S. Ganguly, N. C. Sahoo, and D. Das, Recent advances on power distribution system planning: A state-of-the-art survey. *Energy Syst.*, vol. 4, no. 2, pp. 165–193, 2013.

26. N. S. Rau and Y.-H. Wan, Optimum location of resources in distributed planning. *IEEE Trans. Power Syst.*, vol. 9, no. 4, pp. 2014–2020, 1994.

27. G. Carpinelli, G. Celli, F. Pilo, and A. Russo, Distributed generation siting and sizing under uncertainty. In *Proceedings of the 2001 IEEE PowerTech*, Porto, Portugal, pp. 1–7, September 10–13, 2001.

28. J.-H. Teng, Y.-H. Liu, C.-Y. Chen, and C.-F. Chen, Value-based distributed generator placements for service quality improvements. *Int. J. Electr. Power Energ. Syst.*, vol. 29, no. 3, pp. 268–274, 2007.

29. G. Celli, E. Ghiani, S. Mocci, and F. Pilo, A multiobjective evolutionary algorithm for the sizing and siting of distributed generation. *IEEE Trans. Power Syst.*, vol. 20, no. 2, pp. 750–757, 2005.

30. H. Falaghi and M.-R. Haghifam, ACO based algorithm for distributed generation sources allocation and sizing in distribution systems. In *Proceedings of the 2007 IEEE PowerTech*, Lausanne, Switzerland, pp. 555–560, July 1–5, 2007.

31. Y. M. Atwa, E. F. El-Saadany, M. M. A. Salama, and R. Seethapathy, Optimal renewable resources mix for distribution system energy loss minimization. *IEEE Trans. Power Syst.*, vol. 25, no. 1, pp. 360–370, 2010.

32. M. F. Akorede, H. Hizam, I. Aris, and M. Z. A. Ab Kadir, Effective method for optimal allocation of distributed generation units in meshed electric power systems. *IET Gener. Transm. Distrib.*, vol. 5, no. 2, pp. 276–287, 2011.

33. K. Zou, A. P. Agalgaonkar, K. M. Muttaqi, and S. Perera, Distribution system planning with incorporating DG reactive capability and system uncertainties. *IEEE Trans. Sustain. Energ.*, vol. 3, no. 1, pp. 112–123, 2012.

34. W.-S. Tan, M. Y. Hassan, M. S. Majid, and H. A. Rahman, Optimal distributed renewable generation planning: A review of different approaches. *Renew. Sust. Energ. Rev.*, vol. 18, pp. 626–645, 2013.

35. A. Keane, L. F. Ochoa, C. L. T. Borges, G. W. Ault, A. D. Alarcon-Rodriguez, R. A. F. Currie, F. Pilo, C. Dent, and G. P. Harrison, State-of-the-art techniques and challenges ahead for distributed generation planning and optimization. *IEEE Trans. Power Syst.*, vol. 28, no. 2, pp. 1493–1502, 2013.

36. V. H. Méndez, J. Rivier, J. I. de la Fuente, T. Gómez, J. Arceluz, J. Marín, and A. Madurga, Impact of distributed generation on distribution investment deferral. *Int. J. Electr. Power Energ. Syst.*, vol. 28, no. 4, pp. 244–252, 2006.

37. H. A. Gil and G. Joos, On the quantification of the network capacity deferral value of distributed generation. *IEEE Trans. Power Syst.*, vol. 21, no. 4, pp. 1592–1599, 2006.

38. A. Piccolo and P. Siano, Evaluating the impact of network investment deferral on distributed generation expansion. *IEEE Trans. Power Syst.*, vol. 24, no. 3, pp. 1559–1567, 2009.

39. D. T.-C. Wang, L. F. Ochoa, and G. P. Harrison, DG impact on investment deferral: Network planning and security of supply. *IEEE Trans. Power Syst.*, vol. 25, no. 2, pp. 1134–1141, 2010.

40. M. F. Shaaban, Y. M. Atwa, and E. F. El-Saadany, DG allocation for benefit maximization in distribution networks. *IEEE Trans. Power Syst.*, vol. 28, no. 2, pp. 639–649, 2013.

41. V. V. Thang, D. Q. Thong, and B. Q. Khanh, A new model applied to the planning of distribution systems for competitive electricity markets. In *Proceedings of the 4th International Conference on Electric Utility Deregulation and Restructuring and Power Technologies*, Weihai, People's Republic of China, pp. 631–638, July 6–9, 2011.

42. V. F. Martins and C. L. T. Borges, Active distribution network integrated planning incorporating distributed generation and load response uncertainties. *IEEE Trans. Power Syst.*, vol. 26, no. 4, pp. 2164–2172, 2011.

43. E. Naderi, H. Seifi, and M. S. Sepasian, A dynamic approach for distribution system planning considering distributed generation. *IEEE Trans. Power Deliv.*, vol. 27, no. 3, pp. 1313–1322, 2012.

44. I. Ziari, G. Ledwich, A. Ghosh, and G. Platt, Integrated distribution systems planning to improve reliability under load growth. *IEEE Trans. Power Deliv.*, vol. 27, no. 2, pp. 757–765, 2012.

45. M. E. Samper and A. Vargas, Investment decisions in distribution networks under uncertainty with distributed generation—Part I: Model formulation. *IEEE Trans. Power Syst.*, vol. 28, no. 3, pp. 2331–2340, 2013.

46. M. E. Samper and A. Vargas, Investment decisions in distribution networks under uncertainty with distributed generation—Part II: Implementation and results. *IEEE Trans. Power Syst.*, vol. 28, no. 3, pp. 2341–2351, 2013.

47. A. A. Chowdhury and D. O. Koval, Current practices and customer value-based distribution system reliability planning. *IEEE Trans. Ind. Appl.*, vol. 40, no. 5, pp. 1174–1182, 2004.

48. G. M. Masters, *Renewable and Efficient Electric Power Systems, 2nd ed.* New York, NY: Wiley-Interscience, 2013.

49. E. Lorenzo, G. Araujo, A. Cuevas, M. Egido, J. Miñano, and R. Zilles, *Solar Electricity. Engineering of Photovoltaic Systems*. Sevilla, Spain: Progensa, 1994.

50. L. Baringo and A. J. Conejo, Transmission and wind power investment. *IEEE Trans. Power Syst.*, vol. 27, no. 2, pp. 885–893, 2012.

51. L. Baringo and A. J. Conejo, Correlated wind-power production and electric load scenarios for investment decisions. *Appl. Energy*, vol. 101, pp. 475–482, 2013.

52. S. J. Kazempour, A. J. Conejo, and C. Ruiz, Strategic generation investment using a complementarity approach. *IEEE Trans. Power Syst.*, vol. 26, no. 2, pp. 940–948, 2011.

53. J. R. Birge and F. Louveaux, *Introduction to Stochastic Programming, 2nd ed.* New York, NY: Springer, 2011.

54. The IBM ILOG CPLEX Website, 2014. Available at: http://www-01.ibm.com/software/commerce/optimization/cplex-optimizer.

55. G. L. Nemhauser and L. A. Wolsey, *Integer and Combinatorial Optimization.* New York, NY: Wiley-Interscience, 1999.

56. L. Blank and A. Tarquin, *Engineering Economy, 7th ed.* New York, NY: McGraw-Hill, 2012.

57. S. P. Bradley, A. C. Hax, and T. L. Magnanti, *Applied Mathematical Programming.* Reading, MA: Addison-Wesley, 1977.

58. S. Binato, M. V. F. Pereira, and S. Granville, A new Benders decomposition approach to solve power transmission network design problems. *IEEE Trans. Power Syst.*, vol. 16, no. 2, pp. 235–240, 2001.

59. A. A. Chowdhury and D. O. Koval, *Power Distribution System Reliability. Practical Methods and Applications.* Hoboken, NJ: John Wiley and Sons, Inc., 2009.

60. R. Billinton and J. E. Billinton, Distribution system reliability indices. *IEEE Trans. Power Deliv.*, vol. 4, no. 1, pp. 561–568, 1989.

61. E. Balas and R. Jeroslow, Canonical cuts on the unit hypercube. *SIAM J. Appl. Math.*, vol. 23, no. 1, pp. 61–69, 1972.

62. The GAMS Development Corporation Website, 2014. Available at: http://www.gams.com.

Index

Note: Locators followed by *"f"* and *"t"* denote figures and tables in the text

Printed and bound by CPI Group (UK) Ltd, Croydon, CR0 4YY

24/10/2024

01778281-0014